Computational Intelligence in Biomedical Engineering

Computational Intelligence in Biomedical Engineering

Rezaul Begg
Daniel T. H. Lai
Marimuthu Palaniswami

CRC Press
Taylor & Francis Group
Boca Raton London New York

CRC Press is an imprint of the
Taylor & Francis Group, an **informa** business

CRC Press
Taylor & Francis Group
6000 Broken Sound Parkway NW, Suite 300
Boca Raton, FL 33487-2742

First issued in paperback 2019

© 2008 by Taylor & Francis Group, LLC
CRC Press is an imprint of Taylor & Francis Group, an Informa business

No claim to original U.S. Government works

ISBN-13: 978-0-8493-4080-2 (hbk)
ISBN-13: 978-0-367-38809-6 (pbk)

Library of Congress Cataloging-in-Publication Data

Begg, Rezaul.
    Computational intelligence in biomedical engineering / Rezaul Begg, Daniel T.H. Lai, and Marimuthu Palaniswami.
        p. ; cm.
    "A CRC title."
    Includes bibliographical references and index.
    ISBN 978-0-8493-4080-2 (hardcover : alk. paper) 1. Artificial intelligence--Medical applications. 2. Biomedical engineering--Computer simulation. I. Lai, Daniel T. H. II. Palaniswami, Marimuthu. III. Title.
    [DNLM: 1. Artificial Intelligence. 2. Biomedical Engineering. 3. Medical Informatics Applications. W 26.55.A7 B416c 2008]
    R859.7.A78B44 2008
    610.285'63--dc22   2007032849

Visit the Taylor & Francis Web site at
http://www.taylorandfrancis.com

and the CRC Press Web site at
http://www.crcpress.com

Dedication

This book is dedicated to,

My caring wife Dola, son Rashad and my parents
Rezaul K. Begg

My lovely wife Carennie, my parents and family
Daniel T. H. Lai

My supportive family and the healthcare workers of the world
Marimuthu Palaniswami

Contents

Preface

In recent years, there has been an explosion of interest in computational intelligence (CI) as evidenced by the numerous applications in health, biomedicine, and biomedical engineering. This book presents innovative developments in the emerging field of applied CI and biomedical engineering. We describe state-of-the-art applications that have benefited from the application of CI, including cardiology, electromyography, electroencephalography, movement science, and biomechanics.

Biomedical engineering is the application of engineering concepts and techniques to problems in medicine and healthcare. It is a relatively new domain of research and development consisting of subdisciplines such as bioinstrumentation, bioinformatics, biomechanics, biosensors, signal processing, medical imaging, and computing. Typical applications include design and development in prosthetics and medical instruments, diagnostic software, imaging equipment, and health-monitoring and drug delivery systems.

CI techniques are computing algorithms and learning machines, including artificial neural networks, fuzzy logic, genetic algorithms, and support vector machines. Most CI techniques involve establishing a nonlinear mapping between the inputs and outputs of a model representing the operation of a real-world biomedical system. Although CI concepts were introduced a few decades ago, impressive new developments have only recently taken place about theoretical advances. New theory has, however, led to a mushrooming in applications, in which input–output relations are too complex to be expressed using explicit mathematical formulations.

This book consists of eight chapters each with a specific focus. Following an overview of signal processing and machine-learning approaches, four application-specific chapters are presented illustrating CI's importance in medical diagnosis and healthcare. Chapter 1 is an overview of biomedical signals and systems; it discusses the origin of biomedical signals within the human body, the key elements of a biomedical system, and the sources of noise in recorded signals. The second chapter continues with an extensive overview of signal-processing techniques commonly employed in the analysis of biomedical signals and for improving the signal-to-noise ratio. Chapter 3 presents an overview of the major CI techniques.

Electrical potentials generated by the heart are recorded on the body's surface and represented using an electrocardiogram (ECG). The ECG can be used to detect many cardiac abnormalities. Chapter 4 concerns recent CI techniques that have been applied to the postprocessing of ECG signals for the diagnosis of cardiovascular diseases. This field is already well advanced and almost every new CI technique has been tested on processed ECG data. Chapter 5 focuses on CI applications in electromyography (EMG), pattern recognition, and the diagnosis of neuromuscular pathologies that affect EMG-related signal characteristics. More recently, CI has also been applied to the control of myoelectric prostheses and exoskeletons. Of interest here is the combination of signal-processing and classifier systems to detect the user's intention to move the prosthesis. In Chapter 6, CI applications to bioelectric potentials representing brain activity are outlined based on electroencephalogram (EEG) recordings. CI approaches play a major role in EEG signal processing because of their effectiveness as pattern classifiers. Here we concentrated on several applications including the identification of abnormal EEG activity in patients with neurological diseases (e.g., epilepsy) and in the control of external devices using EEG waveforms, known as brain–computer interfaces (BCI). The BCI has many potential applications in rehabilitation, such as assisting individuals with disabilities to independently operate appliances. Chapter 7 provides an overview of CI applications for the detection and classification of gait types from their kinematic, kinetic, and EMG features. Gait analysis is routinely used for detecting abnormality in lower-limb function and also for evaluating the progress of treatment. Various studies in this area are discussed with a particular focus on CI's potential as a tool for gait diagnostics. Finally, Chapter 8 discusses progress in biomedical engineering, biomedicine, and human health areas with suggestions for future applications.

This book should be of considerable help to a broad readership, including researchers, professionals, academics, and graduate students from a wide range of disciplines who are beginning to look for applications in health care. The text provides a comprehensive account of recent research in this emerging field and we anticipate that the concepts presented here will generate further research in this multidisciplinary field.

Rezaul K. Begg
Daniel T. H. Lai
Marimuthu Palaniswami

Abbreviations

Acronym	Definition
AD	Alzheimer's disease
AEP	Auditory evoked potential
AIC	Akaike's information criterion
ADALINE	Adaptive linear networks
AMI	Acute myocardial infarctions
ANN	Artificial neural networks
ANFIS	Adaptive neurofuzzy system
A/P	Anterior–posterior
AR	Autoregressive model
ARMA	Autoregressive model with moving average
ART	Adaptive resonance theory
BCI	Brain–computer interface
BP	Backpropagation
CI	Computational intelligence
CP	Cerebral palsy
CV	Cross-validation
CVD	Cardiovascular disease
CHF	Congestive heart failure
CNV	Cognitive negative variation
CSSD	Common spatial subspace decomposition
ECG	Electrocardiogram
EMG	Electromyogram
EVP	Evoked potentials
FA	Fuzzy ARTMAP
FDA	Food and Drug Administration
FIR	Finite impulse response
FFT	Fast fourier transform
FPE	Final prediction error
FLD	Fisher's linear discriminant
GA	Genetic algorithm
GRF	Ground reaction forces
HD	Huntington's disease
HMM	Hidden Markov model
HOS	Higher-order statistics
IIR	Infinite impulse response
ICD	Implantable cardioverter–defibrillators
i.i.d	Independent and identically distributed
IIP	Intelligent information processing
KKT	Karush–Kuhn–Tucker optimality conditions
LS	Least squares
LDA	Linear discriminant analysis

LMS	Least mean squares
LS-SVM	Least squares support vector machines
LRM	Likelihood ratio method
LVQ	Learning vector quantization
MI	Myocardial infarction
ML	Maximum likelihood
M/L	Medio-lateral
MDL	Minimum description length
MOE	Mixture of experts
MRI	Magnetic resonance imaging
MTC	Minimum toe clearance
MUAP	Motor unit action potentials
MUAPT	Motor unit action potential train
NCS	Nerve conduction studies
NSR	Normal sinus rhythm
PSD	Power spectral density
PVC	Premature ventricular contraction
PSVM	Proximal support vector machines
QP	Quadratic programming or quadratic program
RLS	Recursive least squares
RBF	Radial basis function
ROC	Receiver operating characteristics
ROM	Range of motion
RNN	Recurrent neural networks
SA	Sinoatrial node
SEP	Somatosensory evoked potential
SMO	Sequential minimal optimization
SRM	Structural risk minimization
SVM	Support vector machines
SVR	Support vector regressor
VT	Ventricular tachycardia
VPB	Ventricular premature beats
VEB	Ventricular ectopic beats
VEP	Visual evoked potential

Acknowledgments

We gratefully acknowledge support from the ARC Research Network on Intelligent Sensors, Sensor Networks and Information Processing (ISSNIP) for this project and also to Victoria University and the University of Melbourne for providing logistic support to undertake this project. ISSNIP has been instrumental in bringing together advancements from many fields including biology, biomedical engineering, computational intelligence, and sensor networks under a single umbrella. The authors would like to thank Dr. Tony Sparrow for taking the time to proofread the manuscript as well as providing helpful comments. We also thank Dr. Ahsan Khandoker and Dr. Alistair Shilton for their comments, especially on two chapters of the book.

The authors would also like to thank the publishing team at the Taylor & Francis Group LLC (CRC Press), who provided continuous help, encouragement, and professional support from the initial proposal stage to the final publication, with special thanks to T. Michael Slaughter, Jill Jurgensen, and Liz Spangenberger. We also thank Darren Lai for assisting in the preparation of the figures of this book.

Finally, we thank our families for their love and support throughout the writing of this book.

Authors

Rezaul K. Begg received his BEng and MScEng degrees in electrical and electronic engineering from Bangladesh University of Engineering and Technology (BUET), Dhaka, Bangladesh, and his PhD degree in biomedical engineering from the University of Aberdeen, United Kingdom. Currently, he is an associate professor within the Biomechanics Unit at Victoria University, Melbourne, Australia. Previously, he worked with Deakin University and BUET. He researches in biomedical engineering, biomechanics, and machine-learning areas, and has published over 120 research papers in these areas. He is a regular reviewer for several international journals, and has been actively involved in organizing a number of major international conferences. He has received several awards for academic excellence, including the ICISIP2005 Best Paper Award, VC's Citation Award for excellence in research at Victoria University, the BUET Gold Medal, and the Chancellor Prize. He is a senior member of the IEEE.

Daniel T.H. Lai received his BE (Hons) degree in electrical and computer systems from Monash University, Melbourne, Australia, in 2002, and his PhD from Monash University in 2006. He is currently a research fellow in the University of Melbourne. His research interests include decomposition techniques for support vector machines, application of signal processing, computational intelligence techniques, and wireless sensor networks to biomedical engineering applications. He has over 20 peer-reviewed publications and is a current reviewer for the *IEEE Transactions of Information Technology and Biomedicine* and *the International Journal of Computational Intelligence and Applications*. He has been actively involved in the organization of several conferences and workshops including ISSNIP 2005 and ISSNIP 2007.

 **Marimuthu Palaniswami** received his BE (Hons) from the University of Madras; ME from the Indian Institute of Science, India; MScEng from the University of Melbourne; and PhD from the University of Newcastle, Australia, before rejoining the University of Melbourne. He has been serving the University of Melbourne for over 16 years. He has published more than 180 refereed papers. His research interests include support vector machines, sensors and sensor networks, machine learning, neural network, pattern recognition, and signal processing and control. He was given a Foreign Specialist Award by the Ministry of Education, Japan, in recognition of his contributions to the field of machine learning. He served as associate editor for journals/transactions including *IEEE Transactions on Neural Networks and Computational Intelligence for Finance.* He is also the subject editor for the *International Journal on Distributed Sensor Networks.*

1

Introduction

With the advent of the twenty-first century, the focus on global health and personal well-being has increased. In 2006, the population of the Earth was estimated to be 6.5 billion,* which is expected to rise steadily in the coming years. Health has become more important with recent advances in genetic research, development of new and improved drugs, diagnosis of disease, and the design of more advanced systems for patient monitoring. Health now ranks as one of the most important needs in modern society as evidenced by increasing government funding around the globe.

In this opening chapter we examine the fundamental components of a biomedical system, ranging from the medical data to the instrumentation that can be used to diagnose a disorder. The vast amount of data collected has resulted in increasingly extensive databases, which cannot always be efficiently interpreted by clinicians or doctors, resulting in excessive waiting times for a patient to be diagnosed. The problem is exacerbated by the shortage of qualified medical staff causing unnecessary delays, which can be exceptionally fatal to patients with serious disorders. Consequently, recent research issues have focused on new and improved diagnostic methods to alleviate the risk of life loss due to inadequate diagnostic capacity.

A new paradigm in the field of medicine is computational intelligence (CI) for diagnostic systems. Computational intelligence is a combination of advances in artificial learning and computers, which was introduced more than five decades ago. These techniques are extremely valuable for interpreting large volumes of data and have been successfully applied to pattern recognition and function estimation problems. In short, CI techniques provide a possible solution to the growing need for automated diagnostic systems. In the following chapters, we will examine the fundamentals of automated biomedical diagnostic systems and highlight the areas in which CI techniques have been successfully employed.

---

*http://geography.about.com/od/obtainpopulationdata/a/worldpopulation.htm

1.1   Biomedical Systems

1.1.1   Medical Data

Many types of medical data such as seen in Figure 1.1 are collected, interpreted, and archived at a hospital. Alphanumeric data are usually stored in the hospital databases and consist of details such as the patient's biodata, laboratory test results, doctors' notes, and medical histories. Image data include copies of x-rays, magnetic resonance imaging (MRI) scans, computerized tomography (CT) scans, and ultrasound images. Physiological data are also kept, such as recordings of the electrocardiogram (ECG), electroencephalogram (EEG), electromyogram (EMG), and blood pressure readings. Medical data can be stored in several forms such as binary data, which consist of simple yes/no, male/female type answers and categorical data (Hudson and Cohen, 1999), which contain answers from many options; for example, regarding a cardiac drug, the categories could be calcium channel blocker, beta blocker, or arrhythmic agent. Integer data may involve blood pressure readings whereas continuous data consists of real value measurements. Fuzzy data are collected from instruments that have error in their measurements, for example, $4.2\,\mathrm{mm}^3 \pm 0.1$. Extensive types of data include temporal data from patient histories, time series data from instrumentation recordings, and digitized image data such as ultrasound scans. Image data are composed of pixels that can be easily stored as TIFF, JPEG, or BMP file types for future analysis and diagnosis.

   Medical records have increased in size as databases of quantitative data continually grow. Computer databases for medical records have been around for the last 40 years, beginning with the first database COSTAR (computer based medical record system) (Barnett et al., 1979). Since then, other large databases followed such as PROMIS (problem oriented medical system) in the University of Vermont during 1970s. The arthritis, rheumatism and

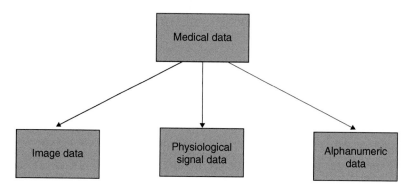

FIGURE 1.1
Categories of medical data that can be collected.

aging medical information systems (ARAMIS) system developed at Stanford University in 1972 was also based on the PROMIS structure but was designed to collect time oriented data to record a patient's recovery progress. Continued developments in technology since then have now allowed medical records to be made accessible online and to be carried around in personal digital assistants (PDAs) or smart cards.

1.1.2 Medical Instrumentation

There are several categories of medical instruments ranging from those used in surgery to the ECG for monitoring the cardiovascular system. In this book, we will be mostly concerned with medical instruments that record physiological signals including the ECG, EMG, and EEG that measure the voltage potentials on the skin. These measurements are done by placing electrodes on specific locations of the body. Nonevasive electrodes are surface electrodes, which can be placed on the skin whereas needle electrodes are evasive in the sense that they must pierce the area to obtain measurements. Besides electrodes, other sensors are also available to transduce quantities such as body temperature and blood pressure.

Detected electrical signals can be processed using a computer or simply a digital signal processor (DSP) board. The processor receives the measurement signals and then performs specific tasks such as administering intravenous drips or alerting the doctor. Other instruments such as the heart pacemaker are surgically implanted. Modern pacemakers have wireless communication capabilities, which enable them to display operating parameters or vital signs on nearby monitors. The device parameters such as amplitude of heart stimulation may also be adjusted in a similar fashion.

A major source of noise in medical instruments is due to the polarizable electrodes. When these electrodes are in direct contact with the skin, a two-layer charge develops between the electrode and the gel surface. Movement of the electrode against the electrolyte results in a change of the half-cell potential and causes distortions to the recordings made between two moving electrodes. This is known as a motion artifact, which is the result of a superimposed biosignal and is a serious measurement noise source. Some researchers have shown that the outer skin layer, stratum corneum, possesses dielectric or piezoelectric properties that increases the probability of motion artifacts (Togawa et al., 1997). Unfortunately, removal of this skin layer is not advisable due to risk of infection.

1.1.3 Medical Systems

Biomedical systems are designed to collect, process, and interpret medical data. In the early 1980s, computer programs were designed to model physical systems of the body such as the cardiovascular system and biological processes such as respiration. Unfortunately, more complex models such as the

central nervous system proved extremely difficult due to the lack of understanding in the individual workings of the subsystems. It was realized that
newer techniques, such as classifiers in pattern recognition, could contribute
greatly toward understanding these subsystems and research began in integrating CI into systems modelling.

Traditional modelling approaches had a number of weaknesses; for example, strict algorithms were inadequate in modelling biological systems since
they required precise knowledge, which was unavailable then. Lack of understanding resulted in a shortage of deterministic models, and other approaches
such as knowledge-based modelling were also inefficient. This type of method
required opinions of a small number of experts to make a diagnosis. Serious
problems arose when these experts disagreed on the diagnosis. Further, it
was discovered that generalization to other cases were inadequate since they
depended on data instances and patient physiological characteristics tended
to vary from person to person.

Nevertheless, biomedical systems have been created using several other
approaches. The symbolic approach constructs a knowledge-based system
using a group of experts. The numerical approach relies on computational
techniques such as neural networks for pattern recognition whereas hybrid
systems are birthed from two or more techniques combined to solve a single
problem. A biomedical system incorporating these intelligent techniques is
depicted in Figure 1.2.

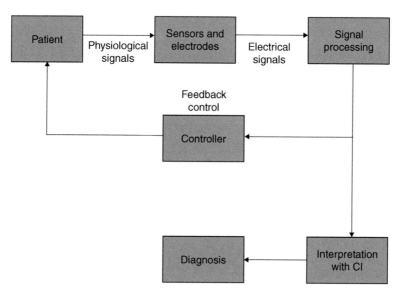

FIGURE 1.2

Flowchart diagram detailing the fundamental components in a biomedical
monitoring system.

1.2 Bioelectric Signals and Electrode Theory

Many quantities of the human body can be used to infer the state of health. These include pressure either of the direct or indirect type, flow and velocity, motion and force, temperature and evaporation, and bioelectric and biomagnetic measurements. The early chapters of this book are devoted to the processing and interpretation of bioelectric signals from the major organs in the body such as the heart, muscle, and brain. Recording and interpretation of these kind of signals require signal processing and recently developed CI methods.

The majority of bioelectric events in the human body have amplitudes less than 1 V, for example, the resting membrane potential of a cell can be as high as 0.05–0.1 V, whereas the voltages due to muscle activity are in the order of microvolts (Togawa et al., 1997). Table 1.1 depicts the voltage and frequency ranges of several commonly measured bioelectric signals. In contrast, biomagnetic signals recorded from organs are extremely weak, reaching a maximum of approximately 10^{-10} Tesla (T).

The bioelectric signal is generally measured by electrodes, which are made of various materials, for example, silver, platinum alloys. The type of electrode used varies according to the biosignal being measured as described in later chapters. For now, we will examine the electrical processes in obtaining the bioelectric signal.

The primary source of electrical signals in the body is the movement of *ions*, which are positively or negatively charged particles. The amount of ions in a unit solution is measured by its *concentration*. If two solutions of different concentration are separated by a barrier, say a semipermeable membrane, a *concentration gradient* forms, which forces ions to move from the higher concentration to lower concentration solution. Flow of ions across a membrane along the concentration gradient is referred to as *diffusion*. If diffusion is allowed to occur uninterrupted, the concentration gradient in both solutions

TABLE 1.1

Bioelectric Measurement Techniques and the Respective Voltage and Frequency Ranges

Bioelectric Measurements	Signal Range (mV)	Frequency Range (Hz)
Electrocardiography (ECG)	0.5–4	0.01–250
Electromyography (EMG)	Variable	d.c.–10,000
Electroencephalography (EEG)	0.001–0.1	d.c.–150
Nerve potential	0.01–3	d.c.–10,000
Electrooculography (EOG)	0.005–0.2	d.c.–50

Source: Adapted from Togawa, T., Tamura, T., and Oberg, P.A. in *Biomedical Transducers and Instruments*. CRC Press, Boca Raton, FL, 1997.

finally acquires the same concentration and is said to be in *equilibrium*. Ion flow against the concentration gradient can be achieved by an active pump, which requires energy, a process known as *active transport*. Ions being charged particles are also influenced by electric potentials, that is, positive potentials attract negative ions and vice versa. The resultant ion concentration (M^+) is therefore a function of both charge and concentration gradient and can be described by the Nernst potential:

$$E_M = \frac{RT}{nF} \ln(M^+) + E_0 \qquad (1.1)$$

where R is the universal gas constant, T the absolute temperature, n the ion valence, F the Faraday constant, and E_0 some constant. An ionic solution is sometimes referred to as an *electrolyte* because it is used to conduct charges between two electrodes.

Ions in the human body can be divided into majority and minority carriers. Majority carriers are responsible for carrying most of the electrical charge and include ions such as sodium (Na^+), chloride (Cl^-), and potassium (K^+) (usually at higher frequencies, such as in the muscle). These ions are present in high concentrations in the human body. Minority carriers are other ions present in lower concentration and do not contribute significantly to carrying charges. Ions can form interfaces between metal electrodes, between electrodes and electrolytes, and between electrolytes and membranes. The electrode conducts charges from the ionic solution to a metal and can then be recorded and stored.

When a metal is placed in an electrolyte, double layers are formed near the electrode surface, a net current results, which passes from the electrode to the electrolyte. The current consists of electrons and negative ions (anions) moving in opposite direction to the current as well as positive ions (cations) moving in the same direction of the current. These can be written as the following chemical equations:

$$C \leftrightarrows C^{n+} + ne^-$$
$$A \leftrightarrows A^{m-} - me^-$$

where n is the valence of C and m the valence of A. Assuming that the electrode is made of the same material as the cations in the electrolyte, the metal atoms at the surface of the electrode will give off electrons and form cations. The anions in the interface will acquire these electrons to form neutral atoms. The process occurs even if no current is present albeit at a lower rate and is reversible if external current is applied. The local concentration of the cations in the interface now changes affecting the anion concentrations. This difference in concentrations results in the electrode possessing a different electrical potential from the rest of the solution. The potential difference between the metal and the solution is then known as the *half-cell potential*. It is not possible to measure the potential of a single electrode with respect to the solution. Potential difference measurements are generally made with respect to the standard hydrogen electrode comprising a platinum black electrode in contact with a solution of hydrogen ions in dissolved hydrogen gas. Most electrodes for measuring bioelectric signals are designed based on this principle.

1.3 Signal Processing and Feature Extraction

Biosignals form the raw data that are collected from medical instrumentation such as ECG, EEG, and EMG. These analog signals are often converted into digital signals for storage. In cases where the signals are small in amplitude, some form of amplification is performed using differential amplifiers. Filtering is then applied to retrieve a particular signal frequency band and to reduce noise. Currently, fundamental signal preprocessing and digitization of these analog signals are automatically performed by the measuring instruments. Digitization is also a step toward transforming the information in the signal to a form, which can be manipulated by the computer.

The next phase in interpretation of the recorded biosignal is feature extraction, which is used to identify important and special characteristics in the digital signal. These "features" are associated with a particular pathology, motion, or state of the body. However, note that it is very common to have a feature shared by two or more disease types. A natural question, which arises then, is how to select the best features, the answer to which is as yet not definitive. Research in all fields of medicine currently seeks better feature extraction methods with the goal of obtaining better data representations of the subject of interest. As a result, we now possess a plethora of feature extraction methods, which have emerged from various efforts in the signal-processing field. In this book, we will try to highlight the common feature extraction methods and compare them to illustrate the importance of selecting the correct features for detection and diagnosis of human diseases.

Feature extraction methods focus on different properties of the signal in time and frequency domain. The signal waveform can be easily described and graphed using any plotting software, for example, Matlab. The shape or "morphology" of the waveform is the first step in finding clues of the pathology involved. If the waveform is periodic and of constant morphology, that is, approximately the same shape and repetitive, one may decide to measure morphological details such as waveform spike height, period, number of spikes and troughs, and distance between spikes. Signal-processing techniques such as Fourier series decomposition and power spectra analysis are more advanced techniques to extract information that cannot be otherwise measured directly. Comparison of these properties between healthy people and patients is sometimes adequate to determine the type of disease if not at least the presence of it. More complex signals with varying frequencies can be investigated using the popular wavelet decomposition technique.

1.4   Computational Intelligence Techniques

In practical applications, the extracted features of the problem are rarely able
to fully describe the pathology. This problem arises particularly from using
a suboptimal or redundant set of features to describe the problem. Instead
of looking for better features, we assume that the extracted features are nonlinearly related to the inputs and outputs of a biomedical system. For example, automatic diagnostic systems for detecting heartbeat irregularities using
ECG waveforms use the processed ECG data as inputs. The aim is to learn
the relationship between this information and the corresponding pathology,
so that when new data are available, the system would recognize the correct
pathology. Such tasks can be accomplished by using CI techniques.

There are three disciplines, which form the core of today's applied CI technology, although many other interesting CI concepts such as unsupervised
learning, reinforcement learning, symbolic learning, and cognition sciences
(Shavlik and Dieterich, 1990) are still being researched on. Figure 1.3 depicts
the relationship between the three core disciplines, which emerged around
the same era; supervised learning beginning with the perceptron (Rosenblatt,
1958), fuzzy logic (Zadeh, 1965, 1973; Zadeh et al., 2002), and genetic algorithms (Barricelli, 1954; Fraser, 1957).

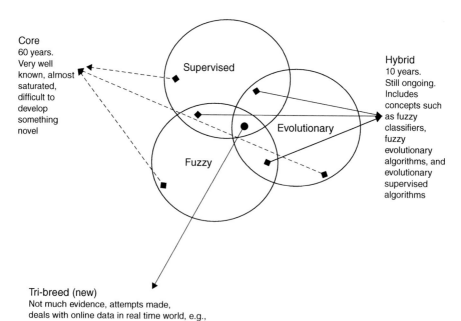

FIGURE 1.3

The main computational intelligence paradigms and the state of current
research.

1.4.1 Supervised Learning

The most common CI technique in use is supervised learning, which involves constructing a classifier or regressor from a set of data. This construction process is commonly known as *training* and the data set is known as the *training set*. In supervised learning, a supervisor designates class labels for the data and presents it to the machine to train it. For example, in the diagnosis of myocardial infarctions (heart attacks), the supervisor would be a cardiologist who examined ECG readings and labeled them as healthy or belonging to someone suffering from possible myocardial infarction. This form of learning is commonly used in pattern recognition applications and is the most widely researched and understood area, examples being neural networks and support vector machines.

1.4.2 Fuzzy Logic

In standard mathematics, values in a set have fixed and defined numbers. For example, a set $\mathbf{A} = \{x, y, z\}$ in the normal sense contains three elements x, y, and z, which take specific values such as $x = 1, y = 2$, and $z = 8$. In fuzzy set theory, the concept of defined values for a set is changed to accommodate elements with uncertainties. Quantities such as x, y, z could belong to two or more sets with their degree of belongingness defined by a membership function. An element x will then have different membership values indicating the closeness of the element to a particular set and is referred to as a fuzzy element. In fuzzy logic, variables need not be distinct values but can take on abstract measurements described in linguistic terms. For example, rather than stating the patient's heart beat is 60 beats/min, we can assign terms such as the heart beat is *fast, moderate,* or *slow.* The heart beat is now treated as a linguistic variable described in these linguistic terms. Fuzzy logic is thus an extension of hard rules, which use thresholding, where rather than the familiar "If $x = 5$, then $y = 10$," we use rules such as "If heart beat is fast, then disorder is ventricular tachycardia."

1.4.3 Evolutionary Algorithms

Evolutionary algorithms refer primarily to optimization algorithms that are based on the evolution of biological systems and include algorithms such as genetic algorithms, particle swarm optimization, and ant colony optimization algorithms. Genetic algorithms perform optimization based on genetic evolution. The idea here is to divide the main problem into smaller problems and solve them individually. The subproblems or subsolutions are treated as genes and the iterative step toward the main objective is made by retaining only the good genes while discarding the suboptimal solutions. These genes are either combined according to a certain fitness function or randomly to simulate evolution. Other evolutionary elements such as "mutation" are introduced to avoid the algorithm terminating at local optimum points. Particle

swarm and ant swarm optimization are similar in concept; they use observations on how swarms work collectively to achieve a global objective. In
these techniques, a global fitness function or objective is defined along with
several local objectives. These local objectives depict members of a swarm
and are continuously updated according to certain criteria so that the swarm
eventually discovers the optimal solution. One must define the fitness function, local objectives, local update rules, and elements of the swarm. The
main drawback, however, is algorithm stability and convergence, which is
difficult to guarantee at times. In Chapter 3, we describe several CI techniques in detail but have not included the topic of evolutionary algorithms
because there are very few biomedical applications that have applied them.
There is nevertheless scope for their application, and the interested reader
is referred instead to some of the more authoritative references (Fogel, 1995;
Sarker et al., 2002) for applying this CI technology to problems in biomedical
engineering.

1.5   Chapter Overview

Chapter 2. This chapter provides an introductory background and begins with
the definition of signals before going on to review the theory of underlying
signal-processing techniques. Of particular importance are digital signals now
commonly used in CI because they are obtained from analog biosignals via
sampling. Incorrect sampling can lead to the introduction of noise and aliasing errors. Noise is undesirable because it corrupts the signal and results
in information loss whereas aliasing errors result in erroneous digital representations of the original signal. Noise can be removed by filters, which
enhance the signal and attenuate the noise amplitudes. Aliasing errors can
be circumvented by using appropriate sampling rates following the Nyquist
sampling theorem. Standard analog filters use passive components such as
resistors whereas digital filters can be implemented using computer algorithms. If noise exists in the same frequency band as the desired signal,
adaptive filter technology can be employed to filter the noise. Signal information can also be extracted using signal transforms, which fundamentally
represent the signal as a sum or a series of basis functions. Some of the popular transforms, we look at include the Fourier transform, wavelet transform,
and discrete cosine transform. These transforms are important in providing
characteristic information of a signal and can also be employed as compression methods allowing for a more concise signal representation and signal
reconstruction.

Chapter 3. In this chapter, we review the CI techniques used in several biomedical applications. These techniques include neural network models such as the
multilayered perceptron and self-organizing map, support vector machines
(SVM), hidden Markov models (HMM), and fuzzy theory. CI techniques are

primarily employed for solving problems such as function estimation where no direct mathematical model can be built to mimic the observed data. Recent biomedical problems usually use classifiers to assist in diagnosis of a particular disease or disorder. Classifiers generally require training data, which are composed of labeled examples from which a separating or decision surface is constructed to classify unseen data. In addition, we examine the theory behind a selection of techniques, which have been chosen for the diversity of their application. Along the way, we will mention the various issues that are pertinent to the successful implementation of these techniques, the algorithms that can be used to train them, and references to software implementations of these techniques.

Chapter 4. One of the major organs of the human body is the heart, which preserves us by pumping oxygen carrying blood to the organs. The heart forms the core of the cardiovascular system that can be monitored using the ECG. ECGs measure the net action potentials generated from the mechanical pumping action of the heart during every cardiac cycle. Clinicians have used this close relationship between ECG and cardiac action to monitor the heart's health. Cardiovascular disorders such as arrhythmias (abnormal heart beats) and myocardial infarctions (heart attacks) have been detected by cardiologists from ECG readings. A contemporary research is devising methods to automatically detect heart diseases from ECG inputs using CI techniques. We will show how the ECG is recorded, preprocessed, and used in conjunction with CI techniques such as neural networks to diagnose certain cardiovascular disorders. Of special interest is the accuracy in correctly identifying the disorder and proper feature selection for improved classification. An inherent problem is the considerable amount of ECG variation observed between patient age groups in waveform characteristics such as amplitude and period. This is not surprising, given that children have generally faster heart beats of smaller amplitude whereas the elderly have slower heart rates. This means that better processing methods for feature extraction are required to obtain more independent and robust features in the design of an automated diagnostic system.

Chapter 5. Our ability to move is largely due to the neuromuscular system, which is primarily responsible for providing control nerve signals and actuation. Disorders in this system range from motor neuron diseases to diseases of the neuromuscular junction and diseases within the muscle fibers. The motor unit action potential (MUAP) is generated by muscle fibers during contraction and can be measured by an EMG. One serious complication is that each muscle group contains potentially thousands of muscle fibers each generating action potentials. Depending on the type of EMG collected (needle, wire, or surface), the final waveform will be a superposition of MUAPs with some suffering dispersion effects as they propagate through fat or tissue although others are affected by electrode movements that cause motion artifacts. Automated diagnostic systems face an enormous task of first decomposing the EMG into individual MUAP waveforms or simpler waveforms. This process requires considerable filtering and signal processing to acquire waveforms characteristic of the disorder. Of late, another approach has been adopted where signal processing is minimally done and classifiers used to classify the composite waveform itself. This direction poses several intricate challenges, the primary one being that the superimposed MUAP waveforms are not unique to the disorder, and using them as such will lead to erroneous diagnoses. This chapter will examine how the composite MUAPs are decomposed into simpler waveforms, how feature extraction and information processing is done, and also how classification is undertaken. In addition, CI application in the development of prostheses and exoskeletons are also examined. Here, EMG signals are used as primary input signals to a classifier to detect the intention of a particular user to move.

Chapter 6. The brain is the main control centre of our central nervous system where generated control signals are used to execute bodily functions. The electrical activity relating to the brain's function can be picked up by placing electrodes on the scalp, which is commonly known as EEG. EEG signals vary according to changes in brain functions, for example, due to neurological conditions. CI techniques play a major role in the processing and analysis of EEG signals and in Chapter 6, we focus on EEG signal acquisition and analysis, and in particular CI-assisted major applications involving the detection and classification of abnormalities in EEG signals. Furthermore, we provide an overview of EEG-based communication, widely known as BCI or the brain–computer interface, which holds many potential applications such as helping the disabled to restore functional activities.

Chapter 7. It focuses on gait- and movement-related signals, commonly used biomechanical and motion analysis approaches, and the key features that are extracted from movement patterns for characterizing gait. We briefly discuss some of the significant application areas that have benefited as a result of the usage of CI methods, such as the detection of tripping falls in the elderly. Analysis of gait or human movement offers interesting insights into the types of information that can help the diagnosis and treatment of ambulatory disorders. In analyzing gait, both statistical and machine-learning techniques are used by researchers to assess movement patterns. The applications related to motion analysis are extensive: ranging from clinical, rehabilitation and health improvements to sports performance.

Chapter 8. In the final chapter, we review briefly the major areas in biomedical engineering, which have benefited from CI applications and the significant developments to date. Much of the research in CI for biomedical engineering has focused on classifying disorders based on pathological characteristics. CI techniques are gradually becoming more and more effective in solving a variety of problems, in particular those in which human intervention might be replaced by a machine-based automated response. However, to bring CI to the forefront of healthcare technology, several fundamental challenges remain to be addressed. Some of these challenges would require contributions from the machine-learning community as these are regarded as the core technologies that drive these techniques. Finally, we discuss the future developments in biomedical engineering and how CI could play a role in them. Some examples include the development of wireless healthcare systems and the use of sensor networks and intelligent information techniques.

1.6 Book Usage

This book can be used in a variety of ways. Chapters 2 and 3 contain background material that would be useful for comprehending the following chapters. Selected topics in these chapters may be used together with material from one or more other chapters for undergraduate and postgraduate courses in biomedical engineering. For biomedical researchers, we hope that this book provides a review of the advances and remaining challenges in the application of CI to problems in the biomedical field. We have tried to present the material such that each chapter can be independently read and we hope that the readers will find it useful and enjoy the material to read as much as we have enjoyed presenting it.

References

Barnett, G., N. Justice, M. Somand, J. Adams, B. Waxman, P. Beaman, M. Parent, F. Vandeusen, and J. Greenlie (1979). COSTAR—a computer-based medical information system for ambulatory care. *Proceedings of the IEEE 67(9)*, 1226–1237.

Barricelli, N. A. (1954). Esempi numerici di processi di evoluzione. *Methodos*, 45–68.

Fogel, D. (1995). *Evolutionary Computation: Toward a New Philosophy of Machine Intelligence*. New York: IEEE Press.

Fraser, A. (1957). Simulation of genetic systems by automatic digital computers I. Introduction. *Australian Journal of Biological Sciences 10*, 484–491.

Hudson, D. L. and M. E. Cohen (1999). *Neural Networks and Artificial Intelligence for Biomedical Engineering*. IEEE Press Series in Biomedical Engineering. New York: Institute of Electrical and Electronics Engineers.

Rosenblatt, F. (1958). The perceptron: a probabilistic model for information storage and organization in the brain. *Psychological Review 65*, 386–408.

Sarker, R., M. Mohammadian, and X. Yao (Eds) (2002). *Evolutionary Optimization*. Boston, MA: Kluwer Academic.

Shavlik, J. and T. Dietterich (1990). *Readings in Machine Learning*. The Morgan Kaufmann series in machine learning. San Mateo, CA: Morgan Kaufmann Publishers.

Togawa, T., T. Tamura, and P. A. Oberg (1997). *Biomedical Transducers and Instruments*. Boca Raton, FL: CRC Press.

Zadeh, L. (1965). Fuzzy sets. *Information and Control 8*, 338–353.

Zadeh, L. (1973). Outline of a new approach to the analysis of complex systems and decision processes. *IEEE Transactions on Systems, Man, and Cybernetics 8*, 28–44.

Zadeh, L., T. Lin, and Y. Yao (2002). *Data Mining, Rough Sets, and Granular Computing*. Studies in Fuzziness and Soft Computing, vol. 95. Heidelberg: Physica-Verlag.

2

Biomedical Signal Processing

2.1 Introduction

An important aspect of biomedical engineering is measuring and interpreting biosignals originating from almost every system in the body, from the nervous system to the endocrine system and the brain. These signals are electrical signals that result from action potentials in our cells, the nervous system being the primary example. It is also possible to measure quantities such as blood pressure or frequency of a heart valve opening and closing, and treat them as biosignals. The correct measurement and interpretation of signals is vital in assisting physicians with diagnosis, patient monitoring, and treatment, for example, heart bypass surgery requires stringent monitoring of blood oxygenation levels to avoid tissue damage. As such, signal-processing techniques have become important in biomedical engineering to interpret the information and to present it effectively.

In this chapter, we review the various aspects of signal processing, beginning with the definition of a signal followed by signal transforms, basic power spectral estimation, filters, and parametric models. These topics are covered in greater detail in other sources (Akay, 1994; Banks, 1990; Clarkson, 1993; Oppenheim et al., 1999; Stein, 2000; Stergiopoulos, 2001). We have limited the mathematics because most of the techniques described later in the chapter have been implemented in commercially available software or can be downloaded from the Internet.

2.2 Signals and Signal Systems

A signal can be represented in both time and frequency domains. By this, we mean that we can specify the signal as a progression of values through time or a variation in the frequency range. An analog signal is usually continuous in time and represented by a smooth variation in voltage, for example, ECG, EEG, and electrooculogram (EOG). A digital signal is discrete in time and represented by a sequence of quantized voltage levels. The levels correspond to discrete intervals of time and often obtained by sampling an analog signal.

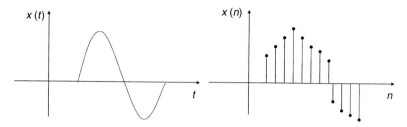

FIGURE 2.1

Left: A time continuous analog signal, $x(t)$. *Right*: A discrete sequence or digital signal, $x(n)$.

Digital signals are important for storage and transfer of information from an analog signal to some form of hardware or computer.

An analog signal is often written as $x(t)$, where the function denotes a smooth variation of voltage with time t. A digital signal is represented as a sequence $x(n)$, where $-\infty < n < \infty$ and n is an integer, that is, $n \in Z$. These two types of signal are depicted in Figure 2.1 where it should be noted that one type of signal can be derived from the other. Periodic signals such as heart rate, for example, have a constant variation in voltage over fixed intervals and are said to be *deterministic*. A signal that does not have a fixed period is random, for example, signal activity in the spinal cord. A purely random signal is indeterministic because the voltage variations cannot be predicted with absolute accuracy.

2.2.1   Linear Shift Invariant Systems

There are many forms of deterministic or periodic signals; the two most common are the unit impulse function and the unit step function. The unit impulse function (Figure 2.2) is defined as

$$\delta(n) = \begin{cases} 1 & \text{for } n = 0 \\ 0 & \text{for } n \neq 0 \end{cases} \tag{2.1}$$

The unit step function (Figure 2.3) is defined as

$$h(n) = \begin{cases} 1 & \text{for } n \geq 0 \\ 0 & \text{for } n < 0 \end{cases} \tag{2.2}$$

The energy of a signal is estimated by

$$E = \sum_{n=-\infty}^{\infty} |x(n)|^2 \tag{2.3}$$

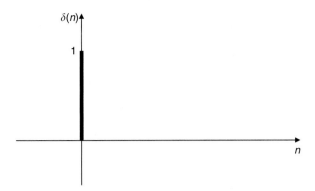

FIGURE 2.2
A unit impulse function, $\delta(n)$.

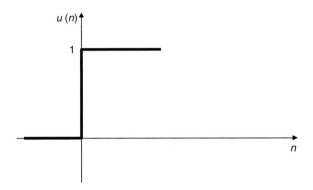

FIGURE 2.3
A unit step function, $u(n)$.

whereas the mean energy or mean power of a digital signal having N sequences is

$$E_{\text{mean}} = \lim_{n \to \infty} \frac{1}{N} \sum_{n=-\frac{N}{2}}^{\frac{N}{2}} |x(n)|^2 \qquad (2.4)$$

It is usual to define a transformation operator by which an input signal $x_1(n)$ is mapped to an output signal $x_2(n)$. The transformation operator $\phi(\cdot)$ defines the transfer characteristics of the system and is commonly known as the transfer function. Figure 2.4 depicts system components and the transformation of inputs to outputs through this transformation operator. Mathematically, we have

$$x_2(n) = \phi[x_1(n)] \qquad (2.5)$$

A system is said to be a *linear* system if the following principle of superposition is satisfied:

$$\phi[ax_1(n) + bx_2(n)] = a\phi[x_1(n)] + b\phi[x_2(n)] \qquad (2.6)$$

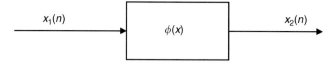

FIGURE 2.4
A signal system consisting of one input signal, which is transformed into the output.

where $x_1(n)$ and $x_2(n)$ are different inputs to the system. A *shift invariant* system is a digital system, which maps $x_1(n)$ to $x_1(n-d)$, and $x_2(n)$ to $x_2(n-d)$, where d is a delay or shift in the digital sequence. Since n is the indices for the sequence number, it can be thought of as time for a discrete or digital system. The analog version of this system is the *linear time invariant* system. Now let the impulse response of a system be written as

$$h = \phi(\delta_0) \tag{2.7}$$

which is the output of the system due to an impulse excitation. Any output signal $x_2(n)$ can be written in terms of the input signal $x_1(n)$ as follows:

$$x_2(n) = \sum_{k=-\infty}^{\infty} x_1(k)\delta_0(n-k) \tag{2.8}$$

Then the output of a general linear shift invariant system can be represented by

$$x_2(n) = \sum_{k=-\infty}^{\infty} x_1(k)h(n-k) \tag{2.9}$$

that is, the linear shift invariant system is characterized completely by its impulse response. Equation 2.9 can also be written as a *discrete convolution* sum:

$$x_2(n) = x_1(n) \times h(n) \tag{2.10}$$

Convolution is a technique used to represent the output of a signal in terms of the input and transformation operator. The convolution operator is commutative, associative, and distributive, that is, the following properties are satisfied:

$$x_1(n) \times [x_2(n) \times x_3(n)] = [x_1(n) \times x_2(n)] \times x_3(n)$$
$$x_1(n) \times [x_2(n) + x_3(n)] = [x_1(n) \times x_2(n)] + [x_1(n) \times x_3(n)]$$

Further details on how to compute the convolution for simple signals can be found in Akay (1994).

2.2.2 Sampling and Analog Reconstruction

Many digital systems are obtained by sampling an analog system to obtain discrete values at fixed time intervals, usually denoted by T. The motivation for digitization includes greater functionality and reproducibility, more modularity and flexibility, better performance (Stein, 2000), and convenient storage. Digitization requires proper signal representation so that minimal signal information is lost, hence the goal of sampling is to obtain a digital representation of the analog signal accurate enough to reconstruct the original. Sampling can be performed by passing the analog signal $x(t)$ through a switch that allows certain points of the analog waveform through when turned on. The sampling frequency can be controlled by a control signal (Figure 2.5), which controls the frequency of the switch operation. Since analog signals can have a range of frequencies the sampling has to be performed carefully. For example, if the analog signal has low frequency, then we would sample it at larger time intervals and vice versa. Unfortunately, the disadvantage of digitization is the introduction of noise or high-frequency components in the power spectra of the signal.

To minimize this, the input analog signal $x(t)$ is bandlimited with cutoff frequency f_C. The sampling theorem then states that the sampling frequency, f_{SR}, should be at least twice the cutoff or maximum frequency, f_C, that is,

$$f_{SR} \geq 2f_C \qquad (2.11)$$

The minimum sampling rate is known as the Nyquist rate, where the Nyquist frequency is defined as

$$f_{NR} = \frac{f_{SR}}{2} \qquad (2.12)$$

so that the Nyquist intervals for a signal frequency f are then uniformly defined as

$$\frac{f - f_{SR}}{2} < f < f + \frac{f_{SR}}{2} \qquad (2.13)$$

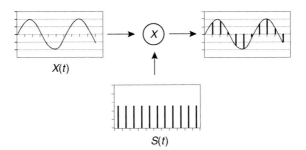

$X(t)$

$S(t)$

FIGURE 2.5
Analog to digital signal conversion using sampling where switch X is turned on and off by the control signal $S(t)$. (Adapted from Winter, D.A. and Palta, A.E., *Signal Processing and Linear Systems for the Movement Sciences*, Waterloo Biomechanics, Waterloo, Canada, 1997.)

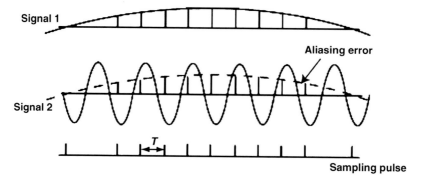

FIGURE 2.6

Aliasing error due to low-sampling frequency. (Adapted from Winter, D.A. and Palta, A.E., *Signal Processing and Linear Systems for the Movement Sciences*, Waterloo Biomechanics, Waterloo, Canada, 1997.)

This interval is important because if the sampling rate is too low, interference effects known as *aliasing* or frequency folding begin to happen. Figure 2.6 shows the case of two signals with different frequencies being sampled at the same rate. Signal 1 is adequately sampled at more than twice the signal frequency and the samples are representative of the signal but in signal 2, the sampling rate is less than half the signal frequency, which causes important characteristics to be missed. As a result, digital signal 2 ends up with a lower frequency than the original analog signal 2. This is known as *aliasing error* and occurs when insufficient samples of the analog signal are taken, causing erroneous digital representation. To avoid these effects, the analog signal is preprocessed using a low-pass filter with cutoff $f_{SR}/2$, so that higher-order frequencies are removed; this may also be viewed as sampling a subband frequency of the intended signal.

Analog reconstruction is the opposite of digitization with the aim to recover the original time continuous signal given a sequence of discrete values. To do this, the digital signal is low-pass filtered with cutoff frequency equal to the Nyquist frequency. This removes the higher frequency components and smoothens the gaps between digital samples. Using Shannon's sampling theorem, the reconstructed analog signal can be expressed in terms of the impulse response of the reconstructor $h(nT)$ and the digital sequence $x_1(nT)$ as

$$x_1(t) = \sum_{n=-\infty}^{\infty} x_1(nT)h(t - nT) \tag{2.14}$$

The ideal reconstructor may be difficult to implement in practice because its impulse response is not entirely causal.

2.2.3   Causality and Stability

A system is said to be *causal* if the output at any time depends only on inputs up to that time and not after. We can expand the linear shift invariant system of Equation 2.9 as follows:

$$x_2(n) = \cdots + x_1(1)h(n-1) + x_1(2)h(n-2) + \cdots$$

and note that this system is causal if and only if

$$h(n-k) = 0 \quad \text{for } k < n$$

In other words, the system is causal if the unit step response is zero for $k < n$. A signal is bounded if for some $\alpha \in \Re$ and for all $n \in \mathbb{Z}$ we have

$$|x(n)| \leq \alpha \tag{2.15}$$

and α is independent of the time n. The linear time invariant system is stable (proof see, for example, Banks, 1990) if and only if

$$\sum_{k=-\infty}^{\infty} |h(k)| < \infty \tag{2.16}$$

This implies that the outputs of the system are bounded for all bounded input sequences.

2.2.4 Noise

Most signal processing involves extracting the desired signal from some noisy signal, where noise is regarded as an unwanted signal corrupting the desired signal. In biomedical engineering, the main source of noise is the 50/60 Hz AC mains power supply that interferes with signals such as those from surface electrodes. Knowledge of the type of noise is important because we may decide to remove it by filtering or even use it for simulations.

Noise is often regarded as unwanted signals, which occur randomly. This is known as *pure noise*, which is entirely composed of probabilistic elements that are completely random and cannot be predicted accurately. Most signals, which we label as noise are not pure noise, for example, background radiation is pure noise but the mains power is not because its frequency is known. Noise signals are "stationary" if the probabilistic characteristics of the signal do not change with time, for example, constant mean or variance, otherwise it is "nonstationary." *White noise* refers to signals with a flat spectrum, which is constant and independent of frequency; independence implies that given the complete history of the signal, we cannot predict the future signal characteristics. If this noise follows a Gaussian distribution, it is referred to as Gaussian white noise; the converse is *coloured noise*, a signal whose spectrum is confined to part of the frequency range. An example of coloured noise is a bandpassed signal with frequencies outside the passband removed. It may be more useful to consider noise as consisting of several classes of unpredictable signals, which include pseudorandom signals, incompletely known signals, chaotic signals, and genuinely stochastic signals (Stein, 2000).

Pseudorandom signals are generated using some complex and unknown algorithm, so that the generated signals do not appear uniform or even periodic. Pseudorandom signals are deterministic and are usually generated to behave as noise in simulated systems, for example, simulation of reliability in telecommunications channels. One popular algorithm for pseudorandom generation is the linear recursion method due to D.H. Lehmer in 1951 (Stein, 2000). The algorithm performs the following integer recursion starting with some randomly selected x_0:

$$x(n + 1) = (\lambda x(n) + \gamma) \bmod a \tag{2.17}$$

where λ is some large number and both γ and a are taken to be prime numbers. If these parameters are not chosen correctly, then the algorithm becomes ineffective. It can be seen that this algorithm generates a periodic sequence due to a being finite. Another pseudorandom generator requires extra memory and is given as

$$x(n + 1) = (x(n - j) + x(n - k)) \bmod a \tag{2.18}$$

where j, k, and a are carefully selected. The values j and k are maintained in a buffer and retrieved when needed. These algorithms are used for generating uniformly distributed noise, but other applications may require noise drawn from a particular distribution, for example, Gaussian distribution. Gaussian noise can be generated by the Box–Muller algorithm, which first defines a random point on a circle and then generates the random numbers. The second type of noise is incompletely known signals, which though are completely deterministic arise from an algorithm with variables not fully known.

Chaotic signals in contrast are generated by an algorithm, which is completely known, but as the variables of the algorithm change, the signal approaches a region where it cannot be predicted precisely. This could occur, for example, due to the numerical sensitivity of the algorithm, which results in loss of information because of rounding errors. The study of chaotic signals is more concentrated toward explaining the variable ranges, which cause the chaotic behavior and how this behavior persists. The final type of noise is pure noise or genuinely stochastic noise, which includes signals from background radiation or radioactive decay, which are not generated by any algorithm, and are completely indeterministic.

2.3   Signal Transforms

In this section, we examine several signal transforms that have been used mainly for improving correlation, convolution, and other spectral methods. The Karhunen–Loeve transform is ideal in being easily computed, having compression capabilities, and minimum square error. Unfortunately it is not practically realizable, hence the need for other algorithms, which approximate it or perform the transform differently. In the following sections, we examine several different transforms, which have been well used for this.

2.3.1 The Fourier Transform

In previous sections we represented the signal, whether analog $x(t)$ or digital $x(n)$, in terms of values, which occur at different integer intervals. This is known as the *time-domain* representation of the signal. The Fourier transform is a way of representing the signal in terms of frequencies, thereby transforming the signal into the *frequency-domain* representation. The crux of the Fourier transform is that any signal can be represented as a sum of sinusoids of different frequencies as long as the signal is of finite amplitude.

For a continuous time signal $x(t)$, the Fourier transform is given by the integral

$$X(j\omega) = \int_{-\infty}^{\infty} x(t)e^{-j\omega t}dt \tag{2.19}$$

and $x(t)$ can be recovered by the inverse Fourier transform

$$x(t) = \frac{1}{2\pi} \int_{-\infty}^{\infty} X(j\omega)e^{j\omega t}d\omega \tag{2.20}$$

The discrete Fourier transform (DFT) for digital signals $x(n)$ is similar to the transform for the continuous analog signal $x(t)$, except that the angular frequency ω is divided into N equal parts on the interval $[0, \omega]$. The transform for the digital sequence is then a summation of N parts written as

$$X(k) = \sum_{n=1}^{N} x(n)e^{-j\omega_k n} \tag{2.21}$$

where the angular frequency $\omega_k = 2\pi k/N$ for $k, n = 1, \ldots, N$. The inverse discrete Fourier transform is just

$$x(n) = \frac{1}{N} \sum_{k=1}^{N} X(k)e^{j\omega_k n} \tag{2.22}$$

Note here that both $x(n)$ and $X(k)$ are complex digital signals and the transform is sometimes known as the N-point DFT due to the number of frequencies. Since there are N samples $X(k)$ each involving N summations, calculating the DFT requires N^2 operations, which becomes an increasingly slow operation as the number of samples increases.

2.3.2 Fast Fourier Transform

One method to increase the efficiency in calculating the DFT is the fast Fourier transform (FFT) algorithm described by Cooley and Tukey (1965).

Their algorithm makes use of the fact that $e^{-2\pi j/N}$ is symmetrical and the number of computations can be reduced from N^2 to $N \log_2 N$. The original algorithm is applicable only when N is an integer power of 2. There are, however, more recent algorithms, which improve on the original algorithm (Stein, 2000; Oppenheim et al., 1999) by decomposing the DFT into smaller Fourier transforms and are known as *decimation in time* and *decimation in frequency* algorithms. We describe the decimation in time algorithm here because the decimation in frequency shares the same principle with the exception that it is in frequency domain. Further details can be found in references such as Oppenheim et al. (1999) and Stein (2000).

The decimation in time algorithm decomposes the time sequence $x(n)$ into two terms, one containing the "even" terms and the other the "odd" terms corresponding to values of n. The DFT sequence is usually rewritten as

$$X(k) = \sum_{n=1}^{N} x(n) W_N^{nk} \tag{2.23}$$

for $n = k = 1, \ldots, N$ and $W_N^{nk} = e^{-j\omega_k n}$. Now if $N = 2^p$, we can rewrite Equation 2.23 as

$$X(k) = \sum_{n=2j}^{N} x(n) W_N^{nk} + \sum_{n=2j+1}^{N} x(n) W_N^{nk} \tag{2.24}$$

for $j = 1, \ldots, \frac{N}{2}$ or equivalently as

$$X(k) = \sum_{n=1}^{\frac{N}{2}} x(2n) W_N^{2nk} + \sum_{n=1}^{\frac{N}{2}} x(2n+1) W_N^{(2n+1)k}$$

$$= \sum_{n=1}^{\frac{N}{2}} x(2n) W_N^{2nk} + W_N^k \sum_{n=1}^{\frac{N}{2}} x(2n+1) W_N^{2nk} \tag{2.25}$$

This scheme works principally for even sequence lengths but if the sequence length is odd, an augmenting zero term is usually added to make it even. Equation 2.25 can be further reduced to

$$X(k) = X_e(k) + W_N^k X_o(k) \tag{2.26}$$

where $X_e(k)$ is the DFT of the even terms and $X_o(k)$ the DFT of the odd terms. This algorithm is also known as the Cooley–Tukey algorithm and is best illustrated by a sample computation of DFT for eight samples.

Let the discrete signal X contain the following samples:

$$X(k) = \{x_0, x_1, x_2, x_3, x_4, x_5, x_6, x_7\}$$

Then form the DFT as in Equation 2.26 to obtain

$$X(k) = X_{e1}(k) + W_N^k X_{o1}(k) \tag{2.27}$$

where we have

$$X_{e1}(k) = \text{DFT} \quad \text{for terms } x_0, x_2, x_4, x_6$$
$$X_{o1}(k) = \text{DFT} \quad \text{for terms } x_1, x_3, x_5, x_7$$

Note now that the 8-point DFT has been decomposed to computing two 4-point DFTs, which we can write as follows:

$$X_{e1}(k) = X_{e2}(k) + W_{\frac{N}{2}}^{k} X_{o2}(k)$$
$$X_{o1}(k) = X_{e3}(k) + W_{\frac{N}{2}}^{k} X_{o3}(k)$$

where now

$$X_{e2}(k) = \text{DFT} \quad \text{for terms } x_2, x_6$$
$$X_{o2}(k) = \text{DFT} \quad \text{for terms } x_0, x_4$$
$$X_{e3}(k) = \text{DFT} \quad \text{for terms } x_1, x_5$$
$$X_{o3}(k) = \text{DFT} \quad \text{for terms } x_3, x_7$$

These four terms now represent 2-point DFTs and can be decomposed as before. We show one decomposition here for the term $X_{e2}(k)$:

$$X_{e2}(k) = x_2 + W_{\frac{N}{4}}^{k} x_6 \tag{2.28}$$

Hence, the full 8-point DFT can be computed by decomposing the main DFT into smaller DFTs and using them as recomposition equations.

2.3.3   The Wavelet Transform

The wavelet transform (WT) is essentially a correlation between a certain function $\psi(t)$ and signal $x(t)$ at different frequencies (Mallat, 1999). The function $\psi(t)$ is known as the *mother wavelet*, which can be written generally as

$$\psi(t) = \frac{1}{\sqrt{s}} \psi\left(\frac{t-u}{s}\right) \tag{2.29}$$

where s represents a scale parameter and u a translation parameter. Observe that varying these parameters produces functions with different frequencies and center locations in time, which are referred to as *baby wavelets* or wavelet atoms. Correlations at different frequencies are obtained by a form of convolution of the signal $x(t)$ with these wavelets as follows:

$$Wx(u,s) = \frac{1}{\sqrt{s}} \int_{-\infty}^{\infty} x(t)\psi^*\left(\frac{t-u}{s}\right) dt \tag{2.30}$$

This is known as a WT of $x(t)$ at scale s and position u, whereas the factor $1/\sqrt{s}$ is a normalization factor to preserve the energy in the transform. The quantity $Wx(u,s)$ is sometimes known as the WT coefficient. A large correlation or coefficient value at some scale s shows that the signal $x(t)$ is closely related to the wavelet function. Conversely, small coefficients indicate a lower correlation of the signal with the wavelet function. Using the position u, we can also determine the position of the strongest correlation in the signal, then using the WT coefficients to determine the type of frequencies in the signal and where in the signal train they occur, thereby giving us localization information of the signal $x(t)$ in time and frequency. The transform (Equation 2.30) is known as the continuous wavelet transform (CWT) because the signal $x(t)$ is assumed to be continuous on the time domain t.

A wavelet function is usually chosen based on whether it satisfies several properties, some of which are listed in the following:

a.   The wavelet ψ is a function of zero average (Mallat, 1999):

$$\int_{-\infty}^{\infty} \psi(t)\mathrm{d}t = 0 \qquad (2.31)$$

b.   Property (a) can be generalized to the following *zero moments* property:

$$\int_{-\infty}^{\infty} t^n \psi(t)\mathrm{d}t = 0 \qquad (2.32)$$

for $n = 0, 1, \ldots, N-1$. This property describes the polynomial degree of the wavelet function, which depicts the relationship of the wavelet to the slope of the bandpass filter characteristics in the frequency domain.

c.   The wavelet function can be smooth or discontinuous. Smoothness is indicated by the regularity of the function where it is p regular if it is p times differentiable.

d.   The wavelet function is symmetrical about u if

$$\psi(t + u) = \psi(-t + u) \qquad (2.33)$$

This property is required if the wavelet coefficients are to be used to reconstruct the original signal.

In the case of discrete signals $x(n)$, the discrete wavelet transform (DWT) is used. A dyadic grid for the scale s and translation u parameters is frequently employed and is defined as follows:

$$(s_j, u_k) = (2^j, k2^j : j, k \in \mathbb{Z}) \qquad (2.34)$$

If, for example, we have 512 digital samples, that is, $N = 512$ and select $j = 3$, we will examine the correlations for $s = 8$ at positions $u = 8, 16, 32, \ldots, 512$ of the entire signal. One may use different scales and translation values, but the dyadic grid gives a uniform and systematic way of viewing the signal $x(n)$. The DWT can be calculated using matrix multiplication or by using digital filter banks implemented, for example, using Mallat's (1999) algorithm, a faster and more efficient process.

In both CWT and DWT, the ability of the transform to extract meaningful information about time and frequency characteristics of the signal depends largely on the selection of the mother wavelet $\psi(t)$. Many families of wavelets have been proposed ranging from the *Haar, Daubechies, Coiflet, Symmlet, Mexican Hat* to *Morlet* wavelets, each having different properties depending on the application at hand. The Haar wavelet, for example, is discontinuous whereas the Daubechies wavelet is smooth and symmetrical. There are empirical guidelines for the selection of wavelets, for example, Daubechies 4 (Db4) is known to be suitable for signals that have linear approximation over four samples whereas Db6 is appropriate for quadratic approximations over six samples.

It is also known that if the orthogonal wavelet bases (set of functions) are used, one can reconstruct the original signal $x(t)$ given only the wavelet coefficients and the bases. This idea originated from Haar who proposed that a simple piecewise constant function

$$\psi(t) = \begin{cases} 1 & \text{if } 0 \leq t < \frac{1}{2} \\ -1 & \text{if } \frac{1}{2} \leq t < 1 \\ 0 & \text{otherwise} \end{cases} \tag{2.35}$$

could generate an orthonormal basis defined as

$$\left\{ \psi_{j,n}(t) = \frac{1}{\sqrt{2^j}} \psi\left(\frac{t - 2^j n}{2^j} \right) \right\}_{j,n \in \mathbb{Z}} \tag{2.36}$$

Then any finite energy signal can be decomposed over this basis as follows:

$$x = \sum_{j=-\infty}^{\infty} \sum_{n=-\infty}^{\infty} \langle x, \psi_{j,n} \rangle \psi_{j,n} \tag{2.37}$$

Each partial sum defined by

$$d_j(t) = \sum_{n=-\infty}^{\infty} \langle x, \psi_{j,n} \rangle \psi_{j,n} \tag{2.38}$$

represents a variation at scale 2^j, which can then be used to reconstruct the original signal $x(t)$. In most cases the summation is truncated and only a finite number of scale levels, j are used for the reconstruction. Several researchers later found other piecewise functions, which could provide better approximations such as Mallat (1999) and Daubechies (1988, 1990).

Most biomedical applications, which employ wavelet transforms generally have digitized biosignals treated using DWT. It is important to note that using discrete wavelet bases gives rise to border problems and are not continuous making interpretation of the coefficients not straightforward. The DWT, however, is simple to implement compared to the CWT that requires integration over an infinite number of convolutions. The WT is also useful for detecting transients by a zooming procedure across scales that can be used to detect irregularities in the signal evidenced by very large values of wavelet coefficients (also known in this case as *singularities*).

Most wavelet transform packages today are freely available on the Internet from sites such as http://www.wavelet.org. Matlab-based packages such as WAVELAB are also available at http://www-stat.stanford.edu/~wavelab with versions available for Unix, Linux, Macintosh, and Windows.

2.3.4   The z-Transform

The standard Laplace transform is used for representing a linear time–invariant system as an algebraic equation, which is easier to manipulate. The z-transform is used for the linear shift–invariant system and can be viewed as the discrete transform to the Laplace transform. The two-sided transform for a digital sequence $x(n)$ is defined as

$$X(z) = \sum_{n=-\infty}^{\infty} x(n)z^{-n} \tag{2.39}$$

where z is a complex variable and the transform exists provided that the sequence converges. The one-sided transform is simply defined for only $n > 0$, that is,

$$X(z) = \sum_{n=0}^{\infty} x(n)z^{-n} \tag{2.40}$$

It should be noted that the z-transform is similar to the Fourier transform in that it does not converge for all sequences. Convergence in this sense means that the sequence $x(n)$ satisfies

$$\sum_{n=0}^{\infty} |x(n)z^{-n}| < \infty \tag{2.41}$$

or is bounded. The z-transform for the unit impulse and unit step sequences is respectively

$$Z(\delta(n)) = 1$$

$$Z(u(n)) = \frac{1}{1 - z^{-1}}$$

for $|z| > 1$. The inverse z-transform is defined using Cauchy's theorem (derivation details in Akay, 1994) as

$$x(n) = \frac{1}{2\pi j} \oint_\Gamma X(z)z^{n-1}dz \qquad (2.42)$$

with respect to the z-transform (Equation 2.39). The contour of integration Γ is counterclockwise and encloses the singularities of X assuming that the integral exists. In general it is easier to recover the digital signal $x(n)$ if the transform X has the rational function form:

$$X(z)z^{n-1} = \frac{\phi(z)}{(z-p_1)^{m_1}(z-p_2)^{m_2}\cdots(z-p_k)^{m_k}} \qquad (2.43)$$

where p_i are the poles of the system. Then solving Equation 2.42 we get using the residue theorem

$$x(n) = \sum_{i=1}^{k} \operatorname*{res}_{z=p_i} \{X(z)z^{n-1}\}$$

$$= \sum_{i=1}^{k} \frac{1}{(m_i-1)!} \lim_{z=p_i} \frac{d^{m_i-1}}{d^{m_i-1}}\left[(z-p_i)^{m_i}X(z)z^{n-1}\right] \qquad (2.44)$$

The z-transform has several properties, which we list here. Let $x_1(n)$ and $x_2(n)$ be two arbitrary digital sequences, then:

a. *Linearity.* For $a, b \in \mathbb{C}$, the following holds:

$$Z(ax_1(n) + bx_2(n)) = aZ(x_1(n)) + bZ(x_2(n)) \qquad (2.45)$$

and similarly for its inverse

$$Z^{-1}(ax_1(n) + bx_2(n)) = aZ^{-1}(x_1(n)) + bZ^{-1}(x_2(n)) \qquad (2.46)$$

b. *Shift or Delay.* The z-transform of a delayed causal sequence $x(n-d)$ is

$$Z[x(n-d)] = \sum_{n=0}^{\infty} x(n-d)z^{-n} \qquad (2.47)$$

where $d > 0$ is the sequence delay or shift. Let $n = n - d$ and we can substitute into Equation 2.47 to get

$$Z[x(n-d)] = \sum_{n=0}^{\infty} x(n-d)z^{-(n+d)}$$

$$= z^{-d}\sum_{n=0}^{\infty} x(n)z^{-n} = z^{-d}X(z) \qquad (2.48)$$

c.   *Convolution.* Let the z-transform of the impulse function $h(n)$ be denoted by $H(z)$, that is,

$$H(z) = \sum_{n=0}^{\infty} h(n)z^{-n} \tag{2.49}$$

then the z-transforms of the following convolution sum:

$$x_2(n) = \sum_{k} h(k)\,x\,(n-k)$$

can be simply written as

$$X_2(z) = H(z)X_1(z) \tag{2.50}$$

d.   *Exponential Multiplication.* The z-transform of the sequence $a^n x(n)$ is

$$Z[a^n x(n)] = \sum_{n=0}^{\infty} a^n x(n)z^{-n} = \sum_{n=0}^{\infty} x(n)(a^{-1}z)^{-n} \tag{2.51}$$

which is another transform in terms of the variable $(a^{-1}z)^{-n}$, that is,

$$Z[a^n x(n)] = X(a^{-1}z)$$

e.   *Differentiation.* A series of sequences can be differentiated only in its region of convergence. Differentiation in the z-domain is defined as

$$z^{-1}\frac{\mathrm{d}X}{\mathrm{d}z^{-1}} = \sum_{n=0}^{\infty} nx(n)z^{-n} \tag{2.52}$$

This is also equivalent to the z-transform of the sequence $nx(n)$. The second derivative of $X(z)$ is found by differentiating the above equation to obtain

$$z^{-2}\frac{\mathrm{d}^2 X(z)}{\mathrm{d}(z^{-1})^2} = \sum_{n=0}^{\infty} n(n-1)x(n)z^{-n} = Z[n(n-1)x(n)]$$

The z-transform is useful for solving difference equations and also to derive the transfer function of the system, using Equation 2.50 the transfer function can be written as

$$H(z) = \frac{X_2(z)}{X_1(z)} \tag{2.53}$$

The system is said to be stable if its transfer function $H(z)$ has no poles outside the unit circle $|z| = 1$. This can be seen by considering the transfer function $H(z)$ as follows:

$$H(z) = \left| \sum_{n=0}^{\infty} h(n)z^{-n} \right| \le \sum_{n=0}^{\infty} |h(n)||z^{-n}| \tag{2.54}$$

for which $|z| \geq 1$ simplifies to

$$H(z) \leq \sum_{n=0}^{\infty} |h(n)|$$

If there are no poles outside the unit circle, then the magnitude of transfer function is finite, that is,

$$H(z) \leq \sum_{n=0}^{\infty} |h(n)| < \infty$$

and the system is stable.

2.3.5 Discrete Cosine Transform

The discrete cosine transform (DCT) is often used in signal processing for compression purposes, for example, in image processing. Compression reduces the amount of data to be transmitted, which results in faster data rates, narrower transmission bandwidths, and more efficient use of transmission power. The DCT is a transform similar to the DFT, except that it uses only the real part of the signal. This is possible because the Fourier series of a real and even signal function has only cosine terms. The standard DFT can be written as follows for $k = 1, 2, \ldots, N$:

$$X(k) = \sum_{n=1}^{N} x_n e^{\frac{-j2\pi nk}{N}} \tag{2.55}$$

and the DCT can be obtained by taking the transform of the real part giving

$$X_c(k) = \mathrm{Re}\left[\sum_{n=1}^{N} x_n e^{\frac{-j2\pi nk}{N}} \right] = \sum_{n=1}^{N} x_n \cos\left(\frac{2\pi kn}{N} \right) \tag{2.56}$$

A more common form of DCT is given by the following for all $k = 1, 2, \ldots, N$:

$$X_c(k) = \frac{1}{N} \sum_{n=1}^{N} x_n \cos\left(\frac{\pi k(2n+1)}{2N} \right) \tag{2.57}$$

The inverse of the DCT can be obtained using the following:

$$X_c(k) = \frac{1}{2} x_0 \sum_{n=2}^{N} x_n \cos\left(\frac{\pi n(2k+1)}{2N} \right) \tag{2.58}$$

Several other implementations of DCT can improve the computational speed and algorithm efficiencies (Ifeachor and Jervis, 1993).

2.3.6 The Discrete Walsh Transform

The previously discussed transforms were useful for continuous signals and based on sums of cosine and sine functions of varying frequencies. For waveforms with discontinuities such as image processing or signal coding, the Walsh transform is used. The Walsh transform has rectangular functions known as Walsh functions and rather than waveform frequency, the analogous term known as *sequency* is used. Sequency is defined as half the average number of zero crossings per unit time (Ifeachor and Jervis, 1993). The transform is phase invariant and hence not suitable for fast convolutions or correlations.

Like the Fourier transform, the Walsh transform is composed of even and odd Walsh functions. Any waveform $f(t)$ can be written in terms of sums of the Walsh function series as follows:

$$f(t) = a_0 WAL(0,t) + \sum_{i=1}^{N/2-1} [a_i WAL(2i,t) + b_i WAL(2i+1,t)] \qquad (2.59)$$

where $WAL(2i,t)$ are the even Walsh functions, $WAL(2i+1,t)$ are the odd Walsh functions, and a_i and b_i are scalar coefficients. In general $WAL(i,t) = \pm 1$ and any two Walsh functions are orthogonal, that is,

$$\sum_{t=1}^{N} WAL(p,t)WAL(q,t) = \begin{cases} N & \text{for } p = q \\ 0 & \text{for } p \neq q \end{cases} \qquad (2.60)$$

The transform–inverse pair is written formally as follows:

$$X_k = \frac{1}{N} \sum_{i=1}^{N} x_i WAL(k,i) \qquad (2.61)$$

and the inverse is

$$x_i = \sum_{i=1}^{N} X_k WAL(k,i) \qquad (2.62)$$

which is identical to the transform without the scale factor $1/N$. The discrete Walsh transform can be calculated using matrix multiplication and may be written for the transform of Equation 2.61 as

$$\mathbf{X}_K = \mathbf{x}_i \mathbf{W}_{ki} \qquad (2.63)$$

where $\mathbf{x}_i = [x_1, x_2, \ldots, x_N]$ and $\mathbf{W}_{ki}$ is the $N \times N$ Walsh transform matrix written as

$$\mathbf{W}_{ki} = \begin{pmatrix} W_{11} & \cdots & W_{1N} \\ \vdots & \ddots & \vdots \\ W_{N1} & \cdots & W_{NN} \end{pmatrix}$$

2.3.7   The Hadamard Transform

The Walsh–Hadamard transform is similar to the Walsh transform but has rows of the transform matrix ordered differently. The Hadamard-ordering sequence is derived from the Walsh-ordered sequence by first expressing the Walsh function in binary, bit reversing the values and converting the binary values to Gray code, and finally to real values. It is considered as a generalized class of Fourier transforms.

2.4   Spectral Analysis and Estimation

Random signals are best characterized statistically by averages (Proakis and Dimitris, 1996), for example, fluctuations in the weather of temperature and pressure could be viewed as random events. Information in these kind of signals can be obtained from their power spectrum, where the most relevant measure is the autocorrelation function used to characterize the signal in the time domain. The Fourier transform of this autocorrelation function gives the power density spectrum, which is a transformation to the frequency domain. There are several methods of obtaining estimates of the power spectrum (Proakis and Dimitris, 1996; Oppenheim et al., 1999; Clarkson, 1993).

2.4.1   Autocorrelation and Power Density Spectrum

Consider a digital signal $x(n)$ obtained by sampling some analog signal $x(t)$, if the digital signal is of finite length the power spectrum can only be estimated because complete information of the signal is not available. Given a longer sequence of digital samples, our estimates are more accurate provided that the signal is stationary. If the signal is nonstationary it may be more worthwhile to obtain power spectrum estimates from shorter digital sequences and treat them as estimates of signal segments instead of the full signal. Too short a sequence, however, severely distorts our estimate and the goal is to determine the minimum sequence length providing acceptable estimates.

Recall from Equation 2.3 the energy of an analog signal $x(t)$ as

$$E = \int_{-\infty}^{\infty} |x(t)|^2 \mathrm{d}t < \infty \tag{2.64}$$

If the signal has finite energy, then its Fourier transform exists and is given as

$$X(F) = \int_{-\infty}^{\infty} x(t)\mathrm{e}^{-j2\pi Ft} \mathrm{d}t \tag{2.65}$$

By Parseval's theorem, we find that

$$E = \int_{-\infty}^{\infty} |x(t)|^2 \mathrm{d}t = \int_{-\infty}^{\infty} |X(F)|^2 \mathrm{d}F \qquad (2.66)$$

where $|X(n)|^2$ is the distribution of signal energy, commonly referred to as the *energy density spectrum* of the signal, S_{xx}. Here

$$S_{xx} = |X(n)|^2 \qquad (2.67)$$

and the total energy of the signal is the integral over the frequency range. Similar relationships exist for the digital signals except that the integral becomes a sum over all the available samples.

In time domain, the Fourier transform of the energy density spectrum S_{xx} is called the autocorrelation function and written as

$$R_{xx}(\tau) = \int_{-\infty}^{\infty} x^*(t)x(t+\tau)\mathrm{d}t \qquad (2.68)$$

The autocorrelation function for digital signals $x(n)$ is written as

$$r_{xx}(k) = \sum_{n=-\infty}^{\infty} x^*(n)x(n+k) \qquad (2.69)$$

There are two methods for estimating the energy density spectrum of a signal $x(t)$ from the digital samples $x(n)$. The direct method involves computing the Fourier transform followed by the energy density spectrum S_{xx}, while the indirect method computes the autocorrelation $r_{xx}(\tau)$ for the digital signal, then the Fourier transform. Since the number of digital samples is finite, one is forced to use a window and assume some form of signal periodicity. The window method will be looked at in the following section on nonparametric methods. It should be noted that the window smooths the spectrum in the signal, provided that it is sufficiently long and narrow compared to the signal.

If the signal is a stationary random process, it does not possess finite energy or a Fourier transform. The signal in this case can be characterized by their finite average power measured by computing the *power density spectrum*. A signal $x(t)$, which is a stationary random process has an autocorrelation function given by

$$\gamma_{xx}(\tau) = E[x^*(t)x(t+\tau)] \qquad (2.70)$$

where $E[\cdot]$ is the expectation or statistical average. The power density spectrum is then the Fourier transform of this autocorrelation function written as

$$\Gamma_{xx}(F) = \int_{-\infty}^{\infty} \gamma_{xx}(\tau)\mathrm{e}^{-j2\pi F\tau}\mathrm{d}t \qquad (2.71)$$

A problem that arises with this is that usually only a single realization of this random process is available to estimate its power spectrum. Furthermore, we do not have the true autocorrelation function $\gamma_{xx}(\tau)$ and cannot compute the power density spectrum directly. If we use the single realization alone, it is common to compute the time-average autocorrelation function taken over an interval $2T_0$. This is written mathematically as

$$R_{xx}(\tau) = \frac{1}{2T_0} \int_{-T_0}^{T_0} x^*(t)x(t+\tau)\mathrm{dt} \qquad (2.72)$$

Using this, we can then proceed to estimate the power density spectrum as follows:

$$P_{xx}(\tau) = \frac{1}{2T_0} \int_{-T_0}^{T_0} R_{xx}(\tau)e^{-j2\pi F\tau}\mathrm{d}\tau \qquad (2.73)$$

Note that this is an estimate since the true power density spectrum is obtained in the limit as $T_0 \to \infty$. There are two approaches to computing $P_{xx}(\tau)$ similar to the ones described earlier. If the signal $x(t)$ is sampled (digitized), an equivalent form of the time-averaged autocorrelation function can be derived, that is,

$$r'_{xx}(m) = \frac{1}{N-m} \sum_{n=0}^{N-m-1} x^*(n)x(n+m) \qquad (2.74)$$

for $m = 0, 1, \ldots, N-1$ under the assumption that the $x(t)$ is sampled at a rate of $F_s > 2B$, where B is the highest frequency in the power density spectrum of the random process. It can be shown (Oppenheim et al., 1999) that this estimate of the autocorrelation function is "consistent" and "asymptotically unbiased" in the sense that it approaches the true autocorrelation function as $N \to \infty$. With the estimate of the autocorrelation function, we can now estimate the power density spectrum using

$$P_{xx}(\tau) = \sum_{m=-(N-1)}^{N-1} r_{xx}(m)e^{-j2\pi fm} \qquad (2.75)$$

which can be simplified as

$$P_{xx}(\tau) = \frac{1}{N}\left| \sum_{n=0}^{N-1} x(n)e^{-j2\pi fn} \right|^2 = \frac{1}{N}|X(f)|^2 \qquad (2.76)$$

where $X(f)$ is the Fourier transform of the sample sequence $x(n)$. This estimated form of power density spectrum is known as a *periodogram*, but there are several problems due to the use of this estimate, such as smoothing effects, frequency resolution, and leakage. Furthermore, the periodogram is not a consistent estimate of the power density spectrum. In view of this, better estimation techniques such as nonparametric and parametric estimation models were devised to circumvent the problems.

2.4.2 Nonparametric Estimation Models

The classical nonparametric methods that have been popularly used for estimation of power density spectrum include the Bartlett, Welch, and Blackman and Tukey methods. These methods do not make any assumption on the distribution of the digital signal and hence are nonparametric.

The Bartlett method reduces the variance observed in the periodogram by averaging the periodograms. This is achieved in three steps:

1. Let the digital sequence $x(n)$ have N samples and divide them into K nonoverlapping uniform segments, each of length M.

2. For each segment i, compute its periodogram using

$$P_{xx}^i(f) = \frac{1}{M} \left| \sum_{n=0}^{M-1} x_i(n) e^{-j2\pi fn} \right|^2 \tag{2.77}$$

3. Finally, compute the Bartlett power density spectrum as the average of the K periodograms as follows:

$$P_{xx}^B(f) = \frac{1}{K} \sum_{i=1}^{K} P_{xx}^i(f) \tag{2.78}$$

It is known that the variance of the power spectrum estimate using the Bartlett method is reduced by a factor of K. Bartlett's averaging method was later modified by Welch, who allowed the K segments to overlap and computed a modified periodogram, which used just a selected part of the segment. The Welch method can be summarized as follows:

1. Let the digital sequence $x(n)$ have N samples. Divide the N samples into K segments each of length M defined as follows:

$$x_i(n) = x(n + iO) \tag{2.79}$$

where $n = 0, 1, \ldots, M-1$ and $i = 0, 1, \ldots, K-1$. Here iO is an offset indicating the starting point of the ith sequence.

2. For each segment i, compute a modified periodogram using

$$\hat{P}_{xx}^i(f) = \frac{1}{MU} \left| \sum_{n=0}^{M-1} x_i(n) w(n) e^{-j2\pi fn} \right|^2 \tag{2.80}$$

for $i = 0, 1, \ldots, K - 1$ and $w(n)$ is a window function. The normalization factor U is computed via

$$U = \frac{1}{M} \sum_{n=0}^{M-1} w(n)^2 \qquad (2.81)$$

3. Finally, compute the Welch power density spectrum as the average of the K modified periodograms as follows:

$$P_{xx}^{\mathrm{W}}(f) = \frac{1}{K} \sum_{i=1}^{K} \hat{P}_{xx}^{i}(f) \qquad (2.82)$$

The introduction of the window function results in the variance of the power spectrum density being dependant on the type of window used. In the Blackman and Tukey approach, this problem is addressed differently by first windowing the sample autocorrelation sequence and then obtaining the power spectrum from the Fourier transform. The Blackman–Tukey power spectrum estimate is given by

$$P_{xx}^{\mathrm{BT}}(f) = \sum_{i=-(M-1)}^{M-1} r_{xx}(m)w(n)\mathrm{e}^{-j2\pi f m} \qquad (2.83)$$

where the window function is $2M - 1$ in length and is zero for the case $|m| \geq M$.

It is has been shown that the Welch and Blackman–Tukey power spectrum estimates are more accurate than the Bartlett estimate (Proakis and Dimitris, 1996). Computationally, the Bartlett method requires the fewest multiplications although the Welch method requires the most. However, the difference in accuracy, quality, and computational performance is very small such that any method may be selected depending on the application requirements.

2.4.3 Parametric Estimation Models

The nonparametric methods described previously still suffer from problems of leakage due to the window functions used and the assumption that the autocorrelation is zero for values of signal samples outside the signal length N. One way to improve the power spectrum estimates is to assume a model distribution of the generating data and extrapolate the values of the autocorrelation for values of the signal outside the length N. The model used is based on the output of a linear system having the form

$$H(z) = \frac{B(z)}{A(z)} = \frac{\sum_{k=0}^{q} b_k z^{-k}}{1 + \sum_{k=1}^{p} a_k z^{-k}} \qquad (2.84)$$

which for most intentions and purposes can be viewed as a linear filter with a finite number of poles and zeroes in the z-plane. The inverse z-transform provides the resulting difference equation:

$$x(n) = \sum_{k=0}^{q} b_k w(n-k) - \sum_{k=1}^{p} a_k x(n-k) \qquad (2.85)$$

which is a linear model of the signal $x(n)$. Under the assumption that the sequence $w(n)$ is zero-mean white noise with autocorrelation

$$\gamma_{ww}(m) = E[|w(n)|^2]\delta(m) \qquad (2.86)$$

the power density spectrum of the observed data is

$$\Gamma_{xx}(f) = E[|w(n)|^2]\frac{|B(f)|^2}{|A(f)|^2} \qquad (2.87)$$

To compute this, we first require an estimate of the model parameters $\{a_k\}$ and $\{b_k\}$, given N samples, and then use Equation 2.86 to obtain the power spectrum estimate.

There are several methods for computing the model parameters beginning with the autoregressive moving average (ARMA) model of order (p, q). If one sets $q = 0$ and $b_0 = 1$, the process reduces to an autoregressive (AR) process of order p. Alternatively, if we set $A(z) = 1$ and have $B(z) = H(z)$, the output $x(n)$ is referred to as a moving average (MA) process of order q. The AR model by far is the most popular of the three, mainly because it is useful for representing spectra with narrow peaks, and it is easily implemented requiring only the solution of linear equations. For sinusoidal signals, the spectral peaks in the AR estimate are proportional to the square of the signal power.

2.4.3.1   Autoregressive Model

The AR model parameters can be estimated using several methods such as the Yule–Walker, Burg (1968), covariance, and modified covariance methods. These have been implemented in many software packages, for example, MAT-LAB* *Signal Processing Toolbox* and are now easily accessible. The Yule–Walker method uses a biased form of the autocorrelation estimate to ensure a positive semidefinite autocorrelation matrix. The Burg method on the other hand is a form of order-recursive least squares lattice method that estimates parameters by minimizing forward and backward errors of the linear system. The key advantages of the Burg method are that it provides high-frequency resolution, gives a stable AR model, and is computationally efficient. There are however several disadvantages such as spectral line splitting at high signal to noise ratios and the introduction of spurious peaks at high model orders.

---

*See http://www.mathworks.com/products/matlab/

It is also sensitive to the initial phase of sinusoidal signals, particularly for short digital samples resulting in a frequency bias. Nevertheless, these problems have been addressed with modifications such as a weighting sequence on the squared forward and backward errors or the use of window methods.

An important issue in using the AR model is the selection of the optimal order, p. If the model order is too low, we obtain a smoother representation of the signal and also a highly smoothed spectrum. Conversely, if the model order is too high, we potentially introduce spurious low-level peaks in the spectrum. There have been several criteria to select the optimal AR order; the simplest method being to use the mean squared value of the residual error, which decreases as model order increases and is dependant on the estimation method used. One can then monitor the rate of decrease and use the order where the rate of decrease is no longer significant. Akaike (1969, 1974) proposed two criteria, namely the final prediction criterion (FPE) and the Akaike information criterion (AIC), which could be used to select the optimal AR model order. The FPE selects the optimal order based on minimization of the following performance index

$$\text{FPE}(p) = \sigma_{wp}^2 \left(\frac{N+p+1}{N-p-1} \right) \tag{2.88}$$

where σ_{wp}^2 is the estimated variance of the linear prediction error. The AIC is more often used and is based on selecting the order that minimizes the following:

$$\text{AIC}(p) = \log_e \sigma_{wp}^2 + 2\frac{p}{N} \tag{2.89}$$

Other criteria include Rissanen's (1983) criterion based on the order that minimizes the description length (MDL), where MDL is defined as

$$\text{MDL}(p) = N \log_e \sigma_{wp}^2 + p \ln N \tag{2.90}$$

Parzen (1974) had also proposed a criterion autoregressive transfer function where the optimal model order was found by minimizing the function.

2.4.3.2 Moving Average Model

In the MA model, the coefficients a_k are set to zero and the difference equation for the input–output relationship is given as

$$x(n) = \sum_{k=0}^{q} b_k w(n-k) \tag{2.91}$$

The noise whitening filter for the MA process is regarded as an all-pole filter. The AIC can also be used to estimate the best MA model and has a similar form to that of Equation 2.89. This estimation method is approximately equivalent to the maximum likelihood method under the assumption that the signal is Gaussian.

2.4.3.3   Autoregressive Moving Average Model

In this case the linear filter or system is governed by the full difference equation:

$$x(n) = \sum_{k=0}^{q} b_k w(n-k) - \sum_{k=1}^{p} a_k x(n-k) \qquad (2.92)$$

This model can improve on the AR model by using fewer model parameters and has been used in situations where the signal is corrupted by additive white noise.

2.5   Analog Filters

The primary purpose of a filter is to attenuate certain signal frequency ranges and highlight other frequency regions. This serves many purposes with the most common one being the removal of unwanted signals or noise. Filtering is therefore an important preprocessing step in information processing systems as it removes clutter in the signal and restores information clarity. There are two broad classes of filter; namely analog filters for continuous signals and digital filters, which filter discrete signals. In this section, we describe some analog filters and filter design, whereas digital filters that can be implemented on computers are discussed in the following section.

Analog filters are electronic circuits built from components such as resistors, capacitors, and inductors. The introduction of very large-scale integration (VLSI) technology for manufacturing integrated circuits (IC) has seen filters increase in complexity although reducing greatly in size. These days filters come in IC chips, which can be easily fitted on to a circuit board. The main aspect in filter design is the filter response, which is characterized by its transfer function and plotted on Bode plots. The transfer function is written in terms of magnitude and phase:

$$H(e^{jw}) = |H(e^{jw})| \angle \phi(w) \qquad (2.93)$$

where $|H(e^{jw})|$ is the magnitude or amplitude and $\phi(w)$ represents the phase angle. It can also be written in terms of a rational function in terms of a complex variable, $s = \sigma + jw$ as follows:

$$H(s) = \frac{b_0 + b_1 s + \cdots + b_m s^m}{a_0 + a_1 + \cdots + a_n s^n} \qquad (2.94)$$

where a_i and b_j are scalar coefficients. This rational form is often factorized using partial fractions to give the common pole-zero equation:

$$H(s) = \frac{N(s)}{D(s)} = K\frac{(s-z_1)(s-z_2)\cdots(s-z_m)}{(s-p_1)(s-p_2)\cdots(s-p_n)} \qquad (2.95)$$

where z_i are the zeroes or roots of the equation

$$N(s) = 0 \qquad (2.96)$$

and p_i are the poles or roots of the denominator:

$$D(s) = 0 \qquad (2.97)$$

The gain of the system is denoted by K and since the value of the transfer function is a real number, all poles and zeroes are either real or exist as complex conjugate pairs. Poles and zeroes can also be plotted on a s-plane to give a graphic illustration of their positions, which can be used to infer the stability of the system (Dutton et al., 1997).

The main characteristics of a filter as depicted in Figure 2.7 include special regions such as the passband, stopband, and transition. The passband is the interval or range of frequencies, which are allowed to pass through the filter whereas the stopband is the frequency range where signals are filtered out. This is achieved ideally by attenuating the signal amplitude as much as possible within the stopband. The transition region is the frequency range where the amplitude of the signal is gradually attenuated before the stopband. An ideal filter would have no transition period, that is, the attenuation is instantaneous and sharp (Figure 2.7). Owing to the individual characteristics of

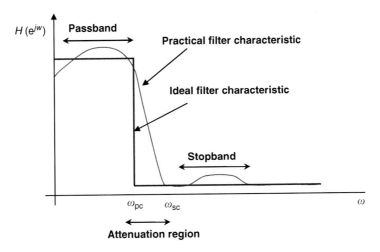

FIGURE 2.7

Filter characteristics and regions indicating the compromise between ideal filters and practical filters.

filter components, it is difficult in practice to design a filter with no transition period. Instead, the compromise is to ensure this attenuation happens as quickly as possible. The allowable design error in an analog filter is denoted by Δp for frequencies in the passband region and Δs for frequencies in the stopband region. There are several widely used analog filter configurations starting with the basic low-pass and high-pass filters, Butterworth, Chebyshev, inverse Chebyshev, Bessel filters, and elliptic filters, some of which are examined in the following subsections.

2.5.1 Basic Analog Filters

Three basic analog electronic components used in the construction of analog filters are the resistor (R), capacitor (C), and inductor (I). Using combinations of these, it is possible to design simple passive filters as well as more complex filters by combining these components either in series or in parallel. These configurations give rise to RC, RL, LC, and RLC circuits in which the letters denote the components. In this section, the fundamentals of RC and RL circuits are examined and we show how they can be employed as analog filters.

2.5.1.1 RC Filter

In a RC circuit, a resistor and capacitor are used to design a simple low-pass filter. The impedance of a resistor is measured in ohms whereas the complex impedance of a capacitor is written as

$$Z_c = \frac{1}{sC} \tag{2.98}$$

where C is the capacitance measured in Farads. The unit s is a complex number usually $s = a + j\omega$, where a is some decay constant and ω the frequency of the current or signal passing through the capacitor measured in rad/s or hertz. The analysis and design of this circuit assumes sinusoidal steady state where transient effects due to initial conditions placed on the components have died out. Steady state also means that the circuit behavior will remain deterministic and no decay effects persist, so $a = 0$ and the impedance of the capacitor can be simplified to

$$Z_c = \frac{1}{j\omega C} \tag{2.99}$$

A low-pass filter can be designed from a series RC circuit with the configuration shown in Figure 2.8. The transfer function is defined as the ratio of the output voltage V_o to the input voltage V_i with the output measured across the capacitor. This can be calculated using the voltage divider principle and Ohm's law as follows:

$$\frac{V_o}{V_i} = \frac{1/sC}{R + (1/sC)} = \frac{1}{1 + sRC} \tag{2.100}$$

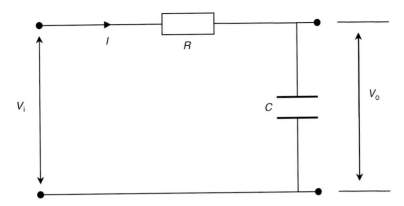

FIGURE 2.8
Circuit diagram for RC low-pass filter.

where the magnitude of the transfer function is computed as

$$|H(\omega)| = \left|\frac{V_o}{V_i}\right| = \frac{1}{\sqrt{1 + (\omega RC)^2}} \qquad (2.101)$$

This magnitude is also known as the *gain* of the circuit because it measures the amplification or attenuation of the output signal with respect to the input. The phase angle of the output voltage taken across the capacitor is given as

$$\phi_c = \tan^{-1}(-\omega RC) \qquad (2.102)$$

and is useful when the input–output phase relationships are important in the filter's design. The filter poles are obtained by setting the denominator of the transfer function (Equation 2.100) to zero and solving for s, which gives

$$s = -\frac{1}{RC} \qquad (2.103)$$

The zeroes of the filter can be obtained by setting the numerator to zero, but in this case our filter has no zeroes. It is known in control theory (Dutton et al., 1997) that negative poles on the left-hand side of the s-plane give rise to a stable system or circuit, hence the RC circuit is stable and no oscillatory behavior should be expected.

Let us now analyze the gain relationship (Equation 2.101) as the input signal frequency ω varies. When ω is small or $\omega \to 0$, we are in the low-frequency region of possible input signals. Substituting $\omega = 0$ into Equation 2.101 gives a gain of unity, that is, $|H(\omega)| = 1$ and as we increase ω, $|H(\omega)|$ decreases slowly or insignificantly. We conclude that the output signal amplitude in relation to the input signal is approximately equal at low frequencies and the RC circuit allows low-frequency signals to pass from input to the output with no attenuation. Now consider the effects of large values of ω or when $\omega \to \infty$. In this case, we are dealing with high-frequency input signals. As ω

increases, the denominator of Equation 2.101 increases and the gain of the circuit decreases faster when ω is larger. When $\omega = \infty$, we have a gain of zero and no output signal. It can be seen that the output signal is increasingly attenuated or "filtered" out as the input signal frequency increases, which is the characteristic of a low-pass filter.

When the input signal frequency is $\omega = 1/RC$, we find the filter gain becomes

$$|H(\omega)| = \frac{1}{\sqrt{2}} \tag{2.104}$$

Taking the $\log_{10}$ of this gives us the gain in decibels or dB as follows:

$$|H(\omega)| = -20 \log_{10} \frac{1}{\sqrt{2}} = -3\,\text{dB} \tag{2.105}$$

We define the frequency at which the filter gain falls by 3 dB as the filter *cutoff frequency*, ω_c. The range of frequencies allowed through is known as the *bandwidth*. Filter characteristics can usually be seen more clearly by plotting the gain of the filter against the range of frequencies. A log scale plot of gain and phase angle against the range of frequencies is known as the Bode plot after Hendrik Wade Bode, a pioneer of modern control theory. An example of a Bode plot for a low-pass filter is given in Figure 2.9.

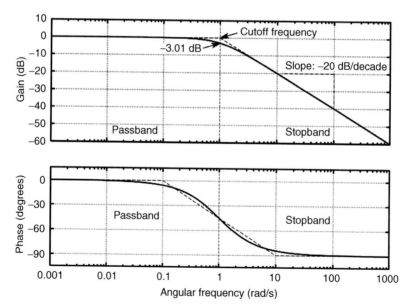

FIGURE 2.9

Bode plot showing magnitude and phase variation over frequency range of input signal for a RC low-pass filter. Note the 3 dB fall in gain at the cutoff frequency, ω_c.

2.5.1.2 RL Filter

We saw how a low-pass filter was designed from two simple components, the resistor and the capacitor. It is possible to design a high-pass filter by taking the output voltage over the resistor but instead we examine how the inductor can be used to design a high-pass filter circuit. The complex impedance of an inductor is given as

$$Z_L = sL \tag{2.106}$$

where L is the inductance measured in Henry. Assuming sinusoidal steady state as before, the inductor impedance can be written as $Z_L = j\omega L$. In a series configuration as depicted in Figure 2.10, the voltage divider rule can be applied as before to obtain the voltage over the inductor. This gives

$$\frac{V_o}{V_i} = \frac{j\omega L}{R + j\omega L} \tag{2.107}$$

and the magnitude of the transfer function is

$$|H(\omega)| = \left| \frac{V_o}{V_i} \right| = \frac{\omega L}{\sqrt{R^2 + \omega^2 L^2}} \tag{2.108}$$

The phase angle of the output voltage taken across the inductor is then

$$\phi_L = \tan^{-1} \left(\frac{R}{\omega L} \right) \tag{2.109}$$

As with the RC low-pass filter, we can analytically determine effects to filter gain for different frequencies. When $\omega \to 0$, the numerator of Equation 2.108 also approaches zero, hence low-frequency signals are attenuated by the RL circuit. Now consider when $\omega \to \infty$, as frequency increases the effect of the constant R in the denominator of Equation 2.108 reduces as the $\omega^2 L^2$ term rapidly grows. The gain in this condition approaches unity and the circuit is passing higher-frequency signals, or is acting as a high-pass filter.

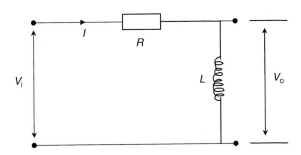

FIGURE 2.10
Circuit diagram for RL high-pass filter.

2.5.2   Butterworth Filters

Butterworth filters are higher-order filters with the special characteristic that as the filter order increases, the filter response approaches that of an ideal filter (Akay, 1994). This can be seen from the transfer characteristics plot in Figure 2.11, where the attenuation region becomes sharper as the filter order increases. When designing a Butterworth filter, two important issues are the filter order, N and the cutoff frequencies, ω_c. We will examine the design of a low-pass Butterworth filter to show how the filter characteristics are determined from the filter specifications. Figure 2.12 shows a Butterworth filter circuit, where R_i and C_i are resistors and capacitors respectively, whereas $UA741$ is an operational amplifier.

The general magnitude of the Butterworth filter response is given as

$$|H(\omega)|^2 = \frac{1}{1 + (\omega/\omega_c)^{2N}} \tag{2.110}$$

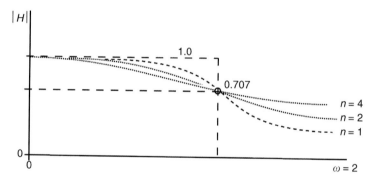

FIGURE 2.11

A typical frequency response of an *nth* order Butterworth filter.

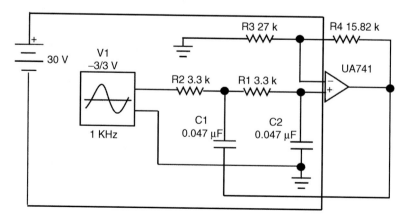

FIGURE 2.12

Circuit diagram of a Butterworth filter.

Note that if $\omega = \omega_c$ then, the response is the 3 dB attenuation or $|H(\omega)| = 1/\sqrt{2}$. After selecting N and ω_c, the Laplace transfer function can be constructed from Equation 2.110 to obtain

$$H(s)\bar{H}(s) = \frac{1}{1 + (s/j\omega_c)^{2N}} \quad (2.111)$$

where $\bar{H}(s)$ is the complement of $H(s)$. Solving this equation by setting the denominator to zero gives the estimate of the roots as

$$s^{2N} = (-1)^{N-1}\omega_c^{2N} \quad (2.112)$$

The poles on the left-hand side of the s-plane correspond to the stable and causal part of the filter and are thus desired in the filter design. Rewriting Equation 2.112 as

$$s_p = \omega_c e^{j\theta_p} \quad (2.113)$$

we obtain the left-hand side s-plane poles by selecting θ_p, such that

$$\theta_p = \frac{\pi}{2N}(N - 1 + 2p) \quad (2.114)$$

for $p = 1, 2, \ldots, N$. It can be seen that this results in $\pi/2 < \theta_p < 3\pi/2$. If the filter order, N, is even, it is possible to arrange the poles as conjugate pairs such that the transfer function denominator can be factorized easily. Thus, a general conjugate pole pair can be written as

$$D_i(s) = \left(1 - \frac{s}{s_i}\right)\left(1 - \frac{s}{\bar{s}_i}\right) \quad (2.115)$$

The Butterworth function can then be easily written as the product of these conjugate pairs, that is,

$$H(s) = \prod_{i=0}^{L}\frac{1}{D_i(s)} = \prod_{i=0}^{L} H_i(s) \quad (2.116)$$

Now if N is even, we usually take $H_0 = 1$ and if N is odd, we set $H_0(s) = 1/1 + (s/\omega_c)$. The filter parameters N and ω_c can now be estimated from the following filter specifications:

$$A_p = 10\log_{10}\left[1 + \left(\frac{\omega_p}{\omega_c}\right)^{2N}\right] \quad (2.117a)$$

$$A_s = 10\log_{10}\left[1 + \left(\frac{\omega_s}{\omega_c}\right)^{2N}\right] \quad (2.117b)$$

where A_p is the maximum passband attenuation, A_s the minimum stopband attenuation, ω_p the passband frequency, and ω_s the stopband frequency. Given these four values, we can compute N and ω_c by rearranging Equations 2.117a and 2.117b. This results in the following:

$$N = \frac{\ln[10^{A_s/10} - 1)/10^{A_p/10} - 1)}{2\ln(\omega_s/\omega_p)}$$

$$\omega_c = \frac{\omega_p}{(10^{A_p/10} - 1)^{1/2N}}$$

This method may be further used to design high-pass and bandpass Butterworth filters.

2.5.3 Chebyshev Filters

Chebyshev filters have steeper attenuation regions than Butterworth filters at the expense of greater rippling effects in the passband region. As with Butterworth filters the Chebyshev filter transfer function comprises poles but unlike the Butterworth poles, which lie on the unit circle of the s-plane, the Chebyshev poles lie on an ellipse. There are two types of Chebyshev filters commonly known as Type-I and Type-II Chebyshev filters. Note the presence of passband ripple in a typical Chebyshev filter response as depicted in Figure 2.13.

Type-I Chebyshev filters are more frequently employed and they possess the following frequency response:

$$|H(\omega)| = \frac{1}{|1 + \epsilon^2 f^2(\omega/\omega_c)|} \tag{2.118}$$

where $\epsilon < 1$ and the gain at the cutoff frequency, $\omega = \omega_c$ is given by

$$|H(\omega_c)| = \frac{1}{|1 + \epsilon^2|} \tag{2.119}$$

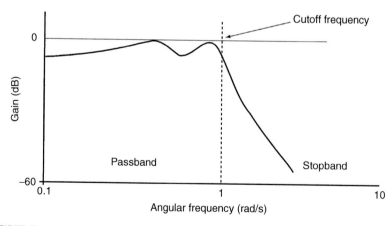

FIGURE 2.13

Frequency response of a fourth order Type-I Chebyshev filter.

Note that for Chebyshev filters the $-3\,\text{dB}$ attenuation, which is the common definition for cutoff frequency, does not hold because the magnitude of the transfer function depends on the value of $f(\omega/\omega_c)$. This function is a Chebyshev polynomial and takes several forms, some of which are outlined in the following:

1. *nth order Chebyshev polynomial.* For $0 \le \omega \le \omega_c$

$$f\left(\frac{\omega}{\omega_c}\right) = \cos\left[n\cos^{-1}\left(\frac{\omega}{\omega_c}\right)\right] \qquad (2.120)$$

For $\omega > \omega_c$

$$f\left(\frac{\omega}{\omega_c}\right) = \cos h\left[n\cos h^{-1}\left(\frac{\omega}{\omega_c}\right)\right] \qquad (2.121)$$

2. *nth order Chebyshev monomial expansions.* For $0 \le \omega \le \omega_c$

$$f\left(\frac{\omega}{\omega_c}\right) = a_0 + a_1\left(\frac{\omega}{\omega_c}\right) + a_2\left(\frac{\omega}{\omega_c}\right)^2 + \cdots + a_n\left(\frac{\omega}{\omega_c}\right)^n \qquad (2.122)$$

For $\omega > \omega_c$

$$f\left(\frac{\omega}{\omega_c}\right) = \frac{\left(\omega/\omega_c\sqrt{(\omega/\omega_c)^2 - 1}\right)^n + \left(\omega/\omega_c\sqrt{(\omega/\omega_c)^2 - 1}\right)^{-n}}{2} \qquad (2.123)$$

The order n of the response determines the number of components such as resistors and capacitors that are required to construct the analog filter.

The less common Chebyshev filter is the inverse Chebyshev or Type-II filter, which possesses the following frequency response:

$$|H(\omega)|^2 = \frac{1}{\sqrt{1 + (1/\epsilon^2 f^2(\omega/\omega_c))}} \qquad (2.124)$$

In this configuration, the $-3\,\text{dB}$ frequency, ω_d is related to the cutoff frequency, ω_c by the following equation:

$$\omega_d = \omega_c\cos h\left(\frac{1}{n}\cos h^{-1}\frac{1}{\epsilon}\right) \qquad (2.125)$$

The Type-II filter has a more gradual attenuation region and less passband ripple but requires more analog components. Interestingly, the passband ripple can be reduced by adding zeroes to the transfer function, but there will be less attenuation at the stopband. The resulting filter due to the addition of zeroes is known as an *elliptic filter*.

2.6   Digital Filters

Digital filters are the counterparts of analog filters used to eliminate noise from digital signals (Tompkins, 1993; Oppenheim et al., 1999) and they do not physically attenuate signal amplitudes rather they modify the coefficients of the digital sequences. Their specifications are usually derived from previously designed and tested analog designs. There are two broad categories of digital filter, the infinite impulse response (IIR) and the finite impulse response (FIR) filters. These filters in turn have a variety of practical realizations such as direct, canonical, and cascade configurations.

2.6.1   Infinite Impulse Response Filters

A linear system can be represented by the following difference equation:

$$\sum_{k=0}^{N} a_k y(n-k) = \sum_{j=0}^{M} b_j x(n-j) \tag{2.126}$$

or its z-transform having the transfer function

$$H(z) = \frac{Y(z)}{X(z)} = \frac{\sum\limits_{j=0}^{M} b_j x(n-j)}{\sum\limits_{k=0}^{N} a_k y(n-k)} \tag{2.127}$$

The output $y(n)$ can be obtained by expressing the system in terms of the previous output and the current and previous input samples giving rise to a recursive formulation, which is characteristic of IIR-type filters with the following form:

$$y(n) = -\sum_{k=0}^{N} \frac{a_k}{a_0} y(n-k) + \sum_{j=0}^{M} \frac{b_j}{a_0} x(n-j) \tag{2.128}$$

There are several IIR filter configurations as discussed in the following sections.

2.6.1.1   Direct IIR Filter

The direct IIR filter is the basic digital filter with the configuration as in Figure 2.14. From Equation 2.126, the direct form of the IIR filter can be achieved by setting $a_0 = 1$, so that Equation 2.126 simplifies to

$$1 + \sum_{k=1}^{N} a_k y(n-k) = \sum_{j=0}^{M} b_j x(n-j) \tag{2.129}$$

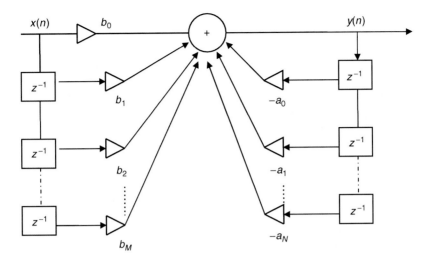

FIGURE 2.14
A direct IIR filter configuration.

Applying the z-transform and rearranging gives the filter transfer function

$$H(z) = \frac{Y(z)}{X(z)} = \frac{\sum\limits_{j=0}^{M} b_j x(n-j)}{1 + \sum\limits_{k=1}^{N} a_k y(n-k)}$$

$$= \frac{b_0 + b_1 z^{-1} + \cdots + b_M z^{-M}}{1 + a_1 z^{-1} + \cdots + a_N z^{-N}} \qquad (2.130)$$

In the time domain this may be written as the difference equation

$$y(n) = -a_1 y(n-1) - a_2 y(n-2) - \cdots - a_M y(n-M) + b_0 x(n)$$
$$+ b_1 x(n-1) + b_N x(n-M)$$

2.6.1.2 Canonical IIR Filter

This filter can be implemented as a system of difference equations where the internal state of the system, s_i can be estimated for a given input $x(n)$ from previous internal states. This may be written as

$$s(n) = x(n) - a_1 s(n-1) - a_2 s(n-2) - \cdots - a_N s(n-N)$$

The output in this case is the summation of the history of internal state values written as

$$y(n) = b_0 s(n) + b_1 s(n-1) + \cdots + b_N s(n-N)$$

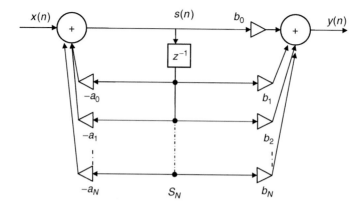

FIGURE 2.15
A canonical IIR digital filter configuration.

with internal states, $s_i(n) = s_{n-i}$ for all $i = 0, 1, \ldots, N$. The states themselves are updated via the equation $s + i(n+1) = s_{i-1}(n)$ for all $i = 0, 1, \ldots, N$. An example of a canonical IIR filter construction is depicted in Figure 2.15.

2.6.1.3 Cascaded IIR Filter

In the cascaded IIR filter, a number of smaller IIR filters are combined or cascaded to form a single filter designed by first decomposing the direct IIR Equation 2.130 into products or partial fractions. The cascaded IIR filter transfer function can be written as a product of the smaller filter transfer functions, that is,

$$H(z) = \prod_{k=1}^{K} H_k(z) = H_1(z)H_2(z)\ldots H_K(z)$$

The cascaded IIR filter in Figure 2.16, for example, has transfer function

$$H(z) = \frac{1 + z^{-1}}{(1 - (1/2)z^{-1})(1 - z^{-1})}$$

2.6.2 Design of IIR Digital Filters

The digital IIR filter can be obtained from the corresponding analog filter that has been designed to specification. There are two common methods to convert the transfer function of the analog IIR filter to their digital counterparts, namely, the *impulse-invariant transformation* and the *bilinear transformation* (Tompkins, 1993). These methods convert the filter pole and zero coefficients to the corresponding poles and zeroes on the unit z-circle.

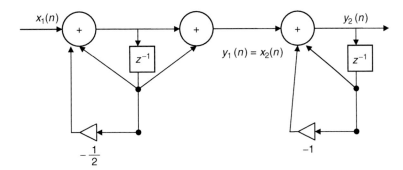

FIGURE 2.16

A cascaded IIR digital filter configuration.

2.6.2.1 Impulse-Invariant Transformation

In this method, the impulse response of the analog filter is first sampled equally to obtain the digital impulse response. This is written as

$$h_D(n) = h_A(nT) \tag{2.131}$$

where T is the sampling period, and $h_D(n)$ the digital response obtained from the analog response $h_A(nT)$. Suppose that the analog transfer function in the time domain is given as

$$h_A(t) = \sum_{p=1}^{N} A_p e^{s_p t} u(t) \tag{2.132}$$

where $u(t)$ is the unit step. This can be transformed to the Laplace equation as

$$H_A(s) = \sum_{p=1}^{N} \frac{A_p}{s - s_p} \tag{2.133}$$

Now using Equation 2.131, we see that the digital response can be similarly obtained as follows:

$$h_D(n) = h_A(nT) = \sum_{p=1}^{N} A_p e^{ns_p T} u(nT) \tag{2.134}$$

The digital transform is written using the z-transform giving

$$H_D(z) = \sum_{p=1}^{N} \frac{A_p}{1 - e^{s_p T} z^{-1}} \tag{2.135}$$

The poles of the analog filter have now been transformed to digital poles through the relationship $z = e^{sT}$. Intuitively, the analog poles on the left of the s-plane have been mapped to poles inside the unit circle where $|z| = 1$.

Poles on the right-handside are mapped outside the circle, which is the region of instability although poles lying on the imaginary axis are transformed to points on the unit circle itself. In summary, the impulse-invariant transformation causes areas of width $2\pi/T$ to be mapped to the unit circle in the z-plane.

Writing the z-transform of the digital transfer function as a Laplace transform, we obtain

$$H(z) = \frac{1}{T} \sum_{-\infty}^{\infty} H_A \left(s + j\frac{2\pi p}{T} \right)$$

$$= \frac{1}{T} \sum_{-\infty}^{\infty} H_A \left(j\frac{2\pi p + \omega}{T} \right) \tag{2.136}$$

where $z = e^{sT}$, $\omega = \Omega T$, and Ω is the analog frequency whereas ω the digital frequency. If the analog filter is bandlimited, the digital filter function takes the form

$$H(z) = \frac{1}{T} H_A \left(j\frac{\omega}{T} \right) \tag{2.137}$$

where $|\omega| \leq \pi$ is the cutoff frequency. At this point we have

$$H_A \left(j\frac{\omega}{T} \right) = 0 \tag{2.138}$$

for $\left| \frac{\omega}{T} \right| \geq \frac{\pi}{T}$.

2.6.2.2 Bilinear Transformation

Another method for deriving digital filters from analog counterparts is the bilinear transformation. This method begins with the assumption that the derivative of the analog signal is obtainable, that is, we have

$$\frac{dy(t)}{dt} = x(t) \tag{2.139}$$

The Laplace transform of this gives us

$$sY(s) = X(s)$$

$$\Rightarrow \frac{Y(s)}{X(s)} = \frac{1}{s} \tag{2.140}$$

Integrating both sides of Equation 2.140 gives us the difference equation as follows:

$$\int_{(n-1)T}^{nT} \frac{dy(t)}{dt} = y(nT) - y[(n-1)T]$$

$$= \int_{(n-1)T}^{nT} x(t) \tag{2.141}$$

This can be approximated using the standard trapezoidal rule

$$y(nT) - y[(n-1)T] = \frac{T}{2}(x(nT) + x[(n-1)T]) \qquad (2.142)$$

This analog difference equation can now be digitized to hold for n discrete samples:

$$y(n) - y(n-1) = \frac{T}{2}[x(n) + x(n-1)] \qquad (2.143)$$

which can be transformed to

$$H(z) = \frac{Y(z)}{X(z)} = \frac{T}{2} \frac{1 + z^{-1}}{1 - z^{-1}} \qquad (2.144)$$

Comparing Equations 2.140 and 2.144 gives us the approximation

$$s = \frac{2}{T} \frac{1 - z^{-1}}{1 + z^{-1}} \qquad (2.145)$$

Alternatively, rearranging Equation 2.145 yields the relationship

$$z = \frac{1 + (T/2)s}{1 - (T/2)s} \qquad (2.146)$$

To design a digital filter from the analog version, we therefore have to replace the s of the analog Laplace transfer function using Equation 2.145 to obtain the digital transfer function in z-transform. Suppose $s = a + bj$, it can be seen that if $|z| > 1$ if $a > 0$ and $|z| < 1$ if $a < 0$. Furthermore, $|z| = 1$ if $a = 0$. This is interpreted as previously where analog poles are mapped to the digital poles within, outside, and on the unit circle. The bilinear transformation is preferable to the previous method because it does not suffer from anti-aliasing effects, though it sometimes results in oscillatory behavior in the resulting digital design. This method is frequently used to derive higher-order filters.

2.6.3 Finite Impulse Response Filters

In FIR filters, the output is determined using only the current and previous input samples. The output then has the following form:

$$y(n) = \sum_{j=0}^{M} \frac{b_j}{a_0} x(n-j) \qquad (2.147)$$

As with IIR digital filters, several configurations of FIR filters are possible as seen in the following sections.

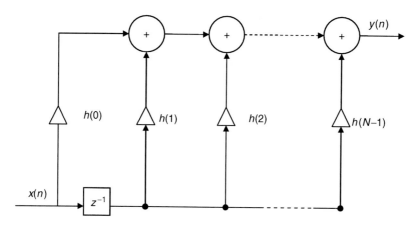

FIGURE 2.17
A direct FIR filter configuration.

2.6.3.1   Direct FIR Filter

This filter configuration (Figure 2.17) is very similar to the IIR system and has the difference equation

$$y(n) = \sum_{j=0}^{M} h(j)x(n-j) \tag{2.148}$$

where $h(j)$ is the impulse response for a given input $x(n)$ and output $y(n)$.

2.6.3.2   Cascaded FIR Filter

This system can be obtained if the transfer function is factored into L-subsystems, each subsystem may be a first- or second-ordered system. The cascaded FIR filter in Figure 2.18 has the following transfer function:

$$H(z) = \left(1 - \frac{1}{2}z^{-1}\right)\left(1 - \frac{2}{3}z^{-1} + z^{-2}\right) \tag{2.149}$$

FIR filters are sometimes preferred over the IIR types due to their stability and ease of implementation, FIR filters also possess linear-phase character-istics where the impulse response $h(n)$ is symmetric about the length of the response. This means that $h(n) = h(N-1-n)$, where N represents the total number of sequences or the length of the response. Unfortunately, FIR filters require long sequences to implement sharper cutoffs in the attenuation region. In the next section, we examine two methods for FIR filter design.

2.6.4   Design of FIR Digital Filters

There are several well-known methods for designing FIR filters such as win-dow, frequency sampling, minmax, and optimal design techniques. We will

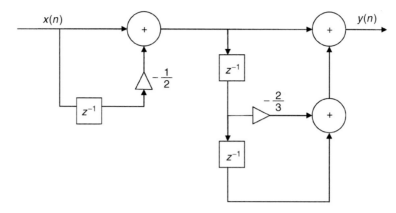

FIGURE 2.18
A cascaded FIR filter configuration.

review just two of these methods, namely, the window and frequency sampling methods.

2.6.4.1 The Window Method

Using this method, an FIR filter can be obtained by selecting a set of infinite duration impulse response sequences. Given an ideal frequency response $H_I(e^{j\omega})$, the corresponding impulse response $h_I(n)$ can be estimated using the following integral:

$$h_I(n) = \frac{1}{2\pi} \int_{-\pi}^{\pi} H_I(e^{j\omega})e^{j\omega n}d\omega \qquad (2.150)$$

where

$$H_I(e^{j\omega}) = \sum_{n=-\infty}^{\infty} h_I(n)e^{-j\omega n} \qquad (2.151)$$

First, the impulse response sequence can be truncated so that only N sequences are selected for consideration. This can be written as follows:

$$h(n) = \begin{cases} h_I(n) & \text{if } 0 \le n \le N-1 \\ 0 & \text{otherwise} \end{cases} \qquad (2.152)$$

where $h_I(n)$ is an impulse response having a sequence of infinite length. Let $h(n)$ be represented by the product of the infinite impulse response $h_I(n)$ and a window $w(n)$, so that

$$h(n) = h_I(n)w(n) \qquad (2.153)$$

It can be seen that Equation 2.152 is achieved for a rectangular window $w(n)$ defined as

$$w(n) = \begin{cases} 1 & \text{if } 0 \leq n \leq N - 1 \\ 0 & \text{otherwise} \end{cases} \tag{2.154}$$

The frequency response can also be obtained using the convolution theorem, which gives

$$H(e^{j\omega}) = \frac{1}{\pi} \int_{-\pi}^{\pi} H_I(e^{j\theta}) e^{j(\omega - \theta)} d\theta \tag{2.155}$$

Now, as the window size $w(n)$ shrinks, the actual filter response $H(e^{j\omega})$ will approach that of the ideal response $H_I(e^{j\omega})$. For a rectangular window, the number of samples N also affects the transfer characteristics of the filter, such as the peak amplitude and oscillatory behavior. There are six popular types of window (Akay, 1994), which are listed here.

1. *Rectangular.* For $0 \leq n \leq N - 1$

$$w(n) = 1$$

2. *Bartlett*

$$w(n) = \begin{cases} \dfrac{2n}{N-1} & 0 \leq n \leq \dfrac{N-1}{2} \\ 2 - \dfrac{2n}{N-1} & \dfrac{N-1}{2} \leq n \leq N-1 \end{cases}$$

3. *Hanning.* For $0 \leq n \leq N - 1$, the window is defined as

$$w(n) = \frac{1}{2} \left[1 - \cos\left(\frac{2\pi n}{N-1}\right) \right]$$

This is also referred to as the raised cosine window because it is symmetric.

4. *Hamming.* For $0 \leq n \leq N - 1$, the window is defined as

$$w(n) = 0.54 - 0.46 \cos\left(\frac{2\pi n}{N-1}\right)$$

This window is also known as the raised cosine platform, which has N nonzero terms.

5. *Blackman.* For $0 \leq n \leq N - 1$, the window function is written as

$$w(n) = 0.42 - 0.5 \cos\left(\frac{2\pi n}{N-1}\right) + 0.08\left(\frac{4\pi n}{N-1}\right)$$

6. *Kaiser.* For $0 \leq n \leq N-1$, the window function is written as

$$w(n) = w_{\mathrm{R}}(n) I_0 \frac{\alpha \sqrt{1-(n/N)^2}}{I_0(\alpha)}$$

where $w_{\mathrm{R}}(n)$ is the rectangular window, α the Kaiser parameter, and I_0 a modified Bessel function of the first kind of order zero, which has the following form:

$$I_0(\alpha) = \sum_{m=0}^{\infty} \left[\frac{\alpha}{m! 2^m} \right]^2$$

One advantage of the Kaiser window is increased design flexibility, the Kaiser parameter α can be used to control the main lobe widths and the side band ripple. This is useful for deriving other window types, for example, if $\alpha = 9$, the Kaiser window reduces to the Blackman window. Generally, window design is iteratively done because it is difficult to know in advance where the attenuation band of the window will affect the desired frequency response. Normally, one would adjust the band edges until the FIR filter fits the required frequency response. Other window methods such as Gaussian and Chebyshev windows can be found in Roberts and Mullis (1987) and more detailed design examples of low-pass, high-pass, and bandpass FIR filters can be found in Akay (1994).

2.6.4.2 Frequency Sampling Method

The frequency sampling method (Oppenheim et al., 1999) is another popular technique for designing FIR filters. First, the discrete Fourier transform is employed to represent the coefficients of the frequency response. Then the impulse response coefficients are determined via the inverse discrete Fourier transform.

The discrete Fourier representation for a frequency response $H(k)$ can be written as

$$H(k) = H_{\mathrm{I}}(e^{j\omega}) \tag{2.156}$$

where the samples have frequency $\omega = 2\pi k/N$ for $k = 0, 1, 2, \dots, N-1$. Then the filter coefficients are derived using the inverse transform via

$$h(n) = \frac{1}{N} \sum_{k=1}^{N-1} H(k) e^{\frac{j 2\pi n k}{N}} \tag{2.157}$$

for $n = 0, 1, 2, \dots, N-1$. Since the FIR coefficients take real values they can be estimated from all the complex terms in the complex conjugate pairs, so that we have

$$h(n) = \frac{1}{N} \left(H(0) + 2 \sum k = 1^{\frac{N-1}{2}} \mathrm{Re} \left[H(k) e^{\frac{j 2\pi n k}{N}} \right] \right) \quad \text{if } N \text{ is odd} \tag{2.158}$$

$$h(n) = \frac{1}{N} \left(H(0) + 2 \sum k = 1^{\frac{N}{2}-1} \mathrm{Re} \left[H(k) e^{\frac{j 2\pi n k}{N}} \right] \right) \quad \text{if } N \text{ is even}$$

$$\tag{2.159}$$

Using the computed $h(n)$ values, the filter response can be obtained in z-transform using the relationship

$$H(z) = \sum_{n=0}^{N-1} h(n)z^{-n} \tag{2.160}$$

The frequency sampling method is therefore a direct method compared to the window technique because it avoids transformations from the time domain to the frequency domain. The large overshoots that usually occur at the attenuation or transition band of the filter response can be minimized by leaving some unconstrained terms in the response function. It is now known that this technique yields more efficient filters than the window method (Tompkins, 1993).

2.6.5   Integer Filters

Integer filters are another form of digital filter that are primarily deployed in environments requiring fast online processing. The previous digital filters can be implemented on computer software, however, the floating point operations performed on the real coefficients of the transfer function limit somewhat the speed of computation. In integer filters, these coefficients are replaced by integers making the computations faster and more efficient by using integer arithmetic operations. These operations require only bit shifting operations rather than the slower floating point unit (FPU) for computations. Such filtering is especially desirable for high-frequency digital signals or when computers have slow microprocessors. The major limitation of the integer filter is that it becomes difficult to obtain sharp cutoff frequencies by using only integer coefficients in the filter transfer function.

2.6.5.1   Design of Integer Filters

The design of integer filters revolves around placement of zeroes on the unit circle in the z-plane followed by positioning of poles to cancel the zeroes, such that frequencies defined by these specific locations are allowed through the passband. Since each point on the unit circle corresponds to a particular frequency, the frequency response of the integer filter can be determined. The transfer function commonly used (Tompkins, 1993) is

$$H_{\mathrm{I}}(z) = \frac{[1 - z^{-m}]^p}{[1 - 2\cos\theta z^{-1} + z^{-2}]^q} \tag{2.161}$$

where m represents the number of zeroes around the unit circle, p and q are integers, which give the order of magnitude for the filter and θ is the angular location of the poles. Raising the order of magnitude of the filter corresponds to cascading these filters. The first step is to place the zeroes using the numerator of the transfer function in Equation 2.161. One may use either of the following factors to determine the location of the zeroes:

$$(1 - z^{-m})$$

$$(1 + z^{-m})$$

It can be seen that m is equal to the number of zeroes placed at a particular location, beginning with $\theta = 0°$ and displaced evenly by $360°/m$. The next step is to ensure that the poles are positioned on the unit circle using complex conjugate factors. This can be seen by noting the Euler relationship for complex numbers, that is,

$$e^{j\theta} = \cos\theta + j\sin\theta \qquad (2.162)$$

and observing that the denominator of Equation 2.161 may be written as

$$(z^{-1} - e^{j\theta})(z^{-1} - e^{-j\theta}) \qquad (2.163)$$

This can be seen by using the fact that

$$\cos\theta = \frac{e^{j\theta} + e^{-j\theta}}{2}$$

and multiplying Equation 2.163 out to obtain

$$1 - 2\cos\theta z^{-1} + z^{-2}$$

The possible positions of the poles are given by the values of θ when $2\cos\theta$ is an integer. This occurs only for $\theta = 0°, \pm0°, \pm60°, \pm90°, \pm120°$, and $\pm180°$ as seen in the plot on Figure 2.19.

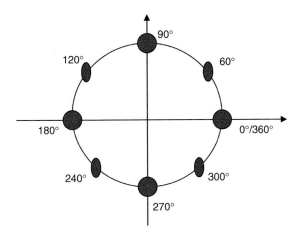

FIGURE 2.19
Possible pole zero placements on the unit circle in the z-plane for integer filter design.

As with the other types of filter, it is possible to design low-pass, high-pass, and even bandpass integer filters. For low-pass integer filters, the transfer function used is usually

$$H(z) = \frac{1 - z^{-m}}{1 - z^{-1}}$$

This results in a lower-frequency lobe larger in magnitude than the high-frequency lobe, which effectively means that the circuit acts as a low-pass filter. Lower cutoff frequencies can be achieved by adding more zeroes to the transfer function or using a lower sampling frequency. The former is preferred due to sharper cutoff frequencies that can be obtained, whereas using a lower sampling frequency is undesirable due to possible information loss.

High-pass integer filters can be designed using the standard high-pass design similar to the low-pass and the *subtraction method*. In the first method, we place a pole at $z = (-1, 0)$ corresponding to $\theta = 180°$, and depending on the number of zeroes, we use the following transfer function:

$$H(z) = \frac{1 - z^{-m}}{1 + z^{-1}} \quad \text{if } m \text{ is even}$$

$$H(z) = \frac{1 + z^{-m}}{1 + z^{-1}} \quad \text{if } m \text{ is odd}$$

The highest input frequency for this technique should be less than 50% of the sampling frequency achieved by designing an analog low-pass filter to eliminate frequencies higher than the sampling frequency. If bandwidths greater than 25% are required, the filter subtraction method must be applied instead and a low-pass filter is combined with an all-pass filter. The all-pass filter acts as a pure delay network with constant gain and has the following transfer function:

$$H(z) = Az^{-m} \tag{2.164}$$

where A is the gain, which is set equal to the gain of the low-pass filter. The resulting filter has the form depicted in Figure 2.20.

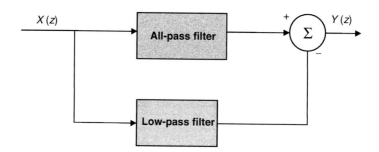

FIGURE 2.20
High-pass integer filter design using a filter subtraction method.

A combination of low-pass and high-pass integer filters can be used to design a bandpass integer filter. This filter shares characteristics of basic integer filters in that the transfer function coefficients are limited to integers, although the poles can only be placed at $\theta = 60°, 90°$, and $120°$. The sampling frequency must be selected so that the passband exists between these three angular locations. As before, the complex conjugate poles $(1 + z^{-m})$ and $(1 - z^{-m})$ are used to cancel the zeroes in the passband frequency. It should be noted that as the number of zeroes is increased, the bandwidth decreases, the amplitude of the side lobes decreases, cutoff steepness increases, and more sample data must be stored to compute the difference equations.

2.7 Adaptive Filters

In some biomedical applications, such as EEG and EMG monitoring, the noise is much greater than the intended signal to be measured. This problem is further exacerbated by the fact that both required signal and noise coexist in the same frequency band and so the noise cannot be selectively filtered out by removing any particular frequency band. In this situation, fixed coefficient filters cannot be used because they would filter out the required signal. In extreme cases, the frequency band in which the noise exists also varies within the band of the required signal. Clearly in such cases, we require a filter that can adjust or adapt to the changing noise.

Adaptive filters are generalized filters with filter coefficients not specified prior to operation but change during the operation in an intelligent manner by adapting to the incoming signal. This is accomplished by using adaptive algorithms such as least mean squares (LMS), recursive least squares (RLS), and Kalman filter-type algorithms. We will review the general adaptive filter and two popular and straightforward adaptive algorithms, namely, the LMS method and the RLS method. The LMS method is preferred due to its stability, but the RLS method has been shown to have superior convergence properties. Kalman filter algorithms occupy a large part of the literature and are well described in references such as Brown (1997), Banks (1990), and Clarkson (1993).

2.7.1 General Adaptive Filter

Adaptive filtering begins with a general filter, which usually has a FIR structure. Other forms such as the IIR or lattice structures can be used (Ifeachor and Jervis, 1993), but the FIR structure as shown in Figure 2.21 is more stable and easily implemented. There are two main components to the filter, the filter coefficients, which are variable and the adaptive algorithm used to change these coefficients.

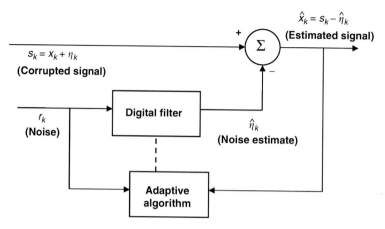

FIGURE 2.21
General adaptive filter structure. Filter coefficients are modified by an adaptive algorithm based on the estimated signal recovered.

The filter receives two input signals, s_k and r_k simultaneously. The composite signal, s_k contains the original signal to be measured x_k, corrupted by noise η_k. The reference noise signal, r_k is correlated in some way with η_k. In this filtering technique, the noise signal η_k is assumed to be independently distributed to the required signal x_k. The aim is to obtain an estimate of the corrupting noise, remove it from the signal to obtain x_k with less noise. The estimated signal is

$$\hat{x}_k = s_k - \hat{\eta}_k = x_k + \eta_k - \hat{\eta}_k \qquad (2.165)$$

The output signal $\hat{x}_k$ is used as an estimate of the desired signal and also as an error signal, which is fed back to the filter. The adaptive algorithms use this error signal to adjust the filter coefficients to adapt to the changing noise.
The FIR filter with N-points is given by the following equation:

$$\hat{\eta}_k = \sum_{i=1}^{N} w_k(i) r_{k-i} \qquad (2.166)$$

where $w_k(i)$ for $i = 1, 2, \ldots, N$ are the filter coefficients, r_{k-i} the input noise reference, and $\hat{\eta}_k$ the optimum estimate of the noise. If multiple input signals are present simultaneously, the equation can be extended to vector notation to denote a system with multiple inputs and a single output.

2.7.2   The LMS Algorithm

Most adaptive algorithms are based on the discrete Wiener filter (Figure 2.22) that we first review to facilitate description of the LMS algorithm. Two signals s_k and r_k are applied to the filter, where s_k contains a component that

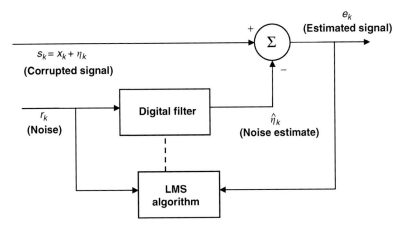

FIGURE 2.22
Adaptive filter with coefficients determined using the least mean squares algorithm.

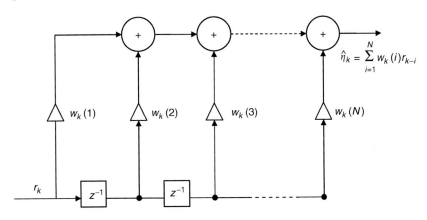

FIGURE 2.23
Structure of an N-point FIR filter with N-filter coefficients.

is correlated with r_k. The Wiener filter then produces an optimal estimate of the correlated part, which is subtracted from s_k. In adaptive filters, the correlated component is the noise to be removed.

For an FIR filter structure (Figure 2.23) with N filter coefficients, the difference between the output of the Wiener filter and the correlated signal is given by

$$e_k = s_k - \sum_{i=1}^{N} w_i r_{k-i} = s_k - \mathbf{w}^{\mathrm{T}} \mathbf{r}_k \tag{2.167}$$

where $\mathbf{w} = w_1, w_2, \ldots, w_N$ is the vector of filter coefficients (sometimes known as weights) and $\mathbf{r} = r_{k+1}, r_k, \ldots, r_N$ is the reference noise signal. The square of the error is obtained by squaring Equation 2.167 giving

$$e_k^2 = s_k^2 - 2s_k \mathbf{w}^{\mathrm{T}} \mathbf{r}_k + \mathbf{w}^{\mathrm{T}} \mathbf{r}_k \mathbf{w}^{\mathrm{T}} \mathbf{r}_k \tag{2.168}$$

Assuming that the signals are jointly stationary, the mean square error (MSE) can be obtained by taking the expectations on both sides of Equation 2.168 giving

$$\begin{aligned} J &= E[e_k^2] \\ &= E[s_k^2] - 2E[s_k \mathbf{w}^{\mathrm{T}} \mathbf{r}_k] + E[\mathbf{w}^{\mathrm{T}} \mathbf{r}_k \mathbf{w}^{\mathrm{T}} \mathbf{r}_k] \\ &= \sigma^2 - 2\mathbf{p}^{\mathrm{T}} \mathbf{w} + \mathbf{w}^{\mathrm{T}} \mathbf{A} \mathbf{w} \end{aligned} \tag{2.169}$$

where $\mathbf{p}$ is the N-dimensional cross-correlation vector, $\mathbf{A}$ a $N \times N$ autocorrelation matrix, and σ^2 the variance of s_k. The gradient of the MSE function is given as

$$\frac{\mathrm{d}J}{\mathrm{d}\mathbf{w}} = -2\mathbf{p}^{\mathrm{T}} + 2\mathbf{A}\mathbf{w} \tag{2.170}$$

and the optimum Wiener filter coefficients are found when the gradient of the MSE function is stationary or zero. This gives us

$$\mathbf{w}^* = \mathbf{A}^{-1}\mathbf{p} \tag{2.171}$$

which is also known as the Wiener–Hopf equation or solution. This filter is difficult to implement because it requires knowledge of $\mathbf{A}$ and $\mathbf{p}$, which are not known *a priori*. In addition, the inversion of the autocorrelation matrix $\mathbf{A}$ becomes slower when the number of filter coefficients increases. Furthermore, if the assumption that the signals are stationary do not hold then $\mathbf{A}$ and $\mathbf{p}$ change continuously in time and $\mathbf{w}^*$ must be constantly recomputed.

The basic LMS algorithm (Widrow et al., 1975a) attempts to overcome these limitations by sequentially solving samples to obtain $\mathbf{w}^*$. The update rule at each sampling instant t is defined as

$$\mathbf{w}^{t+1} = \mathbf{w}^t - \mu \frac{\mathrm{d}J}{\mathrm{d}\mathbf{w}^t} \tag{2.172}$$

or using Equation 2.170

$$\mathbf{w}^{t+1} = \mathbf{w}^t - 2\mu(s_k - \mathbf{w}^{\mathrm{T}} \mathbf{r}_k)\mathbf{r}_k \tag{2.173}$$

This algorithm uses instantaneous estimates of $\mathbf{A}$ and $\mathbf{p}$ and hence the filter coefficients $\mathbf{w}$ obtained are also estimates, but these weights will gradually converge to the optimum value as the filter learns the true chracteristics of the noise. It has been shown that the condition for convergence is

$$\mu > \frac{1}{\lambda_{\mathrm{max}}} > 0 \tag{2.174}$$

where λ_{max} is the largest eigenvalue of the input covariance matrix. In practice, the optimum value $\mathbf{w}$ is never reached but oscillates around the optimal values instead.

The basic LMS algorithm suffers from several weaknesses. First, if the signal environment is nonstationary then the previous assumptions do not hold. This means that the autocorrelation matrix and cross-correlation vector constantly vary resulting in an ever-changing minimum of the MSE function. This creates difficulties to iteratively achieve the minimum using steepest descent algorithms because the curvature of the problem surface is continually changing. The problem can be solved to some extent using the time-sequenced adaptive filter (Ferrara and Widrow, 1981). A further difficulty is that the noise signal is assumed to be uncorrelated to the required signal, but this seldom holds in practice and given a small correlation the recovered signal will be slightly distorted. A bigger problem arises if the original input signal s_k contains no noise causing the introduction of r_k to result in removal of components that completely changes the nature of the recovered signal. In this situation, the filter has destroyed important information rather than filter out the noise. Other problems include computational roundoff errors when programming the filter, which can result in distortion of the recovered signal and slow convergence of the algorithm.

Nevertheless, several further modification improved the basic LMS algorithms giving rise to complex LMS algorithms (Widrow et al., 1975b) for complex data, block LMS algorithms, which improve the computation speed by updating blocks of coefficients rather than just one variable per iteration, and time-sequenced LMS algorithms to deal with nonstationary signal environments.

2.7.3 The RLS Algorithm

The RLS algorithm is an implementation of the well-known least squares method similar to the basic Wiener filter shown in Figure 2.24. The optimum

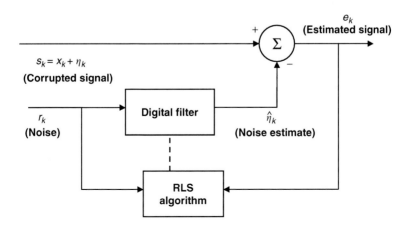

FIGURE 2.24
Adaptive filter with coefficients determined using the recursive least squares algorithm.

estimates in this case are computed for all coefficients in a single iteration and obtained using

$$\mathbf{w}_m = [\mathbf{R}_m^{\mathrm{T}}\mathbf{R}_m]^{-1}\mathbf{R}_m^{\mathrm{T}}\mathbf{s}_m \tag{2.175}$$

where $\mathbf{s}_m = \{s_1, s_2, \ldots, s_m\}$, $\mathbf{w}_m = \{w_1, w_2, \ldots, w_n\}$, and the $m \times m$ matrix $\mathbf{R}_m = \{\mathbf{r}_1^{\mathrm{T}}, \mathbf{r}_2^{\mathrm{T}}, \ldots, \mathbf{r}_m^{\mathrm{T}}\}$. Here the m-dimensional vector $\mathbf{r}_i$ is defined as $\mathbf{r}_i = \{r_i(1), \ldots, r_i(n)\}$ for all $i = 1, 2, \ldots, m$. The least squares Equation 2.175 can be solved using any matrix inversion technique. The filter output is then obtained for $k = 1, 2, \ldots, m$ as follows:

$$\hat{\eta}_k = \sum_{i=1}^{n} \hat{w}_i r_{k-i} \tag{2.176}$$

In the RLS algorithm, the computationally expensive matrix inversions, that is, $[\mathbf{R}_m^{\mathrm{T}}\mathbf{R}_m]^{-1}$ are avoided by sequentially updating the filter coefficients for each new dataset, thereby avoiding direct computation of matrix inverses. The algorithm is obtained by exponentially weighting data to slowly remove the effects of old data in the determination of $\mathbf{w}_m$. This is depicted by the following update rules:

$$\mathbf{w}_k = \mathbf{w}_{k-1} + \mathbf{G}_k \mathbf{e}_k$$

$$\mathbf{P}_k = \frac{1}{\gamma}[\mathbf{P}_{k-1} - \mathbf{G}_k \mathbf{r}^{\mathrm{T}}(k)\mathbf{P}_{k-1}]$$

where

$$\mathbf{G}_k = \frac{\mathbf{P}_{k-1}\mathbf{r}(k)}{\alpha_k}$$

$$e_k = s_k - \mathbf{r}^{\mathrm{T}}(k)\mathbf{w}_{k-1}$$

$$\alpha_k = \gamma + \mathbf{x}^{\mathrm{T}}(k)\mathbf{P}_{k-1}\mathbf{x}(k)$$

The inverse of $\mathbf{R}_m^{\mathrm{T}}\mathbf{R}_m$ is thus computed recursively using $\mathbf{P}_k$ where the indices k refer to the recursive step or sample point and the quantities are defined during that sample point, whereas γ is a forgetting factor, which allows gradual replacement of knowledge from previous signals. The scheme becomes the standard least squares scheme when $\gamma = 1$, although smaller values of γ cause oscillations as the filter attempts to adapt too strongly to the new data. The asymptotic sample length (ASL) is the number of previous samples that still affect the value of the filter coefficients $\mathbf{w}_k$ for a sampling point k. The ASL can also be interpreted as measuring the RLS filter memory and is described by the following relationship:

$$\sum_{k=0}^{\infty} = \frac{1}{1-\gamma} \tag{2.177}$$

Clearly, when $\gamma = 1$, the filter memory becomes infinite, or we obtain the standard least squares method.

Although the RLS method is very efficient, there are problems with direct implementation. If the reference noise signal r_k is zero for some time, the matrix $\mathbf{P}_k$ will increase exponentially due to continuous division by γ, which is less than unity. This no doubt causes overflow errors and further computational errors accumulate. The RLS method is also susceptible to computational roundoff errors, which can give a negative definite matrix $\mathbf{P}_k$ at some stage of the signal causing numerical instability since proper estimation of w requires that $\mathbf{P}_k$ be positive semidefinite. This cannot be guaranteed because of the difference terms in the computation of $\mathbf{P}_k$, but the problem can be alleviated by suitably factorizing the matrix $\mathbf{P}_k$ to avoid explicit use of the difference terms. Two such algorithms that give comparable accuracies to the RLS algorithm are the square root (Ifeachor and Jervis, 1993) and upper diagonal matrix factorization algorithms (Bierman, 1976).

2.8   Concluding Remarks

The signal-processing techniques reviewed in this chapter have been used extensively to process biosignals. In the following chapters, we will investigate how some of these techniques are employed, first to remove noise from the signal and then to extract features containing discriminative information from the raw biosignals. For pattern recognition applications, the coefficients of signal transforms have been commonly used to discriminate between classes. In some cases, these coefficients are sufficient to fully characterize the signal and classification can be easily achieved by setting single thresholds. Unfortunately, in many other cases this is insufficient and instead of spending too much time obtaining fully separable features, we can employ CI techniques. These techniques include advanced classifiers, which are better suited to handling nonseparable data and dealing with nonlinear function estimations, topics that form a major subject of Chapter 3.

References

Akaike, H. (1969). Power spectrum estimation through autoregression model fitting. *Annals of the Institute of Statistical Mathematics 21*, 407–449.

Akaike, H. (1974). A new look at the statistical model identification. *IEEE Transactions on Automatic Control 19*, 716–723.

Akay, M. (1994). *Biomedical Signal Processing*. San Diego, CA: Academic Press.

Banks, S. P. (1990). *Signal Processing, Image Processing, and Pattern Recognition.* Prentice Hall International Series in Acoustics, Speech, and Signal Processing. New York: Prentice Hall.

Bierman, G. (1976). Measurement updating using the UD factorization. *Automatica 12*, 375–382.

Brown, R. (1997). *Introduction to Random Signals and Applied Kalman Filtering: With Matlab Exercises and Solutions.* New York: Wiley.

Burg, J. (1968). *A New Analysis Technique for Time Series Data.* Reprinted in Modern Spectrum Analysis, D. G. Childers (Ed.), IEEE Press New York: NATO Advanced Study Institute on Signal Processing with Emphasis on Underwater Acoustics.

Clarkson, P. M. (1993). *Optimal and Adaptive Signal Processing.* Electronic Engineering Systems Series. Boca Raton, FL: CRC Press.

Cooley, J. and J. Tukey (1965). An algorithm for the machine calculation of complex Fourier series. *Math Computations 19*, 297–301.

Daubechies, I. (1988). Orthonormal bases of compactly supported wavelets. *Communications on Pure and Applied Mathematics 41*, 909–996.

Daubechies, I. (1990). The wavelet transform, time-frequency localization and signal analysis. *IEEE Transactions on Information Theory 36(5)*, 961–1005.

Dutton, K., S. Thompson, and B. Barraclough (1997). *The Art of Control Engineering.* London: Addison and Wesley Longman.

Ferrara, E. and B. Widrow (1981). The time sequenced adaptive filter. *IEEE Transactions on Acoustics, Speech and Signal Processing 29(3)*, 766–770.

Ifeachor, E. C. and B. W. Jervis (1993). *Digital Signal Processing: A Practical Approach.* Electronic Systems Engineering Series. Wokingham, England: Addison-Wesley.

Mallat, S. (1999). *A Wavelet Tour of Signal Processing 2nd Ed.* New York: Academic Press.

Oppenheim, A. V., R. W. Schafer, and J. R. Buck (1999). *Discrete-Time Signal Processing* (2nd ed.). Upper Saddler River, NJ: Prentice Hall.

Parzen, E. (1974). Some recent advances in time series modelling. *IEEE Transactions Automatic Control AC-19*, 723–730.

Proakis, J. and G. M. Dimitris (1996). *Digital Signal Processing: Principles, Algorithms, and Applications.* Englewood Cliffs, NJ: Prentice Hall International.

Rissanen, J. (1983). A universal prior for the integers and estimation by minimum description length. *Annals of Statistics 11*, 417–431.

Roberts, R. and R. Mullis (1987). *Digital Signal Processing.* Reading, MA: Addison Wesley.

Stein, J. (2000). *Digital Signal Processing: A Computer Science Perspective* (1st ed.). Wiley series in telecommunications and signal processing. New York: Wiley Interscience.

Stergiopoulos, S. (2001). *Advanced Signal Processing Handbook: Theory and Implementation for Radar, Sonar, and Medical Imaging Real-Time Systems.* The Electrical Engineering and Signal Processing Series. Boca Raton, FL: CRC Press.

Tompkins, W. J. (1993). *Biomedical Digital Signal Processing: C-language Examples and Laboratory Experiments for the IBM PC.* Englewood Cliffs, NJ: Prentice Hall.

Widrow, B., J. Glover, J. McCool, J. Kaunitz, S. Williams, R. Hearn, J. Zeidler, E. Dibf and R. Goodlin (1975a). Adaptive noise cancelling: principles and applications. *IEEE Proceedings 63*, 1692–1716.

Widrow, B., J. McCool, and M. Ball (1975b). The complex LMS algorithm. *IEEE Proceedings 63*, 719–720.

Winter, D. A. and A. E. Patla (1997). *Signal Processing and Linear Systems for the Movement Sciences.* Waterloo, Canada: Waterloo Biomechanics.

3

Computational Intelligence Techniques

3.1 Computational Intelligence: A Fusion of Paradigms

We have always held a fascination for the mechanisms of learning ever since the time of the Greeks and the great Aristotle (400 BC) himself. In fact, Aristotle was the first to propose that memory was composed of individual elements linked by different mechanisms (Medler, 1998). His ideas later formed the philosophy of connectionism, a study devoted to the connectionist model, which grew richer from further ideas from different schools of thought. There were contributions from philosophers such as Descartes, psychologists such as Spencer (Spencer's connexions), James (James associative memory), and Thorndike (Thorndike's connectionism), and even neuropsychological influences from Lashley and Hebb (Hebbian learning) (Pollack, 1989; Medler, 1998).

Although the philosophy of learning bloomed, other fields of science progressed with equal vigor. Advances in engineering and mathematics saw the development and refinement of calculating devices, which began as the simple abacus and developed to the current computer based on the Von Neumann architecture (stored program computer). Computers soon became an integral part of technology requiring fast calculations and it was inevitable that they would be applied in the study of learning. Prior to this, learning theories had remained cumbersome to test as they required long-term studies of human and animal behavior. If the learning process could be artificially replicated, experimental validation of previous learning theories could be made. Indeed Hull (Hull and Baernstein, 1929) remarked that if the mechanisms of muscle reflexes could be replicated by some inorganic method, then surely powers of reasoning and deduction could be similarly replicated in a machine. It was not until the advent of electronic computers during the 1950s, that connectionist theories could be implemented and tested, and hence the field of machine learning was born. This term later evolved to "computational intelligence" with a given emphasis on computers and machine learning.

There are several machine learning paradigms: supervised learning, unsupervised learning, genetic algorithms, symbolic learning, and others (Shavlik and Dietterich, 1990). Supervised learning has been highly researched having stemmed from the mathematical problem of function estimation. In this formulation, an external teacher or supervisor provides a set of examples attached respectively with class labels or desired outputs. The task is to train a machine to recognize the implicit relationships between the examples and the desired outputs, so that once trained, the output for a previously unseen example could be provided. In some cases, global information may be required or incorporated into the learning process. The main paradigms in supervised learning include risk minimization, reinforcement learning, and stochastic learning.

Unsupervised learning uses only local information and a general set of rules, and is usually referred to as self-organization because the machine self-organizes data presented to the network and detects implicit collective characteristics of the data. Some paradigms of unsupervised learning are Hebbian learning and competitive learning. Genetic algorithms are optimization techniques based on evolutionary principles in biological systems, which include formulations such as genes, mutation, and optimization based on the "survival of the best fitted" principle. Symbolic learning is still very much in its infancy and involves teaching a machine how to learn relationships between symbols and perceptual learning.

This chapter reviews learning techniques that have been successfully applied to the diagnosis of diseases. Techniques such as artificial neural networks (ANN), SVM, HMM, fuzzy logics, and hybrid systems constructed from a combination of these techniques are regarded as core CI techniques. We have omitted the exciting field of evolutionary computation, which includes particle swarm optimization, genetic algorithms, and ant colony optimization since they have only just recently been applied to the field of biomedical engineering. This chapter is meant to provide a general overview of some CI techniques beginning with fundamental concepts of each technique and summarizing the more detailed aspects. Each technique mentioned here has been extensively researched and the reader is encouraged to pursue the references provided for further details of particular techniques of interest.

3.1.1 The Need for Computational Intelligence in Biomedical Engineering

The development of CI beginning with the artificial neuron emerged from the desire to test our understanding of learning concepts. It is now recognized that the computer is a powerful device capable of manipulating numbers and symbols at a higher rate than the average human (Jain et al., 1996). However, the current Von Neumann architecture is inadequate for dealing with problems that require complex perception. For example, we recognize a voice almost instantly and a fleeting glimpse of a face is usually adequate for us to identify it with a name, an address, and a host of other experiences we may have had with that person. However, even the fastest computers today

	Von Neumann Computer	**Biological Computer**
Processor	Complex High speed Single or serveral CPUs	Simple Low speed Many neurons
Memory	Separated from processor Localized Noncontent addressable	Integrated into processor Distributed Content addressable
Computing	Centralized Strictly sequential Stored programs	Distributed Massive parallel Self-learning
Reliability	Vulnerable to electrial failure	Robust with a lot of redundancy
Expertise	Numerical and symbolic manipulations	Perceptual learning, creativity and innovation
Operating environment	Well defined and constrained	Not defined and constraints unknown

FIGURE 3.1
Differences between a standard Von Neumann computer and biological computer, for example, brain. (Adapted from Jain, A.K., Mao, J., and Mohiuddin, K.M., *IEEE Computers*, 29(3), 31–44, 1996, ©IEEE.)

struggle with the most simple task of identification as a result of the many differences the computer has with its biological counterpart, the brain (see Figure 3.1).

The human brain having evolved through millions of years of evolution, possesses many functional properties such as massive parallelism, the ability to learn, generalize, adapt, high reliability and robustness, and low energy consumption. The computer has faster processing and considerable memory capacity with potential to fuse these properties into an efficient information processing system but lacks many other qualities found in the brain. Therefore, to replicate the many powerful characteristics of human intelligence, the Von Neumann architecture must be substantially modified or even completely redesigned. At present, we must however be content to design programs that can perform well with the current computer technology. CI techniques attempt to replicate the human learning processes and represent an initial step toward this goal of simulating human intelligence.

In biomedicine, it has been recognized (Held et al., 1991) that mathematical models of living systems are needed to provide a better visualization of the system. Prior to the 1980s, mathematical models were not sufficiently realistic to model living systems such as the cardiovascular system whereas more complex models were impossible to implement and thus offered little analytical use. With the advent of high-performance computing, these complications have been alleviated and CI plays an increasing role in modelling biological systems. To paraphrase Witten, the computer has become an electronic and virtual laboratory so much so that "what was once done *in* a tube is now being done *on* the tube."*

3.1.2 Problem and Application Domains

Since the 1950s, many computational techniques have been developed, which incorporated various learning theories and concepts from statistics and mathematics. These CI techniques are part of a broader field known as *machine learning*, which focuses on solving several core problems from mathematics and engineering as described in the following:

a. *Classification or pattern recognition.* The most popular application of CI is classification or pattern recognition. In this problem, one is given a set of labelled data and the machine is trained to recognize the relationship between the data structure and the class label. Formally, let $\mathbf{x} = \{(\mathbf{x}_1, y_1), (\mathbf{x}_2, y_2), \ldots, (\mathbf{x}_n, y_n)\}$ be a set of n couples each consisting of a vector $\mathbf{x}_i \in \Re^m$ and $y_i \in \mathbb{N}$ generated by some unknown generating function $f_\theta(\mathbf{x}_i, \boldsymbol{\theta})$. The parameter space $\boldsymbol{\theta} = \{\theta_1, \theta_2, \ldots, \theta_z\}$ is a linear z-dimensional subspace of $\Re^n$. We would like to derive a function, $f_E(\mathbf{x})$, where $f_E(\mathbf{x}) \in \mathbf{F}$ such that $f_E(\mathbf{x})$ approximates the true function $f(\mathbf{x}_i, \boldsymbol{\theta})$ as closely as possible. The space of integers $\mathbb{N}$ indicates that the outputs of the true function take on only discrete values, symbolizing the number of classes involved in the classification problem. The function $f_E(\mathbf{x})$ represents a separating boundary or surface and is referred to as the classifier. Classifiers such as ANNs and SVMs have been successfully applied to a plethora of problems such as text recognition, speech recognition, and video and image recognition. Currently, classification is the most relevant problem in biomedical engineering where classifiers have been developed for data from ECGs, EEGs, x-rays, ultrasound images, and MRI images to assist with the diagnosis of diseases.

b. *Function estimation/regression.* Function estimation is a more general problem than classification where an unknown generating function is estimated using only the corresponding inputs and output responses. More formally, let $\mathbf{x} = \{(\mathbf{x}_1, y_1), (\mathbf{x}_2, y_2), \ldots, (\mathbf{x}_n, y_n)\}$ be a set of n couples, each consisting of a vector $\mathbf{x}_i \in \Re^m$ and $y_i \in \Re$ generated by some unknown generating function $f_\theta(\mathbf{x}_i, \boldsymbol{\theta})$. Note that in this case, the function output y is allowed to be a real number instead of integers as with classification. The parameter space $\boldsymbol{\theta} = \{\theta_1, \theta_2, \ldots, \theta_z\}$ is a linear z-dimensional subspace of $\Re^n$. We are required to derive a function, $f_E(\mathbf{x})$, where $f_E(\mathbf{x}) \in \mathbf{F}$ such that $f_E(\mathbf{x})$ approximates the true function $f(\mathbf{x}_i, \boldsymbol{\theta})$ as closely as possible. The class $\mathbf{F}$ is assumed to contain a broad class of smooth functions and the task would be to determine the space θ as accurately as possible. The function, $f_E(\mathbf{x})$ is known as the *regressor* and several methods have

---

*M. Witten, *High Performance Computational Medicine* in Held et al. (1991).

been used, for example, least squares regression and support vector regression (SVR). Regression is particularly useful for problems in statistics and engineering. Classification and regression is also known as *supervised learning* because a supervisor has already provided the data with the necessary labels or values.

c. *Time series prediction/forecasting.* In this problem, one is presented with a series of data $\{y_1, y_2, \ldots, y_t\}$ generated by some unknown distribution or function. The challenge is to predict the value of the next sample y_{t+1}. This problem often occurs in weather prediction and stock market or financial predictions.

d. *Clustering.* This is a slight variation of the classification problem and more difficult to solve. One is provided with unlabelled data and the task is to find similarities between the individual data and group data. This type of learning is known as *unsupervised learning* or clustering and covers methods such as independent component analysis (ICA), blind source separation (BSS) (Girolami, 1999), and principal component analysis (PCA).

e. *Optimization.* In optimization, the aim is to solve a minimization or maximization problem subject to a set of constraints. The problem is usually modeled mathematically as an *objective function* and the constraint set can be composed of equalities or inequalities, which must be satisfied by the variables. The optimal solution is the set of variables that solve the problem subject to the constraints. Many CI techniques are formulated as optimization problems, which can be solved using techniques such as gradient descent, Newton's method, and conjugate gradients (Chong and Stanislaw, 2001; Fletcher, 1981; Gill et al., 1981).

f. *Content-addressable memory.* In the standard Von Neumann architecture, the memory address is calculated and then accessed. Addresses in memory have no physical meaning and if the incorrect address is calculated, the correct information will not be retrieved. Associative memory is memory, which can be addressed by the content and is more robust to errors or partial addresses than the memory in the Von Neumann architecture. Associative memory can be mimicked by, for example, ANN and is important for applications such as multimedia databases (Jain et al., 1996).

g. *Control.* Control is essential in the design of engineering systems. One has an input–output relationship $\{u(t), y(t)\}$ where $u(t)$ is the input signal, which must be constrained to follow (controlled) the desired output signal $y(t)$ for some time t. Examples of control problems are motor idle speed and water-level control in dams.

In the following sections, we describe the main CI techniques, which are currently more applicable to problems in biomedical engineering.

3.2　Artificial Neural Networks

The ANN is an information-processing paradigm inspired by the structure of biological nervous systems such as the human brain. The primary characteristic of this paradigm is modelling the core information-processing system as a neuron and building a network composed of a large number of highly interconnected neurons working to solve specific problems. ANNs were the earliest attempts to mimic human intelligence with a machine and it was generally accepted that learning in biological systems involved developing interneuron synaptic connections. The key concepts therefore are the characteristics of the neuron, the neuron interconnections, and the neuron activation system consisting of several mathematical functions.

The discovery of the neuron is primarily attributed to Ramon y Cajal (Jain et al., 1996), who postulated that the nervous system was made up of an interconnected network of polarized neurons. This discovery led to the 1906 Nobel Prize in Physiology or Medicine, which he shared with his contemporary Camillo Golgi who had shown that a silver chromate solution could be used to stain brain tissue and reveal the interconnected network of neurons. The major parts of a biological neuron (see Figure 3.2) are the cell body, which contains the nucleus and a branching dendritic tree that collects signals from neighboring neurons. The neuron's output is transmitted via the axon, which forms a connection or synapse with the dendrites of neighboring neurons (see Figure 3.3). The nucleus is the processing center, which integrates the input signals and generates a response that is a nonlinear function of the inputs and the internal state. The neuron alone does not possess considerable information-processing capabilities, but a large network formed from these single neurons gives rise to special processing abilities. We will see later

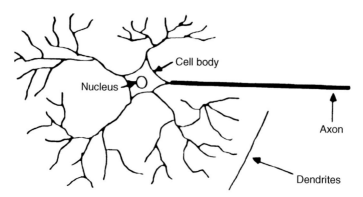

FIGURE 3.2

A neuron collects input signals from nearby neurons through the dendrites, integrates the signals, and produces an output response to other neurons via the axon.

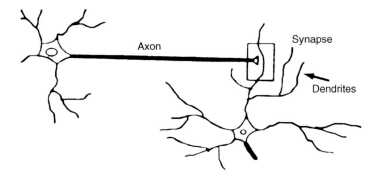

FIGURE 3.3
An axon forms a connection or a synapse with the dendrite tree of a neighboring neuron.

how the neuron is modeled as a mathematical construct and how a network of neurons is represented and trained to perform simple processing tasks.

Although the growth of interest in neural networks is usually attributed to Rosenblatt's discovery of the perceptron in 1958 (Rosenblatt, 1958), a paper by McCulloch and Pitts (1943) is attributed as the source of interest in neural networks. Warren McCulloch, a neurophysiologist, and the logician Walter Pitts designed the first conceptual ANN based on their understanding of neurons. Their paper however was difficult to comprehend and Kleene's (1956) notation is closer to today's standard. Rosenblatt's perceptron represented the neuron as a simple mathematical function, which received many inputs and produced a single output, successfully demonstrating that a learning task, such as pattern classification or object recognition, could be modeled mathematically. It soon became apparent that the early perceptron had several limitations, but the extent of these limitations would only be exposed a decade later. In 1969, Minsky and Papert (Minsky and Papert, 1988/1969) in their initial paper *Perceptrons* pointed out that perceptrons were only suitable for linearly separable classification problems and faced with inseparable problems such as the exclusive-OR (XOR) gate problem; the perceptron was unsuccessful. Several other problems were outlined, serving to further dampen enthusiasm in the connectionist model and neural networks. It was conjectured that the problem could be solved using a multilayered network of perceptrons however the difficulty in training the network remained and research in this field dwindled. This situation changed with the discovery of the backpropagation algorithm first described by Paul Werbos in his 1974 PhD thesis, which was unfortunately not recognized at that time. It was however, later rediscovered independently by Parker and LeCun in 1985 and further popularized by Rumelhart and McClelland (1986), stirring renewed interest in the field. About the same time as the perceptron, Selfridge (1959) proposed a new paradigm for machine learning called "Pandemonium," a learning model that continuously refined and adapted to pattern recognition problems, which it had not previously encountered. Widrow and Hoff (Widrow and Lehr, 1995)

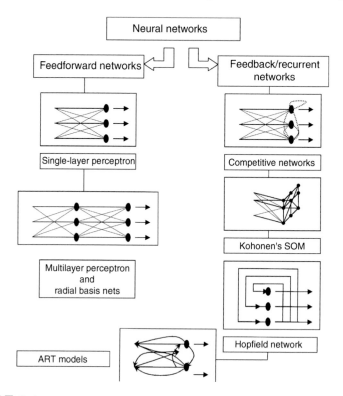

FIGURE 3.4
The various neural network architectures.

in 1960 extended the perceptron to adaptive linear neurons (Adaline) where the output of training a neuron was fed back to the input so that adaptation could be achieved.

There are several different types of neural network architectures ranging from feedforward networks (Haykin, 1999), which include the single-layer and multilayer perceptrons, to recurrent or feedback networks including Hopfield networks and Kohonen self-organizing maps (SOM) (see Figure 3.4).

3.2.1   The Perceptron: An Artificial Neuron

The artificial neuron can be modeled as a system (see Figure 3.5) that receives a quantity of data as inputs and produces a single output. An artificial neuron has many similarities with a biological neuron in that it is referred to as the *perceptron* because it mimics the fundamentals of mammalian visual systems. The perceptron is an example of the simplest feedforward network because it consists of a single neuron, which only allows input signals to travel in a single direction, that is, from input to output. There is no feedback loop, hence the output signal does not affect the neuron further. Feedforward ANNs are straightforward networks that associate inputs with outputs and this bottomup characteristic is very useful for pattern recognition problems.

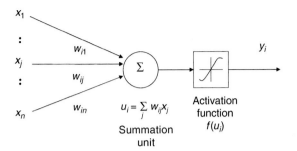

FIGURE 3.5
An artificial neuron or perceptron.

The similarities between the artificial neuron and the biological neuron can be seen in the following:

a. *Inputs.* The scalar x_i represents the electrical input into the neuron. In practice, the input consists of data such as measurements or attributes that mimic the dendrite trees of the biological neuron. The importance of each input is represented by the strength of the synapse connection that is modeled by the scalar weights w_i, which are premultiplied with the individual input x_i.

b. *Summation and activation function.* As in nucleus of a neuron, processing of the weighted inputs and the output response takes place at a summation *node* and is modeled as the following mathematical function:

$$y_i = f\left(\sum_k w_k x_k\right) \qquad (3.1)$$

Here the function f is known as the *activation function* and determines the behavior of the neuron. Mathematically, it is usually taken to be some bounded nondecreasing nonlinear function such as the sigmoid function (see Figure 3.6d)

$$f(u_i) = \frac{1}{1 + e^{-u_i}} \qquad (3.2)$$

Other commonly used functions are the Gaussian, step, piecewise linear (ramp) as seen in Figure 3.6. For linear activation functions, the output activity is directly proportional to the total weighted output. For step or threshold functions, the output is set at one of two levels, depending on whether the total input is greater than or less than some threshold value. Some activation functions are chosen such that very large positive or negative numbers are limited. In this case, they are usually called *squashing functions*, for example, sigmoid and step functions.

Computational Intelligence in Biomedical Engineering

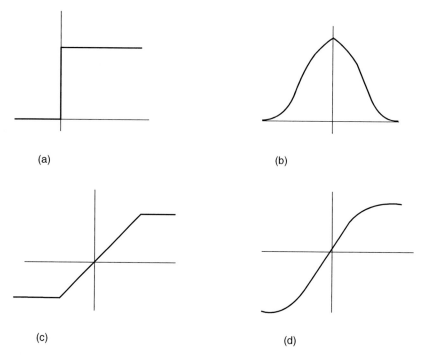

FIGURE 3.6
Examples of activation functions that have been used in artificial neurons.
(a) Step function; (b) Gaussian; (c) piecewise linear; and (d) sigmoid.

   c.   *Output.* The output y_i represents the strength of the electrical impulse travelling along the axon.

   d.   *Feedback.* In some ANNs, there is some provision to feed the output back into the input so that the neural network is adaptive.

    Learning in artificial neurons, perceptrons, or neural networks is by adjusting the neuron weights so that the network output matches the required or true output. In a pattern classification problem (see Section 3.1.2), for example, each training pattern consists of the pair $\{\mathbf{x}_i, y_i\}$ and the aim is to have the neuron learn the complex relationship between the training inputs and the class labels. The error between the output of the neuron, $f(u_i)$ and the actual class label y_i is usually taken to be the difference

$$E_i = y_i - f(u_i) \tag{3.3}$$

If the number of training patterns is n, the sum of errors for a single neuron E_{SN} is written as

$$E_{\text{SN}} = \sum_{i}^{n} E_i = \sum_{i}^{n} y_i - f(u_i) \tag{3.4}$$

The sum of squared errors then is

$$E_{\text{SN}}^{\text{sq}} = \sum_{i}^{n}(y_i - f(u_i))^2 \tag{3.5}$$

Learning or "training" the neuron requires minimizing the sum of errors as far as possible, a process referred to as empirical risk minimization (ERM), where the error is termed *risk* (Vapnik, 2000). If one minimizes the sum of squared errors, this becomes the least squares method. Training is usually done by applying general optimization techniques and designing iterative algorithms that can be easily implemented on a computer. For example, the perceptron learning procedure (Rosenblatt, 1958) can be summarized as follows.

Algorithm 3.1: Perceptron Learning Algorithm.

---

1 Initialize the weights to random values, for example, $0.2 < w_i < 2$.
2 Present the training example $\mathbf{x}_i$ to the perceptron and compute the output $f(u_i)$, where $u_i = \sum_{k} w_k x_k$ as before.

3 Using the sigmoid activation function (Equation 3.2) the weights are updated using

$$w_i^{t+1} = w_i^t - \lambda(y_i - f(u_i))x_i \tag{3.6}$$

where λ is the *learning rate* or *step size*, usually selected from the interval $0.05 < \lambda < 0.75$. Repeat steps 1–3 for all training examples until the sum of errors is either zero or reaches a small predefined threshold.

The random initialization of weights to small values avoids beginning with a large initial error and saturating the activation function. Note that if we start with all weights initialized to zero, then $f(u_i)$ is zero at time $t = 0$. This may be easier to implement in a computer program but could prolong the training times. The weight update rule (Equation 3) is a gradient descent method and leads to slow training times if the selected learning rate is too small. Rosenblatt (1958) showed that if the set of training data was linearly separable, the learning algorithm would converge. This became the famous perceptron convergence theorem and further enhanced the use of the perceptron. Other learning methods such as Boltzmann learning (Wroldsen, 1995; James and Edward, 1988), Hebbian rules (Hebb, 1949), and competitive learning rules such as vector quantization (Kohonen, 1989; Seo and Obermayer, 2003) could also be used to train the perceptron.

3.2.2 Multilayer Perceptrons

The single perceptron can only perform simple classification tasks, specifically those that are linearly separable. If the data is not linearly separable

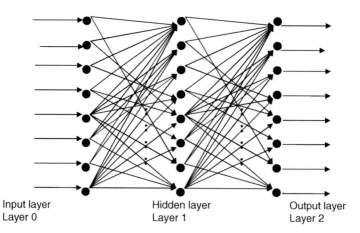

FIGURE 3.7
A double-layer MLP network.

then the learning algorithm would not converge causing problems in further applications. When these perceptrons are connected into a network, however, they could form more complex separating boundaries (Minsky and Papert, 1988/1969). This architecture is the multilayer perceptron (MLP) and use of the backpropagation algorithm (see next section) to train them has made them one of the most popular and powerful neural network architectures. The MLP architecture shown in Figure 3.7 consists of a single layer of inputs connected to a layer of hidden units, which in turn are connected to the output layer. This network is "fully connected" because each node in every layer is connected to every node in the next layer. It is also possible to have a network that is not fully connected. In general, an L-layer MLP network consists of a single input layer, $L-1$ hidden layers and one output layer. There has been some uncertainty concerning how to count the layers in an MLP network (Reed and Marks, 1999), but it is now generally accepted that the input layer is not included because it does not do any processing.

The number of hidden layers determines the complexity of the surface boundary, for example, a single layer is only capable of producing linear separating planes whereas more layers can produce arbitrary separating boundaries depending on the number of nodes in the hidden layers. The sigmoid activation function is also known to produce smooth separating boundaries rather than just piecewise planar boundaries. Several design issues arise from using the MLP structure that can only be empirically answered. These include issues such as the number of hidden layers and nodes per layer to use, the node interconnection topology, and the optimal training data size.

3.2.3 The Back-Propagation Algorithm

When the MLP architecture was introduced, an immediate problem that arose was training it. How could the weights in the hidden layers be adjusted if one was to adapt the learning algorithm for perceptrons to the multilayered network? This problem remained a major hinderance to using the MLP network until the discovery of the backpropagation algorithm. Prior to this algorithm, several attempts were made such as using stepwise functions (Reed and Marks, 1999), but these approaches were not successful.

Backpropagation is composed of two main steps: first, the derivatives of the network training error are computed with respect to the weights through clever application of the derivative chain rule. Second, a gradient descent method is applied to adjust the weights using the error derivatives to minimize the output errors. It can be applied to any feedforward neural network with differentiable activation functions.

Briefly, let w_{ij} denote the weight to node i from node j. Since this is a feedforward network, no backward connections are permitted and w_{ji} has no meaning in the forward direction, that is, $w_{ji} = 0$. If there is no connection from node j, then $w_{ij} = 0$. The network can be biased by adding a bias node with constant activation, for example, $y_{\text{bias}} = 1$. In the first step of the algorithm, the output node values are computed by presenting a training pattern $\mathbf{x}_k$ to the inputs and evaluating each node beginning with the first hidden layer and continuing to the final output node. This procedure can be performed by a computer program that meticulously computes each node output in order. When the outputs at each node are obtained the errors are computed, here the sum of squared errors is used:

$$E_{\text{MLP}}^{\text{sq}} = \frac{1}{2} \sum_k \sum_i (y_{ki} - c_{ki})^2 \tag{3.7}$$

where k indexes the training patterns in the training set, i indexes the output, nodes, y_{ki} the desired output, and c_{ki} the computed network output. The scalar $1/2$ is for convenience when computing the derivative of this error. The error function $E_{\text{MLP}}^{\text{sq}}$ is the sum of individual squared errors of each training pattern presented to the network. This can be seen by writing

$$E_{\text{MLP}}^{\text{sq}} = \sum_k E_k$$

$$E_k = \frac{1}{2} \sum_i (y_{ki} - c_{ki})^2$$

Note that these quantities are independent of one another since they depend on the training pattern being presented. When the errors are obtained, the next step is to compute the derivative of the error with respect to the weights. The derivative of $E_{\text{MLP}}^{\text{sq}}$ is

$$\frac{\partial E_{\text{MLP}}^{\text{sq}}}{\partial w_{ij}} = \sum_k \frac{\partial E_k}{\partial w_{ij}} \tag{3.8}$$

Using the chain rule, the individual pattern derivatives can be written as

$$\frac{\partial E_k}{\partial w_{ij}} = \sum_p \frac{\partial E_k}{\partial u_p} \frac{\partial u_p}{\partial w_{ij}} \tag{3.9}$$

where u_p is the weighted sum input for node p. The contribution of u_i at the node i to the error of a training pattern can be measured by computing the error derivative written as

$$\delta_i = \frac{\partial E_k}{\partial u_i} = \frac{\partial E_k}{\partial c_i}\frac{\partial c_i}{\partial u_i} \qquad (3.10)$$

For output nodes this can be written as

$$\delta_i = -(y_i - c_i)f_i' \qquad (3.11)$$

where f_i' is the derivative of the activation function. This will vary depending on the activation function where, for example, the sigmoid function gives a particularly nice derivative, that is, $f_i' = f(u_i)(1 - f(u_i))$. The tanh function is equally elegant with $f_i' = 1 - f^2(u_i)$.

For nodes in the hidden layers, note that the error at the outputs is influenced only by hidden nodes i, which connect to the output node p. Hence for a hidden node i, we have

$$\delta_i = \frac{\partial E_k}{\partial u_i} = \sum_p \frac{\partial E_k}{\partial u_p}\frac{\partial u_p}{\partial u_i} \qquad (3.12)$$

But the first factor is δ_p of the output node p and so

$$\delta_i = \sum_p \delta_p \frac{\partial u_p}{\partial u_i} \qquad (3.13)$$

The second factor is obtained by observing that node i connects directly to node p and so $(\partial u_p)/(\partial u_i) = f_i'w_{pi}$. This gives

$$\delta_i = f_i' \sum_p w_{pi}\delta_p \qquad (3.14)$$

where δ_i is now the weighted sum of δ_p values from the nodes p, which are connected to it (see Figure 3.8). Since the output values of δ_p must be calculated

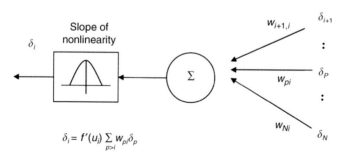

FIGURE 3.8
Backpropagation structure. Node deltas are calculated in the backward direction starting from the output layer.

first, the process begins at the output layer and propagates backward to the hidden layers, hence the term "backpropagation." Generally, it is not necessary to calculate delta values for the input layer but in some applications such as inverse problems, this may be required.

When the derivatives have been obtained, the weights can be updated and the output error iteratively decreased using a gradient descent-type method having the form

$$w_{ij}^{t+1} = w_{ij}^t - \lambda \frac{\partial E_{\text{MLP}}^{\text{sq}}}{\partial w_{ij}} \qquad (3.15)$$

Note the similarity of the update rule for the MLP architecture with the perceptron update (Equation 3) in Algorithm 3.1. The learning rate λ can be selected as before and the weights initialized to some random values as to avoid cycling or immediate saturation of the nonlinear activation functions. Algorithm 3.2 provides a summary of steps for the backpropagation algorithm for an MLP network where the sigmoid function is used for all nodes.

Algorithm 3.2: Backpropagation Learning Algorithm.

---

1 Initialize the weights to random values, for example, $0.2 < w_i < 2$. Select an activation function to use at each node and assume a L-layer MLP.
2 Present a random training example $\mathbf{x}_i$ to the MLP network and propagate the signal through the network to obtain the outputs.
3 Assuming a sigmoid activation function, compute the delta values beginning with the output nodes first, using the following formula

$$\delta_p = -(y_{pi} - c_{pi}) f'(u_k) \qquad (3.16)$$

where $f'(u_k)$ is the derivative of the activation function, u_k the sum of the nodes k in the hidden layer before the output layer, y_{pi} the desired output, and c_{pi} the computed output of the network for the training pattern $\mathbf{x}_i$.
4 Compute the delta values for the preceding layers by propagating the errors backward using

$$\delta_k^l = f'(u_k^l) \sum_i w_{ji}^{l+1} \delta_i^{l+1} \qquad (3.17)$$

where l is the index for layers and $l = L - 1, \ldots, 1$.
5 Update the weights using the computed delta values going forward with

$$w_{ij}^{t+1} = w_{ij}^t - \lambda \delta_j^l y_i^{l-1} \qquad (3.18)$$

The *learning rate* λ is selected from the interval $0.05 < \lambda < 0.75$.
6 Go to step 2 and repeat for all training examples until the sum of errors reaches some small threshold or until a predetermined number of epochs have been completed.

There are two basic weight update variations for the backpropagation algorithm, the batch mode and the online learning mode. In the batch mode, every training pattern is evaluated to obtain the error derivatives and then summed to obtain the total error derivative:

$$\frac{\partial E_{\text{MLP}}^{\text{sq}}}{\partial w} = \sum_k \frac{\partial E_k}{\partial w}$$

Then the weights are updated using this derivative. Each pass through the entire training set is called an *epoch*. Since the derivative of the error is exact, the weight update rule becomes the gradient update rule when the learning rate is small. The disadvantage of this method, however, is that many epochs are required before the MLP network is sufficiently well trained.

The online learning method updates the MLP weights after each training pattern has been presented to the network; this can also be viewed as an incremental method and it may be faster than the batch mode. For a training set with N training examples, the online learning mode would update the weights N times, whereas the batch mode would update them only once, that is, a single epoch. In principle, the disadvantage of the online learning method is that updating the weights based on a single training pattern may result in oscillations during training because the derivative of the error is not computed exactly.

To accelerate the learning of weights in regions where the error derivative $(\partial E_{\text{MLP}}^{\text{sq}})/(\partial w)$ is small, a *momentum* term may be added to the weight update rule. The idea is to stabilize the trajectory of the weight change using a combination of the gradient descent term and a fraction of the previous weight change. The modified weight update rule with momentum is then

$$\mathbf{w}^{t+1} = \mathbf{w}^t - \lambda \frac{\partial E_{\text{MLP}}^{\text{sq}}}{\partial \mathbf{w}} + \alpha(\mathbf{w}^t - \mathbf{w}^{t-1}) \tag{3.19}$$

where typically $0 \leq \alpha \leq 0.9$. This is intended to give the weight vector extra momentum and improve convergence rates.

3.2.4   Recurrent or Feedback Neural Networks

In contrast to feedforward networks such as the MLP, recurrent neural networks allow signals to be fed back to other nodes in the network via feedback loops, hence the term feedback networks. These networks are more complex with more powerful learning capabilities. The feedback network architecture is a dynamic system where network states are changed continuously until a new equilibrium point is found. When a new input pattern is presented, a new equilibrium point is found giving a network that can adapt to data and learn.

The memorization of patterns and response can be grouped into several different mappings. In associative mapping, the network learns to produce a particular pattern when a new pattern is applied to the layer of input units. There are two types of associative mapping:

a. *Autoassociation memory.* An input pattern is associated with the states of input and output units. This is used to provide pattern completion, that is, to produce a pattern whenever a portion of it or a distorted pattern is presented.

b. *Heteroassociation.* In nearest-neighbor recall, an output pattern corresponding to a particular pattern is produced when a similar input pattern is presented. In interpolative recall, the interpolated similarities between patterns are stored and outputted. This variant of associative mapping is closely related to the classification problem where the input patterns are mapped to a fixed set of labels.

In regularity detection, units learn to respond to specific properties of the input patterns and the response of each output unit has special meaning. This type of learning is essential for feature discovery and knowledge representation, which we find in Kohonen's SOM and the Hopfield neural network.

3.2.5   Kohonen's Self-Organizing Map

The SOM developed by Finnish researcher Kohonen (Kohonen, 1989) is an unsupervised method based on ANN. The key characteristic is the property of topology preserving, which mimics aspects of feature maps in animal brains. Topology preserving in this context means that inputs only activate outputs in their neighborhood. The SOM can be used for a variety of unsupervised applications, such as projection of multivariate data (Yin, 2002), density approximation (Kurimo and Somervuo, 1996), and clustering. These have been further applied to problems in speech recognition, image processing, robotics, and control.

The SOM architecture consists of a layer of inputs connected to a two-dimensional (2-D) array of nodes (Figure 3.9). Let there be k input nodes

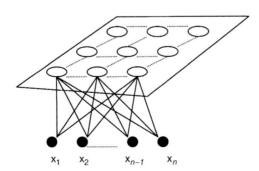

FIGURE 3.9
Kohonen's SOM showing the input layer and the 2-D array of self-organizing nodes.

in the input layer such that the weight $\mathbf{w}_{ij}$ is k-dimensional. In the SOM structure, the weight $\mathbf{w}_{ij}$ is associated with the node having the coordinates (i, j) in the 2-D array. Each neuron or node computes the distance (usually Euclidean) between the input vector $\mathbf{x}$ and the stored weight $\mathbf{w}_{ij}$. Using the Euclidean distance, this may be written as

$$d_i = \|\mathbf{x} - \mathbf{w}_{ij}\| \tag{3.20}$$

The SOM network is an example of competitive learning where sets of neighborhoods $N_{i,j}$ are defined in the 2-D array. The neighborhoods may be circles, squares, or rectangles and are initially set to be half or one third the size of the 2-D array so that a global collection of nodes can be included at the beginning of the iteration. When the neighborhood has been determined, a training pattern is presented to it and propagated through the network as in the MLP network. The neuron with the minimum distance d_i is called the winner and its weight is updated together with those in the neighborhood. The update rule is usually a form of gradient descent with a time-varying learning rate λ_t to avoid oscillation. An exponential decreasing function is used to decrease the size of the neighborhood with the number of iterations. This is similar to competitive learning where the nodes can be seen as competing to be updated as depicted by the SOM algorithm (Algorithm 3.3).

Algorithm 3.3: Self-Organizing Map Learning Algorithm.

---

1 Initialize the weights to small random values, select an initial neighborhood topology and learning rate.
2 Present a random training example $\mathbf{x}_i$ and compute the outputs.
3 Select the *winner* node (i^*, j^*) with the minimum distance, that is,

$$d_{i^*, j^*} = \min_{ij} \|\mathbf{x}_i - \mathbf{w}_{ij}\| \tag{3.21}$$

4 Update only the weights in the neighborhood N_{i^*, j^*}, that is,

$$\mathbf{w}_{ij}^{t+1} = \begin{cases} \mathbf{w}_{ij}^t - \lambda_t [\mathbf{x} - \mathbf{w}_{ij}^t] & \text{if } (i, j) \in N_{i^*, j^*} \\ \mathbf{w}_{ij}^t & \text{otherwise} \end{cases} \tag{3.22}$$

where λ_t is the learning rate.
5 Decrease λ_t and shrink the neighborhood N_{i^*, j^*}.
6 Goto step 2 and repeat for all training examples until the change in the weights is either less than a prespecified threshold or the maximum number of iterations is reached.

With sufficient training time, the SOM would have been able to map the input training patterns to an internal representation in the 2-D array. This mimics a form of memory where information is represented by spatial location

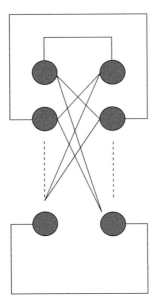

FIGURE 3.10
The Hopfield neural network showing the feedback loops and all nodes are interconnected.

(neighborhood) and weights of the nodes. When an unseen pattern is presented to the SOM, only neurons in the neighborhood associated with the pattern respond, which can be viewed as information retrieval or memory recall.

3.2.6 Hopfield Neural Networks

The Hopfield network shown in Figure 3.10 is a recurrent neural network developed by Hopfield (Hopfield, 1982). These types of networks are content addressable memory systems where the network information is stored as a set of dynamically stable attractors. The Hopfield network is useful for efficiently solving combinatorial problems (Abe, 1997) in that it can search for solutions that minimize an objective function under several constraints.

A Hopfield network consists of a number of processing nodes or units, for example, a three node Hopfield network as seen in Figure 3.11. There are generally two versions of Hopfield networks, binary and continuous valued networks. The states of the nodes are represented as

$$\mathbf{S} = \{S_1, S_2, \ldots, S_N\} \tag{3.23}$$

where $S_i = \{-1, +1\}$ for binary Hopfield networks and $S_i \in \Re$ for continuous valued networks. Each node has a forward and backward connection with symmetrical synapse connections. This implies that between node i and node j we have $w_{ij} = w_{ji}$. Absence of a feedback connection for a node i is represented

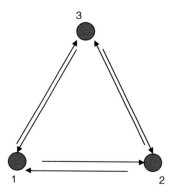

FIGURE 3.11
A different configuration showing a smaller three node Hopfield neural network
with feedback.

by $w_{ii} = 0$. An energy function, which is a combination of an objective function
and constraints is defined for the Hopfield network. The concept is to have the
energy decrease with time and the solution is found by integrating the network
from some starting point. Generally, a state transition table or diagram is used
to derive the energy of each state and the energy function of the network.

For binary Hopfield networks, a node state is represented by

$$S_i = \mathrm{sgn} \left(\sum_j w_{ij} s_j - \theta_i \right) \tag{3.24}$$

where $\mathrm{sgn}(x)$ is the signum function such that $\mathrm{sgn}(x) = +1$ if $x \geq 0$ and
$\mathrm{sgn}(x) = -1$, if $x < 0$. θ_i is some threshold value. For continuous networks,
the node state is represented by a dynamical system

$$\tau_i \frac{\partial u_i}{\partial t} = -u_i + w_{ij} f(u_j) - R_i \theta_i \tag{3.25}$$

where u_i is the net input, f the activation function usually sigmoid, and R_i, τ_i
are constants. The states are represented by $S_i = f(u_i)$ at equilibrium; this is
the solution of Equation 3.25, which can be written as

$$S_i = f \left(\sum_j w_{ij} s_j - \theta_i \right) \tag{3.26}$$

The energy function of a binary Hopfield network in any state $\mathbf{S} = \{S_1, S_2, \ldots, S_N\}$ is

$$E = -\frac{1}{2} \sum_i \sum_j w_{ij} S_i S_j \tag{3.27}$$

But the energy function of a continuous Hopfield network is more complex:

$$E = -\frac{1}{2}\sum_i\sum_j w_{ij}S_iS_j + \sum_i\frac{1}{R_i}\int_0^{S_i} f^{-1}(x)\mathrm{d}x + \sum_i \theta_i S_i \qquad (3.28)$$

There are two methods for updating the weights in the Hopfield network, that is asynchronous and synchronous updates. In synchronous updates all nodes are updated simultaneously at each step whereas in asynchronous updates, one unit is randomly selected to be updated each time. The energy function for both types of networks is dissipative, which means it reduces over time until eventually the network comes to rest at minimum network energy. The points in this equilibrium state are known as *attractors* and can be used to retrieve information on a set of input patterns, hence the Hopfield net is an example of an associative memory construct.

3.3 Support Vector Machines

SVMs were introduced by Vapnik in 1992 as a new supervised machine-learning formulation applied to the well-known problem of function estimation (Vapnik, 2000). It was first demonstrated as a binary classifier based on statistical learning theory and structural risk minimization (SRM) theory developed by Vapnik since the late 1960s to solve pattern recognition problems. It was later extended to general nonlinear problems where an implicit nonlinear mapping mapped data to a feature space. Further work in 1995 extended the framework to regression for estimating real-valued functions. Finally in 1996, it was applied to solving linear operator equations, which included density estimation, conditional and probability density estimations, and time-series predictions (Gestel, 2001; Mukherjee et al., 1997). Since its inception, the SVM has found numerous applications in a diverse range of fields from image and voice recognition (Pirrone and Cascia, 2001; Qi et al., 2001; Ai et al., 2001; Ma et al., 2001), handwriting recognition Maruyama et al., 2002; Bahlmann et al., 2002; Fuentes et al., 2002; Bellili et al., 2001), medicine (Liu et al., 2001; Liang and Lin, 2001; El-Naqa et al., 2002; Chodorowski et al., 2002; Wang et al., 2001; Park et al., 2002; Lee and Grimson, 2002), classification in industry (Rychetsky et al., 1999; Anguita and Boni, 2001; Junfeng et al., 2002), telecommunications (Gong and Kuh, 1999; Chen et al., 2000, 2001; Hasegawa et al., 2001; Gretton et al., 2001; Albu and Martinez, 1999), earth sciences (Zhang et al., 2001; Ramirez et al., 2001), defense (Zhao and Jose, 2001; Li et al., 2001), and finance and economics (Fan and Palaniswami, 2000, 2001; Trafalis and Ince, 2000).

This section summarizes the review contained in the work by Lai (2006) on the basic SVM formulations highlighting the more important concepts to assist in understanding the techniques. We have included the formulations commonly used in the literature when dealing with the generic SVM binary classifier form, and the generic regression form. Classification is also a form of function estimation and for the SVM case, these two forms could be generalized and the notation simplified further. We then describe several training algorithms for solving the SVM optimization problem. The interested reader is referred to the many tutorials (Smola and Schlkopf, 1998; Burges, 1998) and excellent books (Vapnik, 2000; Cristianini and Shawe-Taylor, 2000; Schlkopf et al., 1999; Schlkopf and Smola, 2002) for further technical details and information on other formulations such as SVMs for time series prediction and density estimation problems.

3.3.1   Support Vector Classifiers

The data used to train the support vector classifier is the set Θ where

$$
\begin{aligned}
\Theta &= \{(\mathbf{x_1}, y_1), (\mathbf{x_2}, y_2), \ldots, (\mathbf{x_n}, y_n)\} \\
\mathbf{x}_i &\in \Re^k \\
y_i &\in \{-1, 1\}
\end{aligned}
\tag{3.29}
$$

where $\mathbf{x}_i$ is a feature vector describing the attributes of an example and y_i the corresponding class label. The classification task is to train a machine to learn the nonlinear relationship between the features and their respective labels, that is, $y = f(x)$. By "machine," we mean the mathematical implementation, which is usually a function obtained by solving an optimization problem. The $\mathbf{X}$ vector space is referred to as the *input space* and the dimension k indicates that k different types of measurements of a specific problem are included. In classification, the usual aim is to construct a hyperplane, which separates the two classes of data (Figure 3.12). This form of classification is straightforward because the classes are linearly separable (clearly separable).

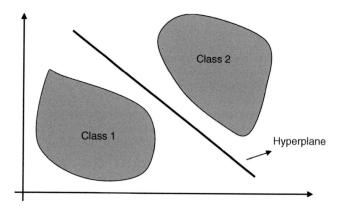

FIGURE 3.12
Two linearly separable classes, which can be clearly separated by an infinite number of hyperplanes.

Figure 3.13 shows a more realistic case where the two classes are nonseparable (overlap) and could confuse the machine.

In the linearly separable case, there are an infinite number of such separating hyperplanes, which can be fitted between the two classes. The margin of the hyperplane is the distance between the hyperplane to the boundaries of the classes.

The principal idea behind the SVMs is to maximize the hyperplane margin to obtain good classifier performance. If our training set is not linearly separable, as in most real-life applications, we define a nonlinear mapping from input space to some (usually) higher dimensional feature space, denoted by

$$\phi(\mathbf{x}) : \mathbf{x} \subset \Re^k \to \Re^m, \quad k << m$$

Figure 3.14 depicts this transformation where we anticipate that the non-separable classes become separable after mapping. We say anticipate because without intricate calculations we have no assurance that the classes will be

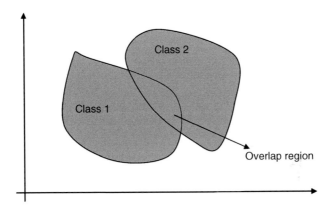

FIGURE 3.13
Two nonlinearly separable classes, which cannot be easily separated.

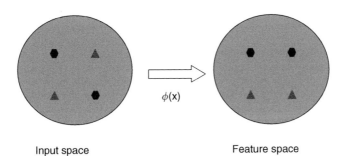

FIGURE 3.14
A 2-D example: the nonlinear mapping $\phi(\mathbf{x})$ transforms inseparable points in input space to feature space where they are now easily separated.

separable following the transformation. The nonlinear map is however never explicitly used in the calculation and does not have to be defined now.

Vapnik (2000) suggested that the form of the hyperplane, $f(\mathbf{x}) \in \mathbf{F}$ be chosen from a family of functions with sufficient capacity. In particular, $\mathbf{F}$ contains functions for the linearly and nonlinearly separable hyperplanes with the following *universal approximator* form

$$f(\mathbf{x}) = \sum_{i=1}^{n} w_i x_i + b \tag{3.30}$$

$$f(\mathbf{x}) = \sum_{i=1}^{m} w_i \phi_i(\mathbf{x}) + b \tag{3.31}$$

Note that each training example $\mathbf{x}_i$ is assigned a weight, w_i, and the hyperplane is a weighted combination of all training examples. For separation in feature space, we would like to obtain the hyperplane with the following properties

$$f(\mathbf{x}) = \sum_{i=1}^{m} w_i \phi_i(\mathbf{x}) + b$$

$$f(\mathbf{x}) > 0 \quad \forall i{:}y_i = +1 \tag{3.32}$$

$$f(\mathbf{x}) < 0 \quad \forall i{:}y_i = -1$$

The conditions in Equation 3.32 can be described by a strict linear discriminant function, so that for each element pair in $\boldsymbol{\Theta}$ we require

$$y_i \left(\sum_{i=1}^{m} w_i \phi_i(\mathbf{x}) + b \right) \geq 1 \tag{3.33}$$

It turns out that only the points closest to the separating hyperplane are important and these points are referred to as *nonbound* support vectors. The distance from the hyperplane to the margin is $1/\|\mathbf{w}\|$ and the distance between the margins of one class to the other class is simply $2/\|\mathbf{w}\|$ by geometry. We noted earlier that even in feature space the two classes may not be linearly separable but this can be addressed by using a soft-margin formulation to relax the strict discriminant in Equation 3.33 through the use of slack variables ξ_i. The SVM soft-margin classifier problem in this case (Cristianini and Shawe-Taylor, 2000) is written as

$$\min_{\text{min}} \Im(\mathbf{w}, \boldsymbol{\xi}) = \frac{1}{2} \sum_{i=1}^{m} w_i^2 + C \sum_{i=1}^{n} \xi_i \tag{3.34}$$

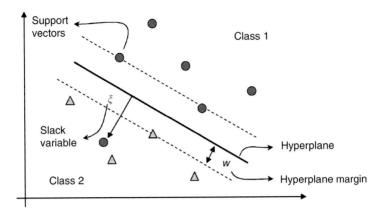

FIGURE 3.15
A 2-D example: the soft-margin support vector classifier.

subject to

$$y_i \left(\sum_{i=1}^{m} w_i \phi_i(\mathbf{x}) + b \right) \geq 1 - \xi_i$$

$$\xi_i > 0$$

$$\forall \, i = 1, \ldots, n$$

The parameter C controls the tradeoff between generalization capability and the number of training misclassifications. Large C reduces the number of classification training errors but gives lower generalization capabilities, whereas small C causes a poor fit with the training data resulting in more training errors. The task is to derive the best hyperplane with the maximum margin between the classes; Figure 3.15 is a simple toy example of data containing two attributes plotted in a $\Re^2$ dimensioned space showing the major components of the SVM classifier.

Lagrangian theory (Bertsekas, 1996) describes optimization of an objective function subject to equality and inequality constraints. Given an objective function $\Im(\mathbf{w})$, equality constraints $G(\mathbf{w})$, and inequality constraints $H(\mathbf{w})$, the Lagrangian function is defined as:

$$L(\mathbf{w}, \boldsymbol{\alpha}, \boldsymbol{\beta}) = \Im(\mathbf{w}) + \sum_{i=1}^{n} \pi_i G_i(\mathbf{w}) + \sum_{j=1}^{n} \alpha_j H_j(\mathbf{w}) \qquad (3.35)$$

Where the variables $\boldsymbol{\alpha}, \boldsymbol{\pi}$ are termed the Lagrangian multipliers. The Lagrangian function has several desirable properties, the primary one being that its solution is equivalent to the original constrained problem. We have also transformed a problem with constraints to a now unconstrained problem, which can be easily dealt with. Furthermore, it can be shown that the

Lagrangian function is a convex function if the objective function $\Im(\mathbf{w})$ is convex and the space defined by the constraint set of $G(\mathbf{w})$ and $H(\mathbf{w})$ is convex. In optimization, convexity is often synonymous with unique and global solutions, which is very desirable because we can easily test for them and then stop the algorithms.

With that in mind, we note the inequality constraint in Equation 3.34 and write the Lagrangian primal problem as follows:

$$\min_{} L_p(\mathbf{w}, \boldsymbol{\alpha}, \boldsymbol{\xi}) = \frac{1}{2}\sum_{i=1}^{m} w_i^2 + C\sum_{i=1}^{n}\xi_i - \sum_{i=1}^{n}\pi_i\xi_i$$

$$- \sum_{i=1}^{n}\alpha_i\left(y_i\left(\sum_{j=1}^{m}w_j\phi_j(\mathbf{x}_i) + b\right) - 1 + \xi_i\right) \qquad (3.36)$$

where $\forall\, i = 1,\ldots,n$

$$\alpha_i, \pi_i \geq 0$$

The primal form of Equation 3.36 offers a dual representation, which is another way of representing the equation in terms of Lagrange multipliers alone. The dual representation is found by differentiating Equation 3.36 with respect to the variables and finding the stationary conditions. The partial gradients with respect to the variables are

$$\frac{\partial L_p}{\partial w_i} = w_i - \sum_{j=1}^{n}\alpha_j y_j\phi_i(\mathbf{x}_j)$$

$$\frac{\partial L_p}{\partial \alpha_i} = y\left(\sum_{j=1}^{m}w_j\phi_j(\mathbf{x}_i) + b\right) - 1 + \xi_i$$

$$\frac{\partial L_p}{\partial b} = \sum_{i=1}^{n}\alpha_i y_i$$

$$\frac{\partial L_p}{\partial \xi_i} = C - \alpha_i - \pi_i$$

Using the stationary conditions of the partial gradients, that is, setting them to zero, we then eliminate the variables by substitution to obtain the Lagrangian dual only in terms of Lagrangian multipliers α_i

$$\min_{} L_D(\boldsymbol{\alpha}) = -\sum_{i=1}^{n}\alpha_i + \frac{1}{2}\sum_{i,j=1}^{n}\alpha_i\alpha_j y_i y_j K(\mathbf{x}_i, \mathbf{x}_j)$$

subject to

$$\sum_{i=1}^{n} \alpha_i y_i = 0$$

$$0 \leq \alpha_i \leq C$$

$$\forall i = 1, \ldots, n$$

Note that the explicit definition of the nonlinear function $\phi(\cdot)$, has been circumvented by the use of a kernel function, defined formally as the dot products of the nonlinear functions

$$K(\mathbf{x}_i, \mathbf{x}_j) = \langle \phi(\mathbf{x}_i), \phi(\mathbf{x}_j) \rangle \tag{3.37}$$

This method is usually attributed to Mercer's theorem (Cristianini and Shawe-Taylor, 2000), which implicitly computes the inner product of the vectors without explicitly defining the mapping of inputs to higher dimensional space since they could be infinitely dimensioned. The use of kernels will not be discussed in detail here, but the interested reader is referred to Schlkopf et al. (1999); Schlkopf and Smola (2002). Many kernels are proposed in the literature, such as the polynomial, radial basis function (RBF), and Gaussian kernels, which will be explicitly defined later. The trained classifier (machine) then has the following form:

$$f(\mathbf{x}) = \text{sign} \left(\sum_{i=1}^{n} \alpha_i y_i K(\mathbf{x}, \mathbf{x}_i) + b \right) \tag{3.38}$$

DEFINITION 3.3.1 (Karush–Kuhn–Tucker Optimality Conditions) *The solution (Vapnik, 2000) to the SVM classification optimization problem is found when the following Karush–Kuhn–Tucker (KKT) conditions are met for all α_i, where $i = 1, \ldots, n$.*

$$y_i f(x_i) > 1 \quad \text{if } \alpha_i = 0$$

$$y_i f(x_i) < 1 \quad \text{if } \alpha_i = C$$

$$y_i f(x_i) = 1 \quad \text{if } 0 < \alpha_i < C$$

The preceding solution is usually *sparse* in that most of the Lagrangian multipliers end up being zero, that is, $\alpha_i = 0$. This causes the corresponding product terms in Equation 3.38 to drop out and so the decision function could be represented solely by nonzero α_i. The training vectors that correspond to nonzero α_i are called the support vectors. Table 3.1 summarizes the training examples according to the values of their multipliers α_i.

3.3.2 Support Vector Regression

In the regression formulation the standard SVM framework is extended to solve the more general function estimation problem. The data set in

TABLE 3.1

The Possible Lagrangian Multiplier Values for SVM Classification
and Their Geometrical Interpretations

α_i	Training Example, $\mathbf{x}_i$	Position about Hyperplane
0	Correctly classified	On the correct side
C	Bound support vector (BSV)	On the wrong side or within the margin of the hyperplane
>0 and $<C$	Nonbound support vector (NBSV)	On margin, i.e., $1/\|\mathbf{w}\|$ from the hyperplane

this setting is

$$\mathbf{\Theta} = \{(\mathbf{x_1}, y_1), (\mathbf{x_2}, y_2), \ldots, (\mathbf{x_n}, y_n)\}$$

$$\mathbf{x}_i \in \Re^k \qquad (3.39)$$

$$y_i \in \Re$$

where instead of just integer labels for y_i, the values of y_i are real, for example, pressure measurements of a hydraulic system. Vapnik proposed the use of a ϵ-insensitive loss function, which for some $\epsilon > 0$ has the form

$$f_L(y, f(\mathbf{x}, \boldsymbol{\alpha})) = f_L(|y - f(\mathbf{x}, \boldsymbol{\alpha})|_\varepsilon) \qquad (3.40)$$

where

$$|y - f(\mathbf{x}, \boldsymbol{\alpha})|_\varepsilon = \begin{cases} 0 & \text{if } |y - f(\mathbf{x}, \boldsymbol{\alpha})| \leq \varepsilon \\ |y - f(\mathbf{x}, \boldsymbol{\alpha})| - \varepsilon & \text{otherwise} \end{cases}$$

This loss function is more robust to outliers in the data due to the introduction of the ϵ-insensitive zone. The effect of the zone is that points within the insensitive zone are considered acceptable and have zero error whereas points outside the tube have losses that will be penalized linearly. In feature space, the estimated function $f_E(\mathbf{x}, \boldsymbol{\alpha})$ then sits inside a tube of width 2ϵ as shown in Figure 3.16. We note that other loss functions can be applied to the SVM framework if some information about the noise distribution in the data set is known *a priori*, for example, polynomial loss and Huber loss (Schlkopf and Smola, 2002) functions.

To obtain the best model, the SVM regression formulation minimizes the following regularized function (linear insensitive loss):

$$\underset{\min}{\Im(\mathbf{w}, \boldsymbol{\xi})} = \frac{1}{2}\|\mathbf{w}\|^2 + C\sum_{i=1}^n (\xi_i^* + \xi_i)$$

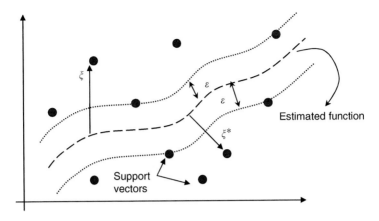

FIGURE 3.16
A 2-D example: the SVR with Vapnik's ϵ-insensitive loss function. The
estimated function $f(\mathbf{x})$ lies in the ϵ-insensitive tube.

subject to

$$\xi_i^* \geq f(\mathbf{x}_i) - y_i - \varepsilon$$

$$\xi_i \geq y_i - f(\mathbf{x}_i) - \varepsilon$$

$$\xi_i^*, \xi_i \geq 0$$

$$i = 1, \ldots, n$$

where the estimated function is chosen from the universal approximator family
(Equation 3.31) as before.

$$f(\mathbf{x}) = \sum_{i=1}^{n} w_i \phi_i(\mathbf{x}) + b$$

The choice of C is either fixed *a priori* or a value found through the use of cross
validation, the concept of cross validation is to train a regressor with different
values of C and select the C giving the best accuracy on a test set. Following
the same methods as for the support vector classification formulation, we first
write the Lagrangian primal problem as follows:

$$L_p(\mathbf{w}, b, \boldsymbol{\xi}^*, \boldsymbol{\xi}, \boldsymbol{\alpha}^*, \boldsymbol{\alpha}) \underset{\min}{=} \frac{1}{2} \sum_{i=1}^{m} w_i^2 + C \sum_{i=1}^{n} (\xi_i^* + \xi_i) - \sum_{i=1}^{n} \alpha_i^* (y_i - f(\mathbf{x}_i) - \varepsilon - \xi_i^*)$$

$$- \sum_{i=1}^{n} \alpha_i (f(\mathbf{x}_i) - y_i - \varepsilon - \xi_i) + \sum_{i=1}^{n} (\pi_i^* \xi_i^* + \pi_i \xi_i)$$

$$(3.41)$$

As before, the stationary partial gradients are computed and used to eliminate
the variables. Substituting into the primal Lagrangian gives us the Lagrangian

dual entirely in terms of $\boldsymbol{\alpha}, \boldsymbol{\alpha}^*$ only.

$$\max_{\substack{L_D(\boldsymbol{\alpha}^*, \boldsymbol{\alpha}) = }} -\frac{1}{2} \sum_{i,j=1}^{n} (\alpha_i - \alpha_i^*)(\alpha_j - \alpha_j^*)K(x_i, x_j)$$

$$-\varepsilon \sum_{i=1}^{n} (\alpha_i + \alpha_i^*) + \sum_{i=1}^{n} y_i(\alpha_i - \alpha_i^*)$$

subject to

$$\sum_{i=1}^{n} y_i(\alpha_i - \alpha_i^*) = 0$$

$$0 \le \alpha_i, \alpha_i^* \le C$$

After optimization, the SVM regressor can be written similarly to the classifier form as

$$f(\mathbf{x}) = \sum_{i=1}^{n} (\alpha_i - \alpha_i^*)K(\mathbf{x}_i, \mathbf{x}) + b \tag{3.42}$$

The dual variables α_i and α_i^* can never be nonzero simultaneously because the constraint set in Equation 3.41 cannot be fully satisfied for every point. This provides an opportunity to use one variable to represent the multipliers in Equation 3.42, but care must be taken when performing the optimization (Schlkopf and Smola, 2002).

3.3.3   Training the Support Vector Machine

The SVM problems Equations 3.36 and 3.41 are constrained quadratic programs in terms of the Lagrangian variables. As such, any optimization technique suitable for solving these types of problems, for example, projected gradient descent, projected Newton method, and conjugate gradients can be used (Fletcher, 1981; Gill et al., 1981). Since data sets can become very large, problems with computational memory invariably surface due to the storage requirements for the dense kernel matrix. To deal with this, the decomposition technique has popularly been applied to solve the SVM by decomposing the main problem into smaller subproblems, which can be solved with the available computer memory. The subproblems are characterized by their variables often referred to as blocks, chunks, or working sets. Decomposition algorithms proceed by sequentially solving subproblems until all KKT conditions of the problem have been fulfilled or until some stopping criterion has been met (Lin, 2002). Some of these algorithms are briefly mentioned as follows:

a. *General chunking and decomposition algorithm.* The general decomposition algorithm operates on part of the problem by taking only a fraction of training points into consideration. The simplest decomposition

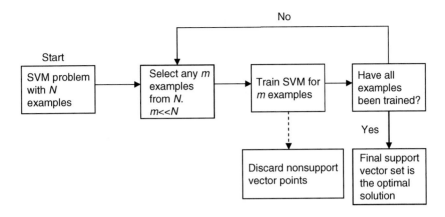

FIGURE 3.17

Flowchart for general chunking or decomposition algorithm.

algorithm randomly selected m training points from the total n training points to solve. The variables, or the working set, at each iteration were then

$$\alpha_p = \{\alpha_1, \alpha_2, \ldots, \alpha_m\}$$

This approach resulted in training a SVM through a series of randomly chosen working sets viewed as chunks and referred to as the "chunking" algorithm (Osuna et al. 1997). An immediate improvement over the random chunking algorithm was to apply a heuristic for the selection of chunks. In particular, one could select points that has the largest KKT violation as suggested by Cristianini and Shawe-Taylor (2000); Schlkopf et al. (1999); Schlkopf and Smola (2002). Once the chunks are selected, a general optimization package could be used to solve the problem. A flowchart depicting the operation of these early chunking algorithms is given in Figure 3.17. The core optimization packages such as LOQO, MINOS (Cristianini and Shawe-Taylor, 2000), and MAT-LAB QP performed the updates on the working set. They were initially programmed for general optimization problems and easily adapted to the SVM problem. But as problem sizes grew, these methods started to slow down significantly and so better algorithms were designed to handle the specific problem structure.

b. *Sequential minimal optimization (SMO) algorithm.* One of the earliest decomposition methods specifically tailored to SVMs was written by Platt (1998) of Microsoft and was called SMO. The initial popularity of SMO was due to the ease of implementing Platt's pseudocode and its improved speed over other chunking methods. The SMO update rule is a special case of the projected Newton method where Platt used a working set containing just two variables and analytically solved the resulting subproblem. The SMO working set consisted of the following couple:

$$\boldsymbol{\alpha}_p = \{\alpha_1, \alpha_2\}$$

Since Platt solved the SVM problem (Equation 3.36), the equality constraint had to be explicitly accommodated to maintain feasibility of the solution. It was observed that if the linear constraint $\boldsymbol{\alpha}^T \mathbf{y} = 0$ was to be maintained at all times during optimization, the following should hold:

$$\alpha_1^t y_1 + \alpha_2^t y_2 = \alpha_1^{t+1} y_1 + \alpha_2^{t+1} y_2 = \text{constant} \qquad (3.43)$$

In other words, the new values of the multipliers should lie on a line constrained in a box due to the inequality constraints, that is, $0 \leq \alpha_1, \alpha_2 \leq C$. Note here that this is only true if one started the algorithm off with all multipliers at zero, that is, $\boldsymbol{\alpha} = 0$ or more generally with the equality constraint holding, that is, $\boldsymbol{\alpha}^T \mathbf{y} = 0$. If this condition does not hold then the following discussion does not apply and a suitable method must be devised, for example, projection that ensures that this constraint holds at some point in the iteration. Assuming that the condition holds, then one multiplier α_2 can first be updated so that,

$$\text{LB} \leq \alpha_2 \leq \text{UB}$$

where the upper-(UB) and lower bounds (LB) are determined as

$$
\begin{aligned}
\text{UB} &= \begin{cases} \min(C, \alpha_1 + \alpha_2) & \text{if } y_1 = y_2 \\ \min(C, C + \alpha_2 - \alpha_1) & \text{if } y_1 \neq y_2 \end{cases} \\[2mm]
\text{LB} &= \begin{cases} \max(0, \alpha_1 + \alpha_2 - C) & \text{if } y_1 = y_2 \\ \max(0, \alpha_2 - \alpha_1) & \text{if } y_1 \neq y_2 \end{cases}
\end{aligned}
\qquad (3.44)
$$

and noting from Equation 3.43 that,

$$s = y_1 y_2$$
$$\alpha_1^{t+1} + s\alpha_2^{t+1} = \alpha_1^t + s\alpha_2^t = \gamma \qquad (3.45)$$

we obtain the value of the second multiplier α_1. To derive the SMO update rule, the training error E_i is defined $\forall i = 1, \ldots, n$ as

$$E_i = f(\mathbf{x}_i) - y_i \qquad (3.46)$$

which is the difference between the trained classifier output and the true label. The minimum of the objective function is first found in the direction of α_2 whereas the other multiplier α_1 is set to a value that maintains the equality constraint. We can rewrite the objective function in Equation 3.37 explicitly in terms of the working set members

$$\Im(\alpha_1, \alpha_2) = -\alpha_1 - \alpha_2 + \frac{1}{2}K_{11}\alpha_1^2 + \frac{1}{2}K_{22}\alpha_2^2 + sK_{12}\alpha_1\alpha_2$$
$$+ y_1\alpha_1 v_1 + y_2\alpha_2 v_2 + \Im_{\text{const}} \qquad (3.47)$$

where

$$v_i = \sum_{j=3}^{n} y_j \alpha_j^t K_{ij} = f(x_1) - y_1 \alpha_1^t K_{1i} - y_1 \alpha_2^t K_{2i}$$

$$\Im_{\text{const}} = -\sum_{i=3}^{n} \alpha_i + \frac{1}{2} \sum_{i,j=3}^{n} \alpha_i \alpha_j y_i y_j K(x_i, x_j)$$

The objective function is then expressed entirely in α_2 and the minimum found in terms of this variable, finally after substitution and rearranging, we obtain the update rule,

$$\alpha_2^{t+1} = \alpha_2^t + y_2 \frac{(E_1 - E_2)}{(K_{11} + K_{22} - 2K_{12})} \qquad (3.48)$$

To ensure that α_2^{t+1} is feasible, we enforce the following:

$$\alpha_2^{t+1,\text{bounded}} = \begin{cases} \text{UB} & \text{if } \alpha_2^{t+1} \geq \text{UB} \\ \alpha_2^{t+1} & \text{if } \text{LB} < \alpha_2^{t+1} < \text{UB} \\ \text{LB} & \alpha_2^{t+1} \leq \text{LB} \end{cases} \qquad (3.49)$$

where the bounds are calculated as before. The update for α_1 is then found by using Equation 3.45 and written explicitly as

$$\alpha_1^{t+1} = \alpha_1^t + s\left(\alpha_2^t - \alpha_2^{t+1,\text{bounded}}\right) \qquad (3.50)$$

The SMO algorithm uses a set of heuristics that would select the working set pair at each iteration; specifically, they approximate the most improvement to the objective function by choosing the *maximal violator* pair. At each iteration, the element with the most positive error, that is, $E_1 > 0$ was chosen and the second element selected to maximize the quantity $|E_1 - E_2|$. This leads to selecting two elements with large error difference or maximum violation of the KKT conditions. Furthermore, this represents mathematically an approximation to the step size (Equation 3.48). If this pair failed to improve the objective function within a certain set tolerance, then two other heuristics were employed. The first heuristic was to pair the first violator with any random non-bounded support vector and try to update the working set. If that also failed, then all possible combinations were tried. In both heuristics, the search was started from some random point. The operation of the SMO algorithm is summarized in the flowchart in Figure 3.18. The randomness introduced by the heuristics caused the original SMO algorithm to have a much slower convergence rate (Keerthi and Gilbert, 2002). Nevertheless, the SMO algorithm remains today as one of the more important SVM training algorithms due to the simplicity and speed of its implementation.

c. **SVM^{light} algorithm.** At approximately the same time as SMO, another SVM training algorithm was proposed by Joachims (1998) following directly from Zoutendijk's (1960) method of feasible directions for solving quadratic programs. The update rule is not explicitly referred to by Joachims but it is generally a steepest search direction method and the novelty of the algorithm lies in Joachims' focus on the working set selection method. Joachims' working set selection heuristic did not have random selection elements and was frequently faster than SMO. A combination of this working set selection and the SMO update rule has been further implemented by Lin in his LibSVM software (Chang and Lin, 2001).

The SVM^{light} method first sorts the points that violate the KKT conditions into an error list. Let z denote the number of violators at

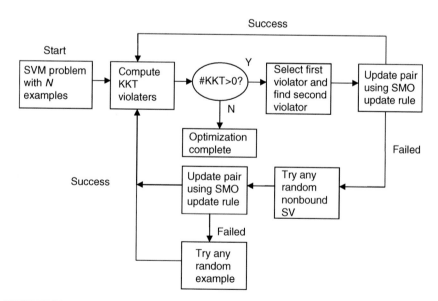

FIGURE 3.18

Flowchart for selecting working sets in the SMO algorithm. #KKT refers to the number of Karush–Kuhn–Tucker violations, that is, points with non-optimal gradients.

the kth iteration and write the violator list as the vector $\mathbf{E}(k)^{\downarrow}$, where $\forall\, k > 0$

$$\mathbf{E}(k)^{\downarrow} = \{y_i v_i | y_1 v_1 \ge y_2 v_2 \ge \cdots \ge y_z v_z, \forall\, i = 1, \ldots, z\}$$

The working set $\boldsymbol{\alpha}_p$ is of size q where q is some even number so that $\forall\, p > 0$, we choose the first $q/2$ elements from $\mathbf{E}(k)^{\downarrow}$ and the last $q/2$ elements, that is,

$$\boldsymbol{\alpha}_p = \{\alpha | \text{first } q/2 \text{ elements from } \mathbf{E}^{\downarrow}, \text{ last } q/2 \text{ elements from } \mathbf{E}^{\downarrow}\} \tag{3.51}$$

This method is a generalization of Platt's method because q variables can now be chosen. Given the ordered error list, the top $q/2$ variables had the largest positive error whereas the bottom $q/2$ variables had the largest negative error. Use of the steepest search direction method implies that gradient descent on q variables could be employed at once, rather than being restricted to just two as in Platt's method. The selection heuristic employed by SVM^{light}, however, was still based on maximal violators as SMO and the step for b was calculated independently. The algorithm flowchart for SVM^{light} is shown in Figure 3.19.

To further decrease the SVM^{light} runtime, Joachims further introduced a heuristic, which he called shrinking. Note that in Figure 3.19, after a selected working set was updated, the gradients of all points were then updated. As convergence was approached, Joachims observed that generally only a subset of multipliers remained to be determined.

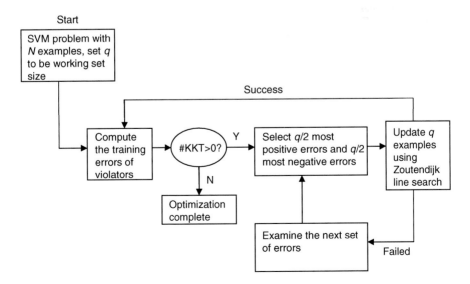

FIGURE 3.19
Flowchart for selecting working sets in the SVM^{light} algorithm.

Multipliers at bounds, that is, $\alpha_i = 0$ and $\alpha_i = C$ tended to remain at bounds and not change further. If it could be known beforehand which multipliers were correctly determined it would not be necessary to update their gradients at every iteration and determine whether they violated the constraints. This would reduce the number of kernel evaluations and increase the algorithm's runtime speed.

Since its inception, several other SVM models have been introduced expanding on Vapnik's original idea. These include Sukyens' least squares SVM (LS–SVM) and its variants (Suykens, 2002; Suykens and Vandewalle, 2000; Suykens et al., 2002), Schlkopf and Smola's (2002) ν-SVM, Weston's adaptive margin SVM (Herbrich and Weston, 1999), and Mangasarian's proximal SVM (PSVM) (Mangasarian and Musicant, 1999; Mangasarian, 2000; Fung and Mangasarian, 2001). Their ideas were mainly motivated by the geometric interpretation of the separating hyperplane resulting in different regularization functions and SVM models (e.g., LS-SVM, PSVM, ν-SVM). These models are variations of the standard SVM, which have exhibited additional useful properties for specific applications. For example, the original LS–SVM could be trained via matrix inversions making its implementation for engineering problems straightforward. In ν-SVM, users could select an upper bound to the number of support vectors by selecting the parameter ν, adjusting the generalization capabilities of the SVM model.

The SVM formulation is powerful for solving function estimation problems and easily implemented. A drawback, however, is that the quadratic optimization problem scales quadratically with the size of the training data, making large-scale problems somewhat slower to solve. Improved optimization algorithms are now being researched to overcome this limitation (see Lai et al., 2004; Dong et al., 2005; Shilton et al., 2005).

3.4 Hidden Markov Models

In signal processing theory, signals can be stationary or nonstationary, pure or noisy, and discrete or continuous. For example, an ECG lead that records heartbeat potentials usually picks up noise from muscle contractions and movement artifacts such that the measured signal is the corrupted version of the actual heart signal. In this case, we aim to build a model of the ECG channel to gain a better understanding of the channel characteristics and to enable removal of the noise from the signal. Such signal models may be either deterministic (known properties) or stochastic (random properties). The HMM is an example of a stochastic signal model used to depict the possible system states. This section contains a summary of the HMM model with further details in excellent tutorials by Rabiner (1989) and Rakesh (Dugad and Desai, 1996).

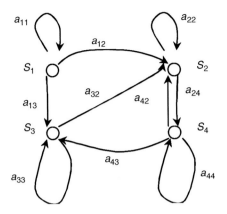

FIGURE 3.20
A 4-state first-order discrete Markov chain with transition probabilities a_{ij}.

3.4.1 The Markov Chain

To begin we review the precursor to the HMMs, the basic discrete Markov chain shown in Figure 3.20. This models a Markov process where all states of a system are observable in that they can be measured. A system that can be modeled by a Markov process usually has several observable states denoted by

$$\mathbf{S} = \{S_1, S_2, \dots, S_N\} \tag{3.52}$$

and is sometimes called an N-state Markov chain. Moving from one state to another is known as a *transition* and the probability of the system moving from state S_i to state S_j is referred to as the transition probability a_{ij}. Given a set of instances or time $T = \{t_1, t_2, \dots, t_k\}$ and denoting the current state of the system at time t as q_t, the state of the system is specified by

$$P[q_t = S_j | q_{t-1} = S_i] \tag{3.53}$$

This is possible provided the system states are time-independent resulting in the transition probabilities

$$a_{ij} = P[q_t = S_j | q_{t-1} = S_i] \tag{3.54}$$

where $1 \leq i, j \leq N$. The state transition coefficients obey the following standard stochastic rules

$$a_{ij} \geq 0$$

$$\sum_{j=1}^{N} a_{ij} = 1$$

3.4.2   The Hidden State

Suppose that you participate in an experiment where someone hidden behind
a curtain draws a ball from a box and announces its color. The ball is either
blue or red and the number of balls in the box is unknown. Assume that we
obtain the observation sequence BBBRBBRB, where B is taken to be blue and
R is red. If the system states are either blue or red, how can the observation
sequence be modeled using a Markov chain? This can be done by assuming
that there were two different colored balls with equal probabilities of drawing
either color and using a 2-state Markov chain to model the sequence. There
were however more blues in the sequence and so it is possible that there were
three balls, two blue and one red and therefore a 3-state Markov system would
be a better model. It is immediately apparent that the previous Markov chain
is too restrictive in this situation due to hidden or unspecified information.
This is the case in many real-world applications with stochastic observations.
One technique to handle this is the HMM, which is an extension of Markov
models to include observations modeled by probabilistic functions.

The HMM model is composed of five key elements described in the following:

a. N, the number of states in the model. Ergodic models have intercon-
     nected states such that each state can be reached from another state
     in a finite number of steps.

b. M, the number of distinct observations in the event. This can be blue
     or red in the ball example, heads or tails for a coin toss, and so on. The
     observations are represented by V_i and the set of distinct observations
     defined as

$$\mathbf{V} = \{V_1, V_2, \ldots, V_M\} \tag{3.55}$$

c.   The set of transition probabilities $A = \{a_{ij}\}$. This set also implicitly
     defines the interconnections between states and also the number of
     states. If there is no direct connection between two states, the transi-
     tion probability is zero.

d.   The set of observations or the observation symbol probability distri-
     bution in state j, $B = \{b_j(k)\}$, where

$$b_j(k) = P[v_k|q_t = S_j] \tag{3.56}$$

     where $1 \leq j \leq N$ and $1 \leq k \leq M$.

e.   The initial state distribution

$$\pi_i = P[q_1 = S_i] \tag{3.57}$$

and $\boldsymbol{\pi} = \{\pi_i\}$.

An HMM can be fully specified by the model parameters $\lambda = \{A, B, \pi\}$ and is used to generate an observation sequence such as

$$\boldsymbol{O} = O_1 O_2 \dots O_T \tag{3.58}$$

This is accomplished by first selecting an initial state according to the state distribution π. Then beginning with $t=1$, select $O_k = V_k$ according to the symbol probability distribution in the state S_i using $b_i(k)$. The new system state $q_{t+1} = S_j$ is obtained by transiting (moving) according to the state transition probability for S_i, that is, using a_{ij}. The model continues until T observations have been obtained.

There are three basic problems to which HMM is applicable.

a. Given an observation sequence $\boldsymbol{O} = O_1 O_2 \dots O_T$ and a model $\lambda = \{A, B, \pi\}$, compute the probability that the model can generate the sequence correctly. This is written as $P(\boldsymbol{O}|\lambda)$ and is actually an evaluation of how well the model can produce the observed sequence. It is useful when we have a number of competing models and wish to select one that produces a sequence to best match the observed sequence.

b. Given an observation sequence $\boldsymbol{O} = O_1 O_2 \dots O_T$ and a model $\lambda = \{A, B, \pi\}$, determine how to select an optimal state sequence $\boldsymbol{Q} = q_1 q_2 \dots q_T$, where optimal reflects how well the generated sequence $\boldsymbol{Q}$ matches the observed sequence $\boldsymbol{O}$. The solution to this problem uncovers the hidden elements of the system but we seldom obtain a HMM model that matches the sequence exactly and an optimality criterion must be used.

c. To adjust the model parameters $\lambda = \{A, B, \pi\}$ to maximize $P(\boldsymbol{O}|\lambda)$, we present the HMM with a set of observation sequences or training data such that the HMM parameters can be adjusted to generate sequences close to the training sequences. Training is important in adaptively adjusting the HMM model to best approximate the real-world application.

3.4.3 Types of Hidden Markov Models

In the previous section, we described the basic ergodic HMM model, such as the commonly used 4-state ergodic model in Figure 3.21a. Several better performing models have been proposed, the left–right model or Bakis model (Jelinek, 1976; Bakis, 1976), for example, is a different model in that connections are only allowed from states with lower indices to states with higher indices. Figure 3.21b shows a 4-state left to right model where the state transitions proceed from left to right. This model is particularly useful for systems with dynamic properties or characteristics that change over time. A definitive characteristic of this model is that the state transition coefficients are defined as

$$a_{ij} = 0 \quad \text{if } j < i$$

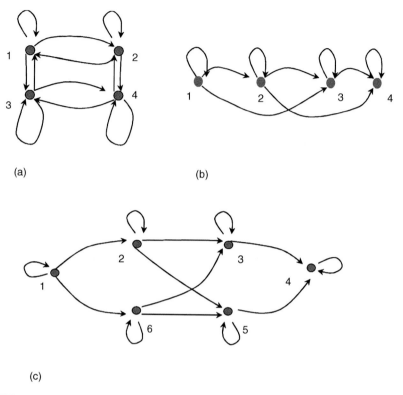

FIGURE 3.21
Diagram depicts three of the common and distinct HMM types. (a) 4-state ergodic model; (b) 4-state left to right model; and (c) 6-state parallel path left to right model.

which implies that backward transitions to lower states (indices) are not permitted. Additional constraints can be placed on the coefficients, for example,

$$a_{ij} = 0 \quad \text{if } j > i + \delta$$

which ensures that large changes in the states are prohibited. If $\delta = 2$, the maximum number of states passed in a single jump (transition) is 2. A slight modification using a cross-coupled connection of two parallel left to right models as in Figure 3.21c adds further flexibility to the previously strict left to right model.

In the models discussed so far, observations have been drawn from some finite alphabet, for example, heads or tails, and red or blue resulting in the discrete probability densities for the model states. In many real-world applications such as telecommunications and speech recognition, continuous signals are digitized using codebooks and mapping signal-quantization levels to code vectors resulting in undesired information loss. This has been addressed using continuous observation densities in HMM models (Liporace, 1982; Juang et al., 1986; Qiang and Chin-Hui, 1997) to be a mixture of densities having the general form

$$b_i(O) = \sum_{m=1}^{M} c_{jm} \Im[\mathbf{O}, \boldsymbol{\mu}_{jm}, \mathbf{U}_{jm}] \tag{3.59}$$

where $1 \le j \le N$. Here $\mathbf{O}$ is the vector of observations to be modeled and c_{jm} is the mixture coefficient for the mth mixture in state j. $\Im$ is some log-concave or symmetric density such as the Gaussian density with mean vector $\boldsymbol{\mu}_{jm}$ and covariance matrix $\boldsymbol{U}_{jm}$ for the mth mixture in state j. The coefficients satisfy the stochastic properties

$$\sum_{m=1}^{M} c_{jm} = 1$$

$$c_{jm} \ge 0$$

and the probability density function (pdf) is normalized as follows:

$$\int_{-\infty}^{\infty} b_j(\mathbf{x}) \mathrm{d}\mathbf{x} = 1$$

where $1 \le j \le N$. The estimation of the model parameters is given by the following formulas:

$$c_{jk} = \frac{\sum\limits_{t=1}^{T} p_t(j, k)}{\sum\limits_{t=1}^{T} \sum\limits_{k=1}^{M} p_t(j, k)}$$

$$\boldsymbol{\mu}_{jk} = \frac{\sum\limits_{t=1}^{T} p_t(j, k) \cdot \boldsymbol{O}_t}{\sum\limits_{t=1}^{T} p_t(j, k)}$$

$$\boldsymbol{U}_{jk} = \frac{\sum\limits_{t=1}^{T} p_t(j, k) \cdot (\boldsymbol{O}_t - \boldsymbol{\mu}_{jk})(\boldsymbol{O}_t - \boldsymbol{\mu}_{jk})^T}{\sum\limits_{t=1}^{T} p_t(j, k)}$$

where $p_t(j, k)$ is the probability of being in state j at time t with the kth mixture component for $\boldsymbol{O}_t$.

There are other variants of HMM models such as the autoregressive HMM models (Poritz, 1982; Juang and Rabiner, 1985) having observation vectors

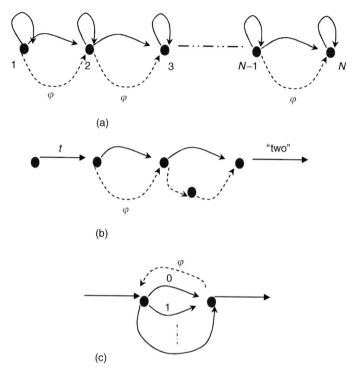

FIGURE 3.22
Diagram depicts three HMM types, which incorporate null transitions.
(a) Left–right model with null transition φ; (b) finite state network;
(c) grammar network.

drawn from an autoregressive process. These models are often employed in
speech recognition (Tisby, 1991). Some variations of the HMM structure
have also been used such as the null-transition and tied-states models (Bahl
et al., 1983). In this model, jumps from one state to another produce no output observation and have been successfully implemented as left–right models
(Figure 3.22a), finite state networks (Figure 3.22b), and grammar networks
(Figure 3.22c). The left–right model has a large number of states and the null
transition allows connectivity between a pair of states but no transitions. The
observation sequences can be generated with one state and still have a path
beginning in state one and ending in state N. The finite state structure in
Figure 3.22b is used primarily to provide a more compact and efficient method
of describing alternate word pronunciations whereas the grammar network
has been employed to generate arbitrary digit sequences of arbitrary length
by returning to the starting state after each digit has been produced.

3.5 Fuzzy Sets and Fuzzy Logic

In the previous sections, we used variables belonging to groups or sets. Set theory is a well-defined mathematical field with applications ranging from engineering to medicine. Consider a collection of elements such as a basket of fruits denoted using set notation as $\mathbf{A}$, and a_i is an arbitrary fruit from the basket. In set theory, we say that "a_i belongs to $\mathbf{A}$" and write $a_i \in \mathbf{A}$ where $\in$ means "in the set." If a_i is not a fruit from the basket, we write $a_i \notin \mathbf{A}$ where $\notin$ means "not belonging to the set" or a_i is not a *member* of the set $\mathbf{A}$. These classical versions of two-valued sets are known as *crisp* sets because an element is either a member or a nonmember of a set.

What happens when we wish to group a set of elements according to abstract variables such as large, hot, tall, and so on? These variables are commonly used in our language and our everyday situations. Such variables are called *linguistic variables* since they originate from our language and usually take vague values. For example, suppose you gather a group of friends for a game of basketball, as the team leader, you would prefer tall team mates to gain an advantage in the game. But the question arises; how tall is tall? Is 170 or 190 cm tall? There appears to be no definitive threshold that defines a person grouped as tall since it varies with individual's perception. For mathematical purposes, we can however denote a linguistic variable as taking a value over a continuous range of truth values normalized to the interval $[0,1]$, rather than just true or false.

The study of such groups forms the crux of fuzzy set theory and the extension to fuzzy logic is a result of applying the fuzzy ideas to Boolean logic. The concept of fuzziness was first described in the 1960s by Lotfi Zadeh from U.C. Berkeley in his seminal papers on fuzzy sets (Zadeh, 1965, 1973). It was extended to fuzzy logic, which is a superset of conventional Boolean logic. Fuzzy logic deals with a concept of partial truths, that is, truth values between "completely true" and "completely false" similar to the way fuzzy sets deal with partial memberships of an element. When applied to real-world problems, fuzzy logic becomes a structured, model-free estimator that approximates a function through linguistic input–output associations. A typical fuzzy system consists of a rule base, membership functions, and an inference procedure. It has been successfully applied to systems that are difficult to model due to the presence of ambiguous elements. It has found many applications, some of which include fuzzy control in robotics, automation, tracking, consumer electronics, fuzzy information systems such as the Internet, information retrieval, fuzzy pattern recognition in image processing, machine vision, and in decision support problems (Bojadziev and Bojadziev, 1997; Kandel and Langholz, 1994; Zadeh et al., 2004b). Pure fuzzy logic itself is seldom directly applicable in devices, perhaps with the exception of the Sony PalmTop that uses a fuzzy logic decision tree algorithm to perform handwritten Kanji character recognition.* In the following section, we describe the basic properties of fuzzy sets and their extension to fuzzy logic applications. Readers interested in further details can consult many other resources on fuzzy theory and applications (Zadeh, 1965, 1973; Zadeh et al., 2002; Brubaker, 1992).

3.5.1   Fuzzy Sets

A fuzzy set is defined as a group or collection of elements with membership values between 0 (completely not in the set) and 1 (completely in the set). The membership values give the property of fuzziness to elements of a fuzzy set in that an element can partially belong to a set. The following definition formalizes the fuzzy set.

DEFINITION 3.5.1   (Fuzzy Set [Zadeh, 1965])   *A fuzzy set is characterized by a membership function, which maps the elements of a domain, space, or an universe of discourse* X *to the unit interval* $[0, 1]$, *written as*

$$A = X \to [0, 1]$$

Hence a fuzzy set A is represented by a set of pairs consisting of the element $x \in X$ *and the value of its membership and written as*

$$A = \{m(x) | x \in X\}$$

A crisp set is our standard set where membership values take either 0 or 1. Fuzzy sets can be defined on finite or infinite universes where when the universe is finite and discrete with cardinality n, the fuzzy set is an n-dimensional vector with values corresponding to the membership grades and the position in the vector corresponding to the set elements. When the universe is continuous, the fuzzy set may be represented by the integral

$$A = \int_x m(x)/x \tag{3.60}$$

where the integral shows the union of elements not the standard summation for integrals. The quantity $m(x)/x$ denotes the element x with membership grade $m(x)$.

An example of a fuzzy set is the set of points on a real line (Pedrycz and Vasilakos, 2001). Their membership functions can be described by a delta function where for $x, p \in \Re$, the membership function is

$$m(x) = \delta(x - p) = \begin{cases} 1 & \text{if } x = p \\ 0 & \text{if } x \neq p \end{cases} \tag{3.61}$$

---

*http://austinlinks.com/Fuzzy/

3.5.2 Fuzzy Membership Functions

Membership functions are used to describe the membership value given to fuzzy set elements. Their use depends on the meaning or linguistic variable attached to the fuzzy set, for example, if the values are taken to denote the linguistic variable "high," then it must be defined with respect to a quantity such as temperature. Examples of membership functions are given in the following:

a. *Triangular membership function.*

$$m(x) = \begin{cases} 0 & \text{if } x \le a \\ \dfrac{x-a}{d-a} & \text{if } x \in [a,d] \\ \dfrac{b-x}{b-d} & \text{if } x \in [d,b] \\ 0 & \text{if } x \ge b \end{cases} \tag{3.62}$$

where d is the modal value and a and b are the upper and lower bounds of $m(x)$, respectively.

b. *Γ-membership function.*

$$m(x) = \begin{cases} 0 & \text{if } x \le a \\ 1 - e^{-k(x-a)^2} & \text{if } x > a \end{cases} \tag{3.63}$$

or

$$m(x) = \begin{cases} 0 & \text{if } x \le a \\ \dfrac{k(x-a)^2}{1+k(x-a)^2} & \text{if } x > a \end{cases} \tag{3.64}$$

where $k > 0$.

c. *S-membership function.*

$$m(x) = \begin{cases} 0 & \text{if } x \le a \\ 2\left(\dfrac{x-a}{b-a}\right)^2 & \text{if } x \in [a,d] \\ 1 - 2\left(\dfrac{x-a}{b-a}\right)^2 & \text{if } x \in [d,b] \\ 1 & \text{if } x \ge b \end{cases} \tag{3.65}$$

Here the point $d = (a+b)/2$ is the crossover of the S-function.

d. *Trapezoidal-membership function.*

$$m(x) = \begin{cases} 0 & \text{if } x < a \\[2mm] \dfrac{x-a}{d-a} & \text{if } x \in [a,d] \\[2mm] 1 & \text{if } x \in [d,e] \\[2mm] \dfrac{b-x}{b-e} & \text{if } x \in [e,b] \\[2mm] 0 & \text{if } x > b \end{cases} \qquad (3.66)$$

e. *Gaussian-membership function.*

$$m(x) = e^{-k(x-d)^2} \qquad (3.67)$$

for $k > 0$.

f. *Exponential-membership functions.*

$$m(x) = \frac{1}{1 + k(x-d)^2} \qquad (3.68)$$

for $k > 1$ or

$$m(x) = \frac{k(x-d)^2}{1 + k(x-d)^2} \qquad (3.69)$$

The membership value assigned to an element in the fuzzy set depends on the application at hand, for example, high temperature in an ore smelting furnace is clearly different from the meaning of high temperature in a house with gas heating. Several methods are used to estimate the proper membership values such as horizontal estimation, vertical estimation, pairwise comparison, and inference. Horizontal estimation uses statistical distributions to estimate the membership value based on responses from a group of observers. Vertical estimation discretely separates the responses of the observer group and sections the elements with respect to the importance placed by the group. In pairwise comparisons, elements are compared with other elements in the fuzzy set **A** and valued according to a suitable ratio scale. Problem inference involves constructing approximating functions to a problem that can be used as membership functions.

3.5.3   Fuzzy Operations

Since a fuzzy set is characterized by its membership function, operations on fuzzy sets manipulate these functions in a similar manner as standard functional analysis and set theory. Some of the more common mathematical notions are outlined in the following.

a. *Cardinality.* For a fuzzy set **A** in a finite universe **X**, the cardinality of the set is

$$\text{card}(\mathbf{A}) = \sum_{x \in \mathbf{X}} m_{\mathbf{A}}(x) \tag{3.70}$$

where the cardinality is a scalar or a sigma count of **A**. If the universe is continuous, the sigma becomes an integral provided that the operation is sensible.

b. *Convexity.* A fuzzy set **A** is convex if its membership function satisfies

$$m_{\mathbf{A}}(\lambda x_1 + (1 - \lambda)x_2) \geq \min[m_{\mathbf{A}}(x_1), m_{\mathbf{A}}(x_2)] \tag{3.71}$$

for any $x_1, x_2 \in \mathbf{X}$ and $\lambda \in [0, 1]$.

c. *Concavity.* A fuzzy set **A** is concave if its membership function satisfies

$$m_{\mathbf{A}}(\lambda x_1 + (1 - \lambda)x_2) \leq \min[m_{\mathbf{A}}(x_1), m_{\mathbf{A}}(x_2)] \tag{3.72}$$

for any $x_1, x_2 \in \mathbf{X}$ and $\lambda \in [0, 1]$.

d. *Support.* A support of a fuzzy set **A** are all elements in the universe **X** belonging to **A** that are nonzero.

$$\text{supp}(\mathbf{A}) = \{x \in \mathbf{X} | m_{\mathbf{A}}(x) > 0\} \tag{3.73}$$

e. *Unimodality.* A fuzzy set **A** is unimodal if the membership function has a single maximum.

f. *Normality and height.* A fuzzy set **A** is normal if the maximum value of its membership grade is 1:

$$\sup_{x} m_{\mathbf{A}} = 1 \tag{3.74}$$

and subnormal if the supremum is less than 1. The supremum is sometimes referred to as the height of **A** and so normal fuzzy sets have height equal to 1.

Several basic set operations for fuzzy sets are given in the following in terms of their membership functions.

a. *Concentration.* The membership functions get smaller and points with higher membership values are clustered closer together, for example,

$$\text{Conc}(\mathbf{A}) = m_{\mathbf{A}}^{p}(x) \tag{3.75}$$

where $p > 1$.

b. *Dilation.* Dilation is the opposite of concentration and can be produced by transforming the membership function as follows:

$$\text{Dil}(\mathbf{A}) = m_{\mathbf{A}}^{\frac{1}{p}}(x) \tag{3.76}$$

where $p > 1$.

c. *Intensification.* The membership values in the fuzzy set are intensified so that greater contrast between set members is achieved. This can be done, for example, by the following membership function modification:

$$\text{Int}(\mathbf{A}) = \begin{cases} 2^{p-1} m_{\mathbf{A}}^{p}(x) & \text{if } 0 \le m_{\mathbf{A}}(x) \le 0.5 \\ 1 - 2^{p-1}(1 - m_{\mathbf{A}}(x))^{p} & \text{otherwise} \end{cases} \tag{3.77}$$

The effect is increased contrast where membership values below 0.5 have a more diminished effect whereas values above 0.5 are intensified.

d. *Fuzzification.* This is the opposite of intensification where the membership function can be modified as follows:

$$\text{Fuzz}(\mathbf{A}) = \begin{cases} \sqrt{m_{\mathbf{A}}(x)/2} & \text{if } m_{\mathbf{A}}(x) \le 0.5 \\ 1 - \sqrt{m_{\mathbf{A}}(x)/2} & \text{otherwise} \end{cases} \tag{3.78}$$

Algebraic operations on fuzzy sets can be viewed as calculation of uncertain numeric quantities. Computing with fuzzy quantities was adapted from Moore (1966), who had developed a branch of mathematics to deal with the calculus of intervals. This resulted in the *extension principle*, which describes how to perform the operations between fuzzy sets.

For example, a fuzzy set $\mathbf{A}$, which undergoes a mapping f forms another fuzzy set $\mathbf{B}$, that is,

$$\mathbf{B} = f(\mathbf{A})$$

In terms of membership functions, this can be written as

$$m_{\mathbf{B}}(b) = \sup_{a \in \mathbf{A}: b = f(a)} [m_{\mathbf{A}}(a)]$$

The supremum over an empty fuzzy set is zero, that is,

$$\sup_{\emptyset}[m_{\mathbf{A}}(a)] = 0$$

The extension principle can be applied provided the following monocity conditions hold:

$$f \le g \Rightarrow f(\mathbf{A}) \subseteq f(\mathbf{B})$$

$$\mathbf{A} \subset \mathbf{A}' \Rightarrow f(\mathbf{A}) \subset f(\mathbf{A}') \tag{3.79}$$

This can also be further generalized to several variables

$$\mathbf{B} = f(\mathbf{A}_1, \ldots, \mathbf{A}_n)$$

and the multivariable membership function is computed as

$$m_{\mathbf{B}}(b) = \sup_{a_1, a_2, \ldots, a_n : b = f(a_1, a_2, \ldots, a_n)} [m_{\mathbf{A}_1}(a_1) \wedge m_{\mathbf{A}_2}(a_2) \wedge, \ldots, \wedge m_{\mathbf{A}_n}(a_n)]$$

It is further known that the extension principle results in a nonlinear optimization problem with some linear or nonlinear constraints. To simplify computations, the triangular membership function is commonly assumed due to its linearity (Pedrycz and Vasilakos, 2001). Further computations on fuzzy sets can be performed using fuzzy rules, which we describe in the following section.

Another aspect of fuzzy sets is the idea of defuzzification (Tomsovic and Chow, 2000), which is the conversion of fuzzy quantities to distinct quantities. This is required because most real-world applications require discrete and certain values. Defuzzification is done by decomposition to find a distinct value that represents the information contained in a fuzzy set. The value obtained is an average or the expectation of the fuzzy set values. There are two commonly used defuzzification methods: the composite moments (centroids) and the composite maximum. In the first method, a fuzzy set centroid is computed by finding the weighted mean of the region, which becomes the center of gravity for the fuzzy set. In the composite maximum method, the aim is to find a point with maximum truth. Several composite maximum methods exist such as the simple maximum, average maximum, and the center of maximums.

3.5.4 Fuzzy Systems: An Application of Fuzzy Logic

In this section, we describe an example of a fuzzy system used in some fuzzy electronic applications. Fuzzy systems operate on a structure of rules that are a formal way of representing strategies, directives, or modes of operation. Systems built on rules derived from a set of facts are called knowledge-based systems. In essence, a rule is a standard conditional statement having the following form:

if condition A then conclusion B

Conditions are sometimes referred to as *antecedents* and conclusions as *consequents*. Rules can be further augmented with quantifiers that add a degree of certainty, for example,

if condition A then conclusion B with certainity λ

Here the quantifier $\lambda \in [0, 1]$ adds a further weight to the rule, when $\lambda = 1$ we immediately have condition B. Rules may also be built to express gradual relationships between condition and conclusion, for example,

the more A the more B

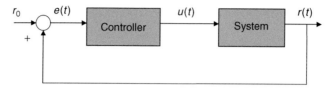

FIGURE 3.23
A general closed loop control system.

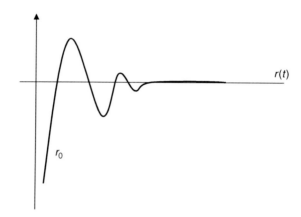

FIGURE 3.24
A desired output response of a good control system.

They can also be extended and generalized to cover a variety of subconditions. A collection of rules such as "if–then" rules can be used to form a rule-based system and in particular act as mappings for fuzzy sets. These are better known as "fuzzy rules" and are used in many applications such as fuzzy controllers, fuzzy schedulers, and fuzzy optimizations. We will describe the fuzzy controller briefly to illustrate how fuzzy rule–based systems are implemented.

The standard control problem can be stated as follows. Given a reference signal $r(t)$, which may be static for all time, design a controller that maps an input r_0 using some control action $u(t)$ so that r_0 achieves $r(t)$ as smoothly and as quickly as possible. The standard controller with feedback is depicted in Figure 3.23 and an example of a desired system response is shown in Figure 3.24. One way to achieve this control is by considering the errors at the system output, which are defined as the difference between the initial signal and the current output signal

$$e(t) = x_0 - x(t)$$

The change in the error can be defined as successive values of the error

$$\triangle e = e(t+1) - e(t)$$

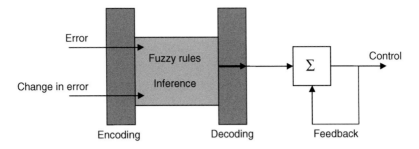

FIGURE 3.25
A basic fuzzy controller structure.

and over time as the derivative of the error. The goal is to design a control signal $u(t)$ such that this error diminishes or the error derivative vanishes.

A fuzzy controller is simply a nonconventional approach to feedback control. It is an example of a commonly used knowledge-based and model-free control paradigm. In this type of controller, a system of rules is applied to control the signal. For example, if the change in error is zero and the error is zero then we would wish to keep the control signal constant. This is aptly denoted by the following rule:

> if error is ZERO and change in error is ZERO, then change in
> control is ZERO

The application of rules should take into account external knowledge such as "common sense" knowledge or knowledge acquired from prior simulation of the control model. An example of a fuzzy controller is depicted in Figure 3.25 comprising an input interface, a processing module, and an output module. The rules and inference form the processing module whereas the encoding and decoding are the interfaces. In this application, the general rule set template has the following form:

> if error is a and change in error is b, then change in control is c

The detailed computations of the control signals can be performed using t-norms or s-norms but will not be discussed here. Scaling coefficients are used to ensure that the same linguistic protocol is applied throughout the system, especially in cases when the inputs and output variables have different ranges. The scaling method has to be further adapted (using moving averages and normalization) because it must be predetermined prior to the start of the system.

This collection of fuzzy rule–based systems can be further employed to construct a fuzzy expert system that uses fuzzy logic instead of Boolean logic. In summary, a fuzzy expert system is a collection of membership functions and rules that are used to interpret data. Unlike conventional expert systems, which are mainly symbolic reasoning engines, fuzzy expert systems are oriented toward numerical processing.

3.6 Hybrid Systems

The previous techniques concerned different problem areas and appeared to be mutually independent of each other. Neural networks are powerful computational tools for function estimation and unsupervised learning whereas SVMs improve on them with enhanced mathematical characteristics and elegance. A signal-processing system with stochastic aspects can be modeled using HMMs, and fuzzy notions provide alternative representations of knowledge-based systems and information. It is thus desirable to design systems that incorporate two or more techniques to solve more complex problems. These systems are called hybrid systems or expert systems (Goonatilake and Khebbal, 1995; Khosla and Dillon, 1997; Ruan, 1997; Ovaska, 2005), and they combine the different paradigms to form more powerful formulations. In this section, we describe two examples of hybrid systems suitable for problems in biomedical engineering, namely, fuzzy neural networks and fuzzy SVMs.

3.6.1 Fuzzy Neural Networks

Fuzzy sets describe knowledge-based aspects of data by allowing varying degrees of uncertainty to be attached to elements in sets thereby giving additional descriptive power to the information that can be represented. But fuzzy sets have poor computing power and are not particularly useful for direct computational purposes. This weakness is inherent in other set theoretic constructs mainly because sets were not designed for direct computation. In retrospect the neural network is a powerful computing technique, which can successfully estimate unknown functions or generate distributions from data. The fuzzy neural system is an extension of the fuzziness notion to the neural network paradigm resulting in a fusion of knowledge-oriented mechanisms. This hybrid system has been further extended to include various fuzzy neural network architectures, neurofuzzy controllers, and neurofuzzy classifiers used in a diverse range of applications such as sensor fusion, logic filtering, and generation of approximate reasoning models.

Attempts have been made to add fuzzy characteristics to various components of the neural network. In training data, fuzzy notions have primarily been used in normalization to allow multiple levels of data granularity and to deal with noisy data. The input–output pairs or training data are allowed a degree of noise insensitivity due to the fuzzification process. The resulting neurofuzzy system therefore possesses a certain degree of robustness with the ability to encode linguistic input terms.

In training algorithms such as the backpropagation algorithm, the update rules are complemented with fuzzy-type rules. Recall the backpropagation update rule, which has the gradient descent form (Equation 3.15; Section 3.2.3):

$$w_{ij}^{t+1} = w_{ij}^t - \lambda \frac{\partial E_{\text{MLP}}^{\text{sq}}}{\partial w_{ij}}$$

Now the larger the learning rate λ, the faster the learning at the expense of overshooting the optimal $E_{\text{MLP}}^{\text{sq}}$. To account for this, it is possible to add *metarules* to the optimization step, an example meta rule being

$$\text{if } \Delta E_{\text{MLP}}^{\text{sq}} \text{ changes then } \lambda \text{ changes}$$

This augmentation is also known as *meta learning* and has shown improved performance compared to the standard update rules found in classical optimization.

Fuzzy perceptrons have also been introduced by Keller and Hunt (1985) to perform data clustering and automatically determine the prototype clusters and class memberships of individual patterns. The update of the perceptron weights is modified to

$$w_j^{t+1} = w_j^t + cx_j \|u_{1j} - u_{2j}\|^p$$

where u_{ij} is the membership grade of the ith training pattern whereas c and p are predetermined scalars, which respectively affect the learning rate and the norm in which the computation takes place.

A general fuzzy neural system (Figure 3.26) usually has inputs consisting of rules where the conditions and consequents are selected *a priori*. Numeric rules such as

$$\text{if } x = 1.2 \quad \text{then } y = 4.5$$

tend to be overly specific and brittle in the sense that they are activated only when the exact condition is met. Alternatively, fuzzy relations are applied over the Cartesian product of input variables such as $\mathbf{A}_1, \mathbf{A}_2, \dots, \mathbf{A}_m$ and output space variables $\mathbf{B}_1, \mathbf{B}_2, \dots, \mathbf{B}_n$ to soften the conditions. The objective now is to discover meaningful rules between the inputs and outputs, that is, $\mathbf{A}_i$ and $\mathbf{B}_j$. In this system the strength of the relationship between the fuzzy input and output variables can be depicted by an association matrix. The input conditions form the m columns of the matrix and the output consequents form the n rows. Each row indicates what the required conclusions are for the given conditions, so if only one dominant element is in the row, then only a single rule is available. If no dominant element exists, the particular rule is insignificant. Besides that, it is also possible to have k dominant elements in

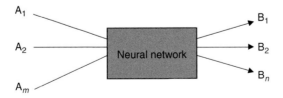

FIGURE 3.26
A basic neural network with fuzzified inputs and outputs. The inputs represent linguistic variables or rules.

a row, which results in k rules governing the fuzzy neural network and the rules have the form

$$\text{if } \mathbf{A}_1 \quad \text{then } \mathbf{B}_{11}$$

$$\vdots$$

$$\text{if } \mathbf{A}_1 \quad \text{then } \mathbf{B}_{1k}$$

This is in fact conflicting because we now have several conclusions for a single conclusion. To alleviate this, we can combine the rules to the following

$$\text{if } \mathbf{A}_1 \quad \text{then } \mathbf{B}_{11} \text{ OR } \mathbf{B}_{12} \dots \text{OR } \mathbf{B}_{1k}$$

With these ideas it is then further possible to construct AND/OR neurons, neurons with inhibitory pathways and feedback fuzzy neurons.

Fuzzy neural systems have also been extended to other neural network architectures such as the fuzzy multilayer perceptron (Figure 3.27), which is the combination of fuzzy neurons similar to the standard MLP network. A further example is the fuzzy SOM with linguistic interpretations attached. The SOM can locate an area of neurons with the greatest weight to simulate associative memory. This corresponds to a substantial amount of outputs as it is in most unsupervised learning cases. For a fuzzy SOM, consider each input variable as having a particular linguistic term defined as a fuzzy set in the corresponding space, for example, $\mathbf{A}_1, \mathbf{A}_2, \dots, \mathbf{A}_n$ for X_1. The activation region indicates how intense or wide the coverage of this input is to the data space and also implicitly its importance in the network. Higher activation levels are interpreted as higher dominance of the linguistic variable or that it is stressed more, for example, the temperature is lower, given X_1 represents coldness. Temporal aspects of network interaction arise when different levels of learning intensity occur.

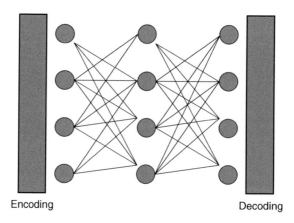

Encoding　　　　　　　　　　　Decoding

FIGURE 3.27

A general fuzzy MLP with fuzzy encoding and decoding.

In summary, there is a variety of hybrid neurofuzzy systems with fuzziness having being extended to almost all neural network architectures. It is therefore important to examine the application so as not to be misled by their variety.

3.6.2 Fuzzy Support Vector Machines

Fuzzy notions have only recently been combined with the SVM and are viewed as a natural extension of fuzzy neural systems. Outliers can seriously affect the training of standard SVM models, in particular the least squares SVM models (Suykens, 2002). In view of this, fuzzifying the training set and applying a SVM classifier was attempted (Inoue and Abe, 2001; Lin and Wang, 2002). Rather than penalizing each training point equally as in the original SVM formulation (Section 3.3.1), a membership grade was assigned to each point with potential outliers having lower membership grades to decrease their influence on the resulting classifier.

Inoue and Abe (2001) and Lin and Wang (2002) modeled the classification problem as

$$\min_{\mathclap{}} \Im(\mathbf{w}, \boldsymbol{\xi}) = \frac{1}{2} \sum_{i=1}^{m} w_i^2 + C \sum_{i=1}^{n} m_i \xi_i \qquad (3.80)$$

subject to

$$y_i \left(\sum_{i=1}^{m} w_i \phi_i(\mathbf{x}) + b \right) \geq 1 - \xi_i$$

$$\xi_i > 0$$

$$\forall \, i = 1, \dots, n$$

where m_i were the membership grades determined prior to the optimization. The simplest membership method was to compute the class centers and scale the Euclidean distances of each training point to their respective class centers (Inoue and Abe, 2001; Lin and Wang, 2002). Another more complex method was to detect outliers using Kohonen's SOM, fuzzy c-means, or some index of fuzziness (Huang and Liu, 2002). This fuzzy SVM model is sometimes referred to as single or unilateral weighted error fuzzy SVM model.

A bilateral method was also proposed where the training set size was doubled (Wang et al., 2005). In this model each training point was assigned a membership grade for both classes in the original dataset

$$\boldsymbol{\Theta} = \{(\mathbf{x_1}, y_1), (\mathbf{x_2}, y_2), \dots, (\mathbf{x_n}, y_n)\}$$

$$\mathbf{x}_i \in \Re^k \qquad (3.81)$$

$$y_i \in \{-1, 1\}$$

was extended to

$$\Theta = \{\mathbf{x}_i, 1, m_i\}, \{\mathbf{x}_i, -1, 1 - m_i\}$$
$$\mathbf{x}_i \in \Re^k \qquad\qquad\qquad (3.82)$$

for $i = 1, \ldots, n$. The result is an optimization problem having the form

$$\min_{\mathbf{w}, \boldsymbol{\xi}} \Im(\mathbf{w}, \boldsymbol{\xi}) = \frac{1}{2} \sum_{i=1}^{m} w_i^2 + C \sum_{i=1}^{n} [m_i \xi_i + (1 - m_i)\eta_i] \qquad (3.83)$$

subject to

$$\sum_{i=1}^{m} w_i \phi_i(\mathbf{x}) + b \geq 1 - \xi_i$$

$$\sum_{i=1}^{m} w_i \phi_i(\mathbf{x}) + b \leq -1 + \eta_i$$

$$\xi_i, \eta_i \geq 0$$

$$\forall i = 1, \ldots, n$$

Both formulations can be solved using the standard Lagrangian approach for dealing with constraints. In the bilateral fuzzy SVM, two sets of Lagrangian multipliers are used to describe the decision surface, which is similar in flavor to the SVM regression case (see Equation 3.42; Section 3.3.2). The bilateral fuzzy SVM decision surface is written as

$$f(\mathbf{x}) = \sum_{i=1}^{n} (\alpha_i - \beta_i) K(\mathbf{x}_i, \mathbf{x}) + b \qquad (3.84)$$

where α_i, β_i are the Lagrangian multipliers.

The generation of membership grades can be done using several types of membership functions such as those used in a credit rating application (Wang et al., 2005). Assuming that some primary credit score s_i had been assigned to a customer i, then the membership functions suggested are:

a. *Linear function.*

$$m_i = \frac{s_i - \min_{1, \ldots, n} s_i}{\max_{1, \ldots, n} s_i - \min_{1, \ldots, n} s_i}$$

b. *Logistic function.*

$$m_i = \frac{e^{as_i + c}}{e^{as_i + c} + 1}$$

for scalars a, c.

c. *Bridge of trapezoidal-type function.*

$$m_i = \begin{cases} 1 & \text{if } s_i > a \\ \dfrac{s_i - c}{a - c} & \text{if } c \le s_i \le a \\ 0 & \text{if } s_i \le c \end{cases} \qquad (3.85)$$

d. *Probit function.*

$$m_i = \phi\left(\frac{s_i - \mu}{\sigma}\right)$$

where $\phi(\cdot)$ is the cumulative normal distribution function with mean μ and standard deviation σ.

Fuzzy SVMs have recently been applied to problems such as content-based image retrieval (Wu and Yap, 2006), biomedical analysis (Chen et al., 2005), and finance (Wang et al., 2005; Bao et al., 2005). The results appear promising but this hybrid system has several drawbacks, which require further research. The selection of the membership function and the allocation of the membership grades have to be carefully done to obtain good classification results. This preprocessing step could make this technique less appealing for online applications since an exhaustive trial of the membership functions has to be done. In addition, the capability of the fuzzy SVM to outperform a normal SVM operating with optimal parameters still remains to be seen. The bilateral fuzzy SVM is slow to train due to the size of the data set and giving only a small improvement to classification accuracy compared to other methods such as logistic regression and neural networks (Wang et al., 2005).

3.7 Concluding Remarks

In this chapter, we examined CI techniques that have been successfully applied to solve various problems. We have concentrated on ANN, SVM, HMMs, and fuzzy logic systems because of their frequent application in the field of biomedical engineering. We introduced these methods and discussed the key issues of their formulation and implementation. In the following chapters, the application of these techniques to the processing of data for diagnosis of various diseases will be described.

References

Abe, S. (1997). *Neural Networks and Fuzzy Systems: Theory and Applications.* Boston, MA: Kluwer Academic.

Ai, H., L. Liang, and G. Xu (2001). Face detection based on template matching and support vector machines. *Proceedings of the International Conference on Image Processing 1*, 1006–1009.

Albu, F. and D. Martinez (1999). The application of support vector machines with Gaussian kernels for overcoming co-channel interference. *Proceedings of the IEEE Signal Processing Society Workshop Neural Networks for Signal Processing IX 1*, 49–57.

Anguita, D. and A. Boni (2001). Towards analog and digital hardware for support vector machines. *Proceedings of the International Joint Conference on Neural Networks*, 2422–2426.

Bahl, L. R., F. Jelinek, and R. L. Mercer (1983). A maximum likelihood approach to continuous speech recognition. *IEEE Transactions on Pattern Analysis and Machine Intelligence 5*, 179–190.

Bahlmann, C., B. Haasdonk, and H. Burkhardt (2002). Online hand-writing recognition with support vector machines—a kernel approach. *Eighth International Workshop on Frontiers in Handwriting Recognition 1*, 49–54.

Bakis, R. (1976). Continuous speech recognition via centisecond acoustic states. *The Journal of the Acoustical Society of America 59*, S97.

Bao, Y. K., Z. T. Liu, L. Guo, and W. Wang (2005). Forecasting stock composite index by fuzzy support vector machines regression. *Proceedings of 2005 International Conference on Machine Learning and Cybernetics 6*, 3535–3540.

Bellili, A., M. Gilloux, and P. Gallinari (2001). An hybrid MLP-SVM hand-written digit recognizer. *Proceedings of the Sixth International Conference on Document Analysis and Recognition 1*, 28–31.

Bertsekas, D. (1996). *Constrained Optimization and Lagrange Multiplier Methods*. Optimization and Neural Computation Series, vol. 4. Belmont, MA: Athena Scientific.

Bojadziev, G. and M. Bojadziev (1997). *Fuzzy Logic for Business, Finance, and Management*. Advances in Fuzzy Systems, vol. 12. River Edge, NJ: World Scientific.

Brubaker, D. (1992). Fuzzy-logic basics: Intuitive rules replace complex math. *EDN Design–Feature*, 111–116.

Burges, C. (1998). A tutorial on support vector machines for pattern recognition. *Data Mining and Knowledge Discovery 2*, 11–167.

Chang, C. and C. Lin (2001). *LIBSVM: A Library for Support Vector Machines*. Software available at http://www.csie.ntu.edu.tw/~cjlin/libsvm.

Chen, S., S. Gunn, and C. Harris (2000). Design of the optimal separating hyperplane for the decision feedback equalizer using support vector machines. *Proceedings of IEEE International Conference on Acoustics, Speech and Signal Processing 5*, 2701–2704.

Chen, X., R. Harrison, and Y.-Q. Zhang (2005). Fuzzy support vector machines for biomedical data analysis. *IEEE International Conference on Granular Computing 1*, 131–134.

Chen, S., A. Samingan, and L. Hanzo (2001). Support vector machine multiuser receiver for DS-CDMA signals in multipath channels. *IEEE Transactions on Neural Networks 12*, 608–611.

Chodorowski, A., T. Gustavsson, and U. Mattsson (2002). Support vector machines for oral lesion classification. *Proceedings of the IEEE International Symposium on Biomedical Imaging 1*, 173–176.

Chong, K. and H. Stanislaw (2001). *An Introduction to Optimization.* Wiley Interscience Series in Discrete Mathematics and Optimization. New York.

Cristianini, N. and J. Shawe-Taylor (2000). *An Introduction to Support Vector Machines: And Other Kernel-based Learning Methods.* New York: Cambridge University Press.

Dong, J., A. Krzyzak, and C. Suen (2005). Fast SVM training algorithm with decomposition on very large data sets. *IEEE Transactions on Pattern Analysis and Machine Intelligence 27*, 603–618.

Dugad, R. and U. B. Desai (1996). *A Tutorial on Hidden Markov Models.* Technical Report SPANN-96-1, Signal Processing and Artificial Neural Networks Laboratory, Department of Electrical Engineering, Indian Institute of Technology. Bombay, India.

El-Naqa, I., Y. Yang, M. Wernick, N. Galatsanos, and R. Nishikawa (2002). Support vector machines learning for detection of microcalcifications in mammograms. *Proceedings of the IEEE International Symposium on Biomedical Imaging 1*, 201–204.

Fan, A. and M. Palaniswami (2000). Selecting bankruptcy predictors using a support vector machine approach. *Proceedings of the International Joint Conference on Neural Networks*, 354–359.

Fan, A. and M. Palaniswami (2001). Stock selection using support vector machines. *Proceedings of the International Joint Conference on Neural Networks*, 511–520.

Fletcher, R. (1981). *Practical Methods of Optimization.* Chichester, England: Wiley.

Fuentes, M., S. Garcia-Salicetti, and B. Dorizzi (2002). Online signature verification: fusion of a hidden Markov model and a neural network via a support vector machine. *Eighth International Workshop on Frontiers in Handwriting Recognition 1*, 253–258.

Fung, G. and O. L. Mangasarian (2001). Proximal support vector machine classifiers. *Proceedings of the Seventh ACM SIGKDD International Conference on Knowledge Discovery and Data Mining*, 77–86.

Gestel, T. (2001). Financial time series prediction using least squares support vector machines within the evidence framework. *IEEE Transactions on Neural Networks 4*(12), 809–821.

Gill, P., W. Murray, and M. Wright (1981). *Practical Optimization.* London: Academic Press.

Girolami, M. (1999). *Self-organising Neural Networks: Independent Component Analysis and Blind Source Separation.* Perspectives in Neural Computing. London: Springer.

Gong, X. and A. Kuh (1999). Support vector machine for multiuser detection in cdma communications. *Conference Record of the Thirty-Third Asilomar Conference on Signals, Systems, and Computers 1*, 680–684.

Goonatilake, S. and S. Khebbal (1995). *Intelligent Hybrid Systems.* Chichester: Wiley.

Gretton, A., M. Davy, A. Doucet, and P. Rayner (2001). Nonstationary signal classification using support vector machines. *Proceedings of the 11th IEEE Workshop on Signal Processing and Statistical Signal Processing*, 305–308.

Hasegawa, M., G. Wu, and M. Mizuno (2001). Applications of nonlinear prediction methods to the Internet traffic. *IEEE International Symposium on Circuits and Systems 2*, 169–172.

Haykin, S. S. (1999). *Neural Networks: A Comprehensive Foundation* (2nd Ed.). Upper Saddle River, NJ: Prentice Hall.

Hebb, D. (1949). *The Organization of Behaviour.* New York: Wiley.

Held, K. D., C. A. Brebbia, R. D. Ciskowski, and C. M. I. S. England (1991). Computers in biomedicine. *Proceedings of the First International Conference, Held in Southampton, UK, 24–26 September 1991.* Southampton, England: Computational Mechanics Publications.

Herbrich, R. and J. Weston (1999). Adaptive margin support vector machines for classification. *Proceedings of the Ninth International Conference on Artificial Neural Networks, ICANN 2*, 880–885.

Hopfield, J. J. (1982). Neural networks and physical systems with emergent collective computational abilities. *Proceedings of the National Academy of Sciences of the USA 79*(8), 2554–2558.

Huang, H. and Y. Liu (2002). Fuzzy support vector machines for pattern recognition and data mining. *International Journal of Fuzzy Systems 4*, 826–835.

Hull, C. L. and H. D. Baernstein. (1929). A mechanical parallel to the conditioned reflex. *Science 70*, 14–15.

Inoue, T. and S. Abe (2001). Fuzzy support vector machines for pattern classification. *Proceedings of International Joint Conference on Neural Networks 2*, 1449–1454.

Jain, A. K., J. Mao, and K. M. Mohiuddin (1996). Artificial neural networks: a tutorial. *IEEE Computer 29*(3), 31–44.

James, A. and R. Edward (1988). *Neurocomputing: Foundations of Research.* Cambridge, MA: MIT Press.

Jelinek, F. (1976). Continuous speech recognition by statistical methods. *Proceedings of the IEEE 64*(4), 532–556.

Joachims, T. (1998). Making large scale support vector machine learning practical. In *Advances in Kernel Methods—Support Vector Learning*, B. Schlkopf, C. J. C. Burges, and A. J. Smola (Eds), Cambridge, MA: MIT Press. pp. 169–184.

Juang, B., S. Levinson, and M. Sondhi (1986). Maximum likelihood estimation for multivariate mixture observations of markov chains. *IEEE Transactions on Information Theory 32*(2), 307–309.

Juang, B. and L. Rabiner (1985). Mixture autoregressive hidden markov models for speech signals. *IEEE Transactions on Acoustics, Speech, and Signal Processing 33*(6), 1404–1413.

Junfeng, G., S. Wengang, T. Jianxun, and F. Zhong (2002). Support vector machines based approach for fault diagnosis of valves in reciprocating pumps. *Proceedings of the IEEE Canadian Conference on Electrical and Computer Engineering 3*, 1622–1627.

Kandel, A. and G. Langholz (1994). *Fuzzy Control Systems*. Boca Raton, FL: CRC Press.

Keerthi, S. S. and E. G. Gilbert (2002). Convergence of a generalized SMO algorithm for SVM classifier design. *Machine Learning 46*(1–3), 351–360.

Keller, J. and D. Hunt (1985). Incorporating fuzzy membership functions into the perceptron algorithm. *IEEE Transactions on Pattern Analysis and Machine Intelligence 7*(6), 693–699.

Khosla, R. and T. S. Dillon (1997). *Engineering Intelligent Hybrid Multi-agent Systems*. Boston, MA: Kluwer Academic Publishers.

Kleene, S. (1956). Representation of events in nerve nets and finite automata. In *Automata Studies*, C. Shannon and J. McCarthy (Eds), Princeton, NJ: Princeton University Press. pp. 3–42.

Kohonen, T. (1982). Self-organized formation of topologically correct feature maps. *Biological Cybernetics 43*, 59–69.

Kohonen, T. (1989). *Self Organization and Associative Memory*. (3rd Ed.). Berlin: Springer.

Kurimo, M. and P. Somervuo (1996). Using the self-organizing map to speed up the probability density estimation for speech recognition with mixture density HMMs. *Proceedings of 4th International Conference on Spoken Language Processing*, pp. 358–361.

Lai, D. (2006). *Subspace Optimization Algorithms for Training Support Vector Machines*. Monash University, Australia: Ph.D. Thesis.

Lai, D., N. Mani, and M. Palaniswami (2004). A new momentum minimization decomposition method for support vector machines. *Proceedings of the International Joint Conference on Neural Networks*, Volume 3, pp. 2001–2006.

Lee, L. and W. Grimson (2002). Gait analysis for recognition and classification. *Proceedings of Fifth IEEE International Conference on Automatic Face and Gesture Recognition 1*, 148–155.

Li, Y., R. Yong, and X. Shan (2001). Radar HRRP classification with support vector machines. *International Conferences on Info-tech and Info-net 1*, 218–222.

Liang, H. and Z. Lin (2001). Detection of delayed gastric emptying from electrogastrograms with support vector machines. *IEEE Transactions on Biomedical Engineering 48*(5), 601–604.

Lin, C. (2002). A formal analysis of stopping criteria of decomposition methods for support vector machines. *IEEE Transactions on Neural Networks 13*(5), 1045–1052.

Lin, C. F. and S. D. Wang (2002). Fuzzy support vector machines. *IEEE Transactions on Neural Networks 13*(2), 464–471.

Liporace, L. (1982). Maximum likelihood estimation for multivariate observations of markov sources. *IEEE Transactions on Information Theory 5*(28), 729–734.

Liu, W., S. Peihua, Y. Qu, and D. Xia (2001). Fast algorithm of support vector machines in lung cancer diagnosis. *Proceedings of the International Workshop on Medical Imaging and Augmented Reality* (1), 188–192.

Ma, C., A. Mark, Randolph, and D. Joe (2001). A support vector machine based rejection technique for speech recognition. *Proceedings of the IEEE International Conference on Acoustics, Speech, and Signal Processing 1*, 381–384.

Mangasarian, O. (2000). Generalized support vector machines. In *Advances in Large Margin Classifiers*, A. J. Smola, B. Schlkopf, P. Bartlett, and D. Schurmans (Eds), Cambridge, MA: MIT Press. pp. 135–146.

Mangasarian, O. and D. Musicant (1999). Successive overrelaxation for support vector machines. *IEEE Transactions on Neural Networks 10*(5), 1032–1037.

Maruyama, K., M. Maruyama, H. Miyao, and Y. Nakano (2002). Handprinted hiragana recognition using support vector machines. *Eighth International Workshop on Frontiers in Handwriting Recognition 1*, 43–48.

McCulloch, W. S. and W. H. Pitts (1943). A logical calculus of the ideas immanent in nervous activity. *Bulletin of Mathematical Biophysics 5*, 115–133.

Medler, D. A. (1998). A brief history of connectionism. *Neural Computing Surveys 1*, 61–101.

Minsky, M. and S. A. Papert (1988/1969). *Perceptrons: An Introduction to Computational Geometry* (expanded edition). Cambridge, MA: MIT Press.

Moore, R. (1966). *Interval Analysis*. Englewood Cliffs, NJ: Prentice Hall.

Mukherjee, S., E. Osuna, and F. Girosi (1997). Nonlinear prediction of chaotic time series using support vector machines. *Proceedings of the IEEE Workshop on Signal Processing*, 511–520.

Osuna, E., R. Freund, and F. Girosi (1997). Training support vector machines: an application to face detection. *Proceedings of IEEE Computer Society Conference on Computer Vision and Pattern Recognition*, 130–136.

Ovaska, S. J. (2005). *Computationally Intelligent Hybrid Systems: The Fusion of Soft Computing and Hard Computing*. IEEE Series on Computational Intelligence. Hoboken, NJ: Wiley.

Park, J., J. Reed, and Q. Zhou (2002). Active feature selection in optic nerve data using support vector machines. *Proceedings of the International Joint Conference on Neural Networks* (1), 1178–1182.

Pedrycz, W. and A. Vasilakos (2001). *Computational Intelligence in Telecommunications Networks*. Boca Raton, FL: CRC Press.

Pirrone, R. and M. Cascia (2001). Texture classification for content-based image retrieval. *Proceedings of 11th International Conference on Image Analysis and Processing 1*, 398–403.

Platt, J. (1998). Fast training of support vector machines using sequential minimal optimization. In *Advances in Kernel Methods-Support Vector Learning*, B. Schlkopf, C. J. C. Burges, and A. J. Smola (Eds), Cambridge, MA: MIT Press. pp. 185–208.

Pollack, J. B. (1989). Connectionism: past, present, and future. *Artificial Intelligence Review 3*, 3–20.

Poritz, A. (1982). Linear predictive hidden Markov models and the speech signal. *IEEE International Conference on Acoustics, Speech, and Signal Processing, ICASSP '82. 7*, 1291–1294.

Qi, Y., D. Doermann, and D. DeMenthon (2001). Hybrid independent component analysis and support vector machine learning scheme for face detection. *Proceedings of IEEE International Conference on Acoustics, Speech, and Signal Processing 3*, 1481–1484.

Qiang, H. and L. Chin-Hui (1997). On-line adaptive learning of the continuous density hidden markov model based on approximate recursive bayes estimate. *IEEE Transactions on Speech and Audio Processing 5*(2), 161–172.

Rabiner, L. R. (1989). A tutorial on hidden Markov models and selected applications in speech recognition. *Proceedings of the IEEE 77*(2), 257–286.

Ramirez, L., W. Pedryez, and N. Pizzi (2001). Severe storm cell classification using support vector machines and radial basis function approaches. *Canadian Conference on Electrical and Computer Engineering* (1), 87–91.

Reed, R. D. and R. J. Marks (1999). *Neural Smithing: Supervised Learning in Feedforward Artificial Neural Networks*. Cambridge, MA: MIT Press.

Rosenblatt, F. (1958). The perceptron: a probabilistic model for information storage and organization in the brain. *Psychological Review 65*, 386–408.

Ruan, D. (1997). *Intelligent Hybrid Systems: Fuzzy Logic, Neural Networks, and Genetic Algorithms*. Boston, MA: Kluwer Academic Publishers.

Rumelhart, D. E. and J. L. McClelland (1986). *Parallel Distributed Processing*. Cambridge, MA: MIT Press.

Rychetsky, M., S. Ortmann, and M. Glesner (1999). Support vector approaches for engine knock detection. *Proceedings of the International Joint Conference on Neural Networks 2*, 969–974.

Schlkopf, B., C. Burges, and A. Smola (1999). *Advances in Kernel Methods: Support Vector Learning*. Cambridge, MA: MIT Press.

Schlkopf, B. and A. Smola (2002). *Learning with Kernels: Support Vector Machines, Regularization, Optimization and Beyond*. Adaptive Computation and Machine Learning. Cambridge, MA: MIT Press.

Selfridge, O. G. (1959). Pandemonium: A Paradigm for Learning. *Proceedings of the Symposium on Mechanisation of Thought Processes*, D. V. Blake and A. M. Uttley, (Eds), pp. 511–529, London: H.M. Stationary Office.

Seo, S. and K. Obermayer (2003). Soft learning vector quantization. *Neural Computation 15*, 1589–1604.

Shavlik, J. and T. Dietterich (1990). *Readings in Machine Learning*. The Morgan Kaufmann Series in Machine Learning. San Mateo, CA: Morgan Kaufmann Publishers.

Shilton, A., M. Palaniswami, D. Ralph, and A. C. Tsoi (2005). Incremental training of support vector machines. *IEEE Transactions on Neural Networks 16*(1), 114–131.

Smola, A. and B. Schlkopf (1998). *A Tutorial on Support Vector Regression*. NeuroCOLT2 Technical Report NC2-TR-1998-030.

Suykens, J. (2002). *Least Squares Support Vector Machines*. River Edge, NJ: World Scientific.

Suykens, J., J. De Brabanter, L. Lukas, and J. Vandewalle (2002). Weighted least squares support vector machines robustness and sparse approximation. *Neurocomputing 48*(1–4), 85–105.

Suykens, J. and J. Vandewalle (2000). Recurrent least squares support vector machines. *IEEE Transactions on Circuits and Systems 47*, 1109–1114.

Tisby, N. (1991). On the application of mixture AR hidden Markov models to textindependent speaker recognition. *IEEE Transactions on Signal Processing 39*(3), 563–570.

Tomsovic, K. and M. Y. Chow (2000). Tutorial on fuzzy logic applications in power systems. *IEEE-PES Winter Meeting, Singapore*, 1–87.

Trafalis, T. and H. Ince (2000). Support vector machines for regression and applications to financial forecasting. *Proceedings of the International Joint Conference on Neural Networks*, 348–353.

Vapnik, V. N. (2000). *The Nature of Statistical Learning Theory* (2nd Ed.). Statistics for Engineering and Information Science. New York: Springer.

Wang, Y., S. Wang, and K. Lai (2005). A new fuzzy support vector machine to evaluate credit risk. *IEEE Transactions on Fuzzy Systems 13*(6), 820–831.

Wang, S., W. Zhu, and Z. Liang (2001). Shape deformation: SVM regression and application to medical image segmentation. *Proceedings of the Eighth IEEE International Conference on Computer Vision* (2), 209–216.

Widrow, B. and M. A. Lehr. (1995). *Perceptrons, adalines, and backpropagation*. In *The Handbook of Brain Theory and Neural Networks*, M. A. Arbib (Ed.), Cambridge, MA: MIT Press.

Wroldsen, J. (1995). Boltzmann learning in a feed-forward neural network. citeseer.ist.psu.edu/wroldsen95boltzmann.html.

Wu, K. and K. H. Yap (2006). Fuzzy SVM for content-based image retrieval: a pseudolabel support vector machine framework. *IEEE Computational Intelligence Magazine 1*(2), 10–16.

Yin, H. (2002). ViSOM—a novel method for multivariate data projection and structure visualization. *IEEE Transactions on Neural Networks 13*(1), 237–243.

Zadeh, L. (1965). Fuzzy sets. *Information and Control 8*, 338–353.

Zadeh, L. (1973). Outline of a new approach to the analysis of complex systems and decision processes. *IEEE Transactions on Systems, Man, and Cybernetics 8*, 28–44.

Zadeh, L., T. Lin, and Y. Yao (2002). *Data Mining, Rough Sets, and Granular Computing.* Studies in Fuzziness and Soft Computing; vol. 95. Heidelberg: Physica-Verlag.

Zadeh, L., M. Nikravesh, and V. Korotkikh (2004a). *Fuzzy Partial Differential Equations and Relational Equations: Reservoir Characterization and Modeling.* Studies in Fuzziness and Soft Computing, vol. 142. Berlin: Springer.

Zadeh, L., M. Nikravesh, and V. Loia (2004b). *Fuzzy logic and the Internet.* Studies in fuzziness and soft computing, vol. 137. Berlin: Springer.

Zhang, J., Y. Zhang, and T. Zhou (2001). Classification of hyperspectral data using support vector machine. *Proceedings of the International Joint Conference on Image Processing 1*, 882–885.

Zhao, Q. and Jose, C. (2001). Support vector machines for SAR automatic target recognition. *IEEE Transactions on Aerospace and Electronic Systems 37*(2), 643–654.

Zoutendijk, G. (1960). *Methods of Feasible Directions: A Study in Linear and Nonlinear Programming.* Amsterdam: Elsevier.

4

Computational Intelligence in Cardiology and Heart Disease Diagnosis

4.1 Introduction

Cardiovascular disease (CVD) is the most widespread cause of death in many countries all over the world. Alarming statistics have been presented by the World Health Organization (WHO) concerning mortality rates due to CVDs. It has been estimated that worldwide, approximately 17 million people die every year as a result of CVDs. This accounts for 30% of deaths with 80% of the cases in developing nations. Demographicwise, in 2002, Russia had the highest death rate from CVD followed by Ukraine, Romania, Hungary, and Poland (Department of Health and Welfare, 2002). Japan has the lowest death rate from CVD followed by France, with Canada being the third lowest. In 2004, Australia was found to have the sixth lowest death rate but with rates up to 1.5 times that of Japan.

CVD is a broad term that encompasses coronary heart diseases such as myocardial infarction (MI), angina, coronary insufficiency, and coronary death. It also includes cerebrovascular diseases such as stroke and transient ischemic attacks, peripheral vascular disease, congestive heart failure (CHF), hypertension, and valvular and congenital heart disease. These specific disorders tend to be commonly recognized by the general public as heart attacks and strokes. Many risk factors that have been identified for CVD include obesity, diabetes, smoking, drinking alcohol, high blood pressure, and high cholesterol levels. Factors such as high blood pressure, smoking, and high cholesterol have been identified as the major causes that increase the risk of heart disease by two- to three folds (WHO, 2005). One way of tackling this major health problem is early diagnosis of cardiac diseases, which can be immediately followed up with appropriate treatments and precautions to minimize fatal events.

Much recent interest has been generated within the research community in the application of CI techniques for detection, classification, and diagnosis of heart diseases. Some examples include diagnosis of MI or "heart attacks," cardiac hypertrophy or myocardium thickening, and myocardial ischemia; all of which generally lead to life-threatening conditions. Early and accurate diagnoses are of paramount importance so that necessary preventive measures or treatment can be administered such as the injection of blood-thinning medicines. Physicians currently use several diagnostic methods ranging from invasive techniques such as heart catheterization and biochemical markers from blood samples, to noninvasive methods such as computerized tomography (CT) scans, MRI, and ECGs. In this chapter, we will mainly focus on the use of ECG due to current availability of the databases and the high probability that most diagnoses can be accurate using CI techniques.

The analysis of ECG signals is very important for monitoring anomalies in the heart function of cardiac patients, mainly because it is fast and ECG monitoring can be continuous. The diagnostic process involves extraction of attributes from ECG waveforms and subsequent comparisons with known diseases to identify any deviation from normal characteristic waveforms. Such a monitoring system should have the capability of detecting heart abnormalities represented by waveshape changes within an ECG cycle. Although clinicians will be required to recognize any deviation in the ECG trace, this is still very tedious and time-consuming for investigating long-term recording waves. CI techniques can automate the process of ECG analysis and the classification between normal and pathological patterns by developing decision surfaces to classify these patterns. Research in this area using neural networks (NNs), fuzzy logic, genetic algorithms, SVMs, and knowledge-based systems is widespread but with various degrees of success. Automatic detection and classification of heartbeats from the ECG waveform using biomedical signal-processing techniques have become a critical aspect of clinical monitoring. The goal of this research is to perfect systems to perform real-time monitoring and alert clinicians when life-threatening conditions begin to surface.

The first section of this chapter is a discussion on the history of cardiology dating back to when humans were beginning to become fascinated with their body. This is followed by a structural account of the heart and the cardiac cycle. We also discuss the early development of the heart in the fetal stage, which sheds light on how biological systems develop. Later, in Section 4.3, the electrical activity of the heart is examined and the major heart signal, namely, the QRS complex is described. We then review several algorithms for QRS peak detection before describing the ECG and ECG monitoring in Section 4.4.4. Several common CVDs that can be detected using ECG readings are considered with the methods used to extract information from the ECG waveforms. In the final sections, we elaborate several CI systems that have been previously proposed to detect these CVDs. Of specific interest is the detection accuracies that can be achieved using a combination of ECG signal-processing and CI techniques.

4.2   The Human Heart

The heart is one of the major organs of the human body, vital to our survival. It is essentially a large pump, whose sole purpose is to maintain blood circulation and keep organs alive. This section presents a brief history of research and medical practice concerning the heart followed by a description of the heart structure and natal development. Most of the material presented here can be found in greater detail in physiology texts such as O'Rourke (2005) and Schlant et al. (1994).

4.2.1   History of Cardiology

William Harvey is usually credited as the first person to begin investigation of the human heart with his discovery of blood circulation in the early seventeenth century. Study of the human heart was given the name cardiology with a history that can be divided into several phases. Cardiology as understood today began with descriptive anatomy and pathology in the seventeenth and eighteenth centuries, followed by interest in auscultation and correlations in the early nineteenth century. In the second half of the nineteenth century, research focused on the causes of cardiac diseases, which led to developments in the field of pathophysiology and surgery in the early twentieth century. Since then, many advances have been made in the diagnosis and treatment of heart diseases with CI being recently employed as a diagnostic tool for the last 20 years.

Despite the reverence given to Harvey's discovery, history of human heart studies can be traced back to ancient civilizations. The heart was initially regarded as a source of heat and the blood vessels as conduits for the life-sustaining spirit known as *pneuma*, which was believed to maintain the vitality of the organs. This idea was strongly advocated by Claudius Galen (AD 130–200), whose erroneous teachings were supported for nearly 1300 years before Andreas Vesalius provided the correct blood vessel anatomy in 1543. In 1616, William Harvey proposed that the heart was a pump meant to circulate blood around the body. When this view was accepted, attention turned toward further observations of the heart. Floyer (1707) invented the 1-min-pulse watch to observe heart rates, whereas suggestions for optimal positions of locating the heart pulse were put forward by Ludwig Traube (1872), Adolf Kussmaul (1873), and James Mackenzie (1902). Auscultation, which had been in use since Hippocrates (460–370 BC) to listen to the heart (by placing the ear to the chest) was also popularized. These bedside methods of observing heart rate were revolutionized in 1816 by Rene Laennec with the invention of the first wooden stethoscope. By 1855, the stethoscope had evolved to a biaural device, and acoustic principles of cardiovascular sound were further understood with the work of Samuel Levine, Rappaport, and Sprague (1940s), Schlant et al. (1994).

Heart monitoring was further improved with the invention of the ECG in 1924 by Willem Einthoven, which won him the Nobel Prize. This work had its beginnings in 1786, when first Galvanis and later Volta discovered that electricity was generated by animal tissue. In 1856, Augustus Waller demonstrated the presence of electrical current in the heart with the first crude capillary electrometer. The device was bulky and inaccurate, limitations that spurred Einthoven's invention in the 1920s. Since then the ECG has been an important instrument for heart diagnosis used in conjunction with x-rays and cardiac fluoroscopy.

More invasive techniques for measuring heart properties and blood pressure were pioneered by Claude Bernard (1844) using a catheter inserted into the hearts of animal and human cadavers. Forssmann, then a 29-year-old surgeon, resident in Germany, was the first to perform a self-catheterization. He confirmed the position of the right atrium as seen on x-rays but was later reprimanded for trying to image his heart by passing sodium iodide through it and discontinued self-experimentation. Only in the 1940s, Andrea Cournand and Dickinson Richards developed and demonstrated a safe and complete right heart catheterization procedure. The heart catheter then became an important instrument for direct measures of pressure, cardiac output, and to collect blood samples and deliver contrasting agents for imaging. Cournand, Richards, and Forssmann shared the 1956 Nobel Prize for this invention, which has since been improved by Zimmerman (1950), Seldinger (1953), and Judkins (1967) to include features such as ultrasound, balloons, and defibrillators.

In the field of imaging, Leornado da Vinci is attributed as the first to attempt imaging of the heart by using a glass cast visualization. In 1895, Konrad Roentgen's discovery of x-rays opened a new world for medical imaging. Moniz and Castellanos separately imaged the heart using sodium iodide, marking them as the first to use a contrasting agent. Herman Blumgart (1927) began nuclear cardiology when he injected radon into the blood to measure circulation time. In 1952, the gamma camera was invented by Hal Anger after the events of World War II. This device had high scanning capabilities and for the first time could image the heart and its chambers. Zaret and Strauss (1973) used potassium 43 as a tracer to detect ischemia. Since then, modern medicine has seen the use of thallium 201 and technetium-99 m as more stable radioactive tracers. In 1954, Inge Edler first used cardiac ultrasound to detect the anterior mitral valve. The ultrasound became an important imaging device, which was expanded from the initial M-echo mode to two-dimensional echo (1974), doppler echo (1975), stress echo (1979), and later, color flow (1982) and the current transesophagael monitoring developed in 1985. Owing to further advances in imaging, other diagnostic methods were developed such as single-photon emission tomography (SPECT, 1963–1981), positron emission tomography (PET, 1975–1987), and MRI (1972–1981).

Work in recognizing heart diseases began in 1768 when William Heberden presented a patient with symptoms of pain in the chest. The patient felt a constant pain akin to strangling in the chest leading Heberden to suspect that this was a new chest disorder. In one of his writings, he referred to this new disease as *angina pectoris*, which when translated, literally meant the strangling of the chest. Little did he know then that these symptoms were related to a coronary artery disease, which was later discovered and confirmed in 1793 after an autopsy by Edward Jenner and Caleb Parry. The lateness of this finding was mainly due to the persistent belief that the heart was immune to diseases and could not fail. However, once clear evidence was presented, traditional views began to change. In 1878, Hammer reported the first case of MI, whereas Johann Bobstein (1833) linked *arteriosclerosis* to coronary disease. Diseases affecting the heart valves were discovered by James Hope (1932), Bertin (1824), Williams (1835), and Fauvel (1843) to name a few. Congenital heart disease, or malformations of the heart at birth, was first presented as cyanotic heart disease. Helen Taussig and Alfred Blalock of the John Hopkins hospital initiated treatment of cyanotic heart disease through shunting, which was successfully performed in 1944 when they saved the life of a blue (lack of oxygen) baby boy.

At the time these advances were taking place, there were important developments in the field of monitoring. Riva-Rocci (1896) was responsible for the invention of the inflatable cuff coupled to a sphygmograph, which measured blood pressure. In 1939, blood recordings were standardized by committees of the American Heart Association (AHA) and the Cardiac Society of Great Britain and Ireland. In the field of surgery, advances were also being made in blood vessel and organ transplants. The Nobel Prize for Physiology or Medicine was awarded to Alexis Carrel in 1912 for his intensive work on vascular sutures and organ or blood vessel transplants. His contributions paved the way for further exploration into heart surgery, which had always been hazardous due to cerebral oxygenation, anesthesia, bleeding, clotting, infection, arrhythmias, and similar problems. In 1953, part of these problems were alleviated when John Gibbons and his wife perfected the pump–oxygenator, which afforded a cardiac surgeon 45 min to repair atrial defects. Alexis Carrel also attempted the first coronary bypass operation and his methods were later perfected by Rene Favoloro and Mason Sones from the Cleveland Clinic.

4.2.2 Structure of the Heart

The heart is a hollow, cone-shaped muscle located between the lungs and behind the sternum (breastbone). It is a four-chambered structure consisting of two parallel independently functioning systems, each consisting of an auricle or atrium (top) and a ventricle (bottom) separated by valves. These systems are separately referred to as the right and left heart. Several major blood vessels connect the heart to other organs in the body, for example, the aorta and superior vena cava, which are two major blood vessels. Figure 4.1 shows an external view of the heart indicating the main structures, which will be described in the following text.

The base of the heart is uppermost and tilts slightly backward toward the spinal column, anatomically, it lies in the *mediastinum*, which is the region between the lungs that holds several organs. Two-thirds of the heart is located to the left of the midline of the body and one-third is to the right, whereas the apex (pointed end) points down and to the left. An adult heart is approximately 5 in. (12 cm) long, 3.5 in. (8–9 cm) wide, and 2.5 in. (6 cm) from front to back, roughly the size of a fist. The average weight of a female human heart is 9 Oz, whereas the male's is slightly heavier, at an average of 10.5 Oz. The heart comprises less than 0.5% of the total body weight and is composed of

Exterior structures of the heart

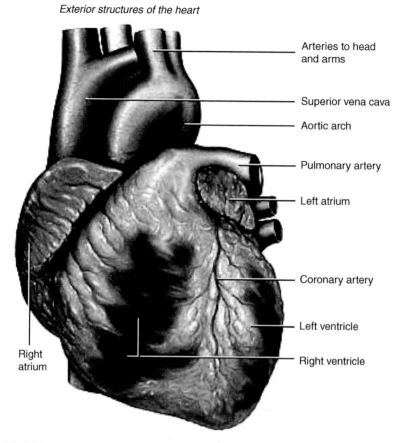

Arteries to head
and arms

Superior vena cava

Aortic arch

Pulmonary artery

Left atrium

Coronary artery

Left ventricle

Right ventricle

Right
atrium

FIGURE 4.1

Diagram of the human heart depicting the major external structures.
(Reprinted from Yale-New-Haven Hospital website, www.ynhh.com.)

three layers: the smooth inside lining of the heart is called the *endocardium*,
whereas the middle layer of heart muscle is called the *myocardium* that is
surrounded by a fluid-filled sac called the *pericardium*. The heart is held in
place primarily by the great arteries and veins and by its confinement in the
pericardium, which is a double-walled sac with one layer enveloping the heart
and the other attached to the breastbone, diaphragm, and the membranes of
the thorax.

   The endocardium lines the inner chambers of the heart and is composed of
three layers, namely, an outer layer in direct contact with the myocardium, a
middle layer, and the innermost layer (see Figure 4.2). The outermost layer
is composed of irregularly arranged collagen fibers that merge with collagen
surrounding the adjacent cardiac muscle fibers. Collagen is a fibrous structural
protein usually found in connective tissues such as ligaments, tendons, and

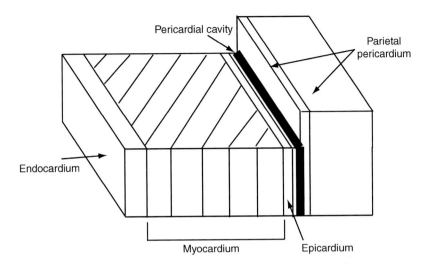

FIGURE 4.2

A sketch of the heart layers showing the epicardium, myocardium, and endocardium.

bones. This layer may contain some Purkinje fibers, which conduct electrical impulses. The middle endocardium layer is the thickest, composed of more regularly arranged collagen fibers containing a variable number of elastic fibers compactly arranged in the deepest part of the layer. Finally, the innermost layer is composed of flat endothelial cells, which form a continuous lining with the endothelial cells lining the blood vessels entering and emerging from the heart. The thickness of the endocardium is variable, being thickest in the atria and thinnest in the ventricles, particularly, the left ventricle. This increased thickness is almost entirely due to a thicker fibroelastic middle layer.

The myocardium layer forms the bulk of the heart and consists of contractile elements composed of specialized striated muscle fibers called *cardiac muscle*. The cardiac muscle is slightly different from the skeletal muscle. Similar to skeletal muscle, the cardiac muscle is striated, has dark Z lines, and possesses myofibrils, which contain actin and myosin filaments (see Chapter 5). These filaments slide over one another using a similar *cross-bridge* action observed in skeletal muscle contraction. In addition to this, the cardiac muscle has several unique properties such as the capability to contract without nerve impulse stimulation (intrinsic contraction). Another striking microscopic feature of these cells is branching patterns and the presence of more than one nuclei in the muscle cell. The individual cells are separated from one another at the ends by intercalated discs, which are specialized cell junctions between cardiac muscle cells. This natural arrangement provides a low-resistance layer, which allows a faster conduction of action potentials along the cells. The resistance is so low that ions move freely through the permeable cell junction allowing the entire atrial or ventricular chamber to function as a single cell. This single functional unit characteristic is referred to as a functional *syncytium*, which means that if an individual muscle cell is stimulated, a contraction in all other muscle cells will be observed. Owing to differences between the action potential conduction properties, however, the cardiac muscle contracts at a slower rate than the skeletal muscle, which is important for natural synchronization of the heart function, as will be seen later. It has also been found that the amount of myocardium and the diameter of muscle fibers in the chambers of the heart differ according to the workload of the specific chamber.

The pericardium sac is composed of two layers separated by a space called the pericardial cavity: the outer layer is called the parietal pericardium and the inner layer is the visceral pericardium. The parietal pericardium consists of an outer layer of thick, fibrous connective tissue and an inner serous layer. The serous layer, consisting largely of mesothelium cells, together with a small amount of connective tissue forms a simple squamous epithelium, which secretes approximately 25–35 mL of fluid. The fluid layer lubricates the surfaces and provides friction-free movement of the heart within the pericardium during muscular contractions. The fibrous layer of the parietal pericardium is attached to the diaphragm and fuses with the outer wall of the great blood vessels entering and leaving the heart. It functions to provide a strong protective sac for the heart and also serves to anchor it within the mediastinum.

The visceral pericardium is also known as the epicardium and as such comprises the outermost layer of the heart proper. The epicardium forms the outer covering of the heart and has an external layer of flat mesothelial cells. These cells lie on a stroma of fibrocollagenous support tissue, which contains elastic fibers as well as the large arteries and venous tributaries carrying blood from the heart wall. The fibrocollagenous tissue is mainly located at the heart valves and are composed of a central fibrous body. These extend to form the valve rings, which support the base of each valve. The valve rings on the left-hand side of the heart surround the mitral and aortic valves and are thicker than those on the right-hand side, which surround the tricuspid and pulmonary valves. A downward extension of the fibrocollagenous tissue of the aortic valve ring called the septum divides the right and left ventricles into two chambers. The walls between the atria are divided by the interatrial septum, whereas the two ventricles are separated by the interventricular septum. The fibrocollagenous tissue or skeleton is important because it provides attachment for the cardiac muscle and lends support to the atrial and ventricular valves. Near the front of the interatrial septum is a characteristic depression known as the *fossa ovalis*. This was the position of a previous opening known as the *foramen ovale* between the two sides of the heart in the fetal stage of development. This opening shunted blood directly from the right atrium into the left atrium, thereby diverting blood from the fetus' still developing lungs.

Each atrium chamber is separated from the ventricle chamber by atrioventricular (AV) valves as shown in Figure 4.3. The AV valves consist of flaps or fibrous tissue, which project from the heart wall into the opening between the atrium and the ventricle. These cusps are sheathed by a layer of endothelium

Interior structures of the heart

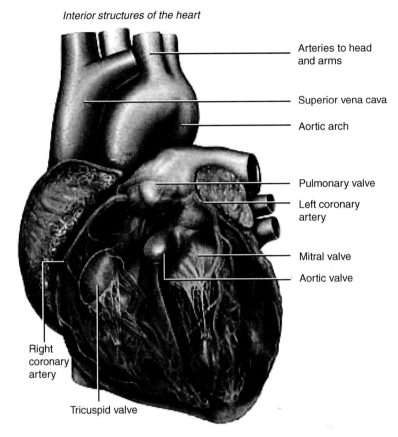

Arteries to head and arms

Superior vena cava

Aortic arch

Pulmonary valve

Left coronary artery

Mitral valve

Aortic valve

Right coronary artery

Tricuspid valve

FIGURE 4.3
Diagram of the human heart depicting the major internal structures. (Reprinted from Yale-New-Haven Hospital website, www.ynhh.com.)

that is continuous with the endocardium of the heart. The right AV valve is called the tricuspid valve because it has three cusps: anterior, posterior, and septal. The left AV valve has only two cusps and is termed the bicuspid or mitral valve. The exits of the ventricles are also guarded by aortic and pulmonic valves with the aortic valve situated in the left ventricle, whereas the pulmonic valve sits in the right ventricle. Each of these valves consists of three cusps, which have the shape of a half-moon. The free end of each cusp is positioned upward into the corresponding blood vessel so that blood flowing in the opposite direction will exert pressure on it and force the cusps together hence closing the valve and preventing back flow.

The large coronary arteries and veins originate above the aortic valve ring in the aorta. They pass over the surface of the heart in the epicardium, sending branches deep into the myocardium. The superficial location of these arteries is important to keep the heart muscle oxygenated and have been used for grafting during coronary bypass surgery. The fibrocollagenous tissue also separates the atrial chamber from the ventricular chamber and therefore, an impulse from the atrium must pass through the AV node before triggering the ventricle.

4.2.3   Early Stages of Heart Development

The heart being an integral part of the human body, begins its growth immediately after conception. The adult human heart is the size of a human fist and after birth grows at approximately the same rate as the fist until adulthood. However, this linear growth rate is not the same for infants in the womb. In this section, we look into some detail at how the heart develops from conception to birth. The interested reader can find further details in McNulty and Wellman (1996).

After conception, the fetal heart occupies a major portion of the midsection for the first few weeks. During this stage, the heart size to body size ratio is nine times greater than in a newborn infant. Initially, the fetal heart lies high in the chest and later slowly moves down to the chest cavity. The fetal heart's development can be divided into several phases as follows:

a. Formation of the heart tube

b. Partitioning of the atria

c. Formation of the right atrium and the left atrium

d. Formation of AV canals

e. Formation of the ventricles

f. Partitioning of the outflow tract

The heart begins growing from a portion of the embryo known as the angiogenic cell cluster located in the cardiogenic plate (see Figure 4.4a). By the twenty-first day, the angiogenic cell clusters have completely fused into the heart tube. The newly formed heart tube may be divided into regions (see Figure 4.4b) beginning with the sinus venosus from the posterior and then the primitive atria, followed by the primitive ventricle. The primitive atria will soon fuse to become a single atria, whereas the primitive ventricle expands to become the left ventricle. The AV sulcus partitions the atria and the primitive ventricle. This is then followed by the bulbous cordis, which is composed of the proximal bulbous cordis, conus cordis, and truncus arteriosus. Finally, the aortic sac and the dorsal aorta are located at the anterior end of the heart tube.

In the beginning, the heart tube grows so fast that it is forced to bend and twist backward forming the familiar fist shape of the heart. During this time, the two atria are already partly separated and there is only one big ventricle (see Figure 4.5). The conus cordis starts enlarging, whereas the truncus arteriosus shows some indentation, which indicates where septation (partitioning)

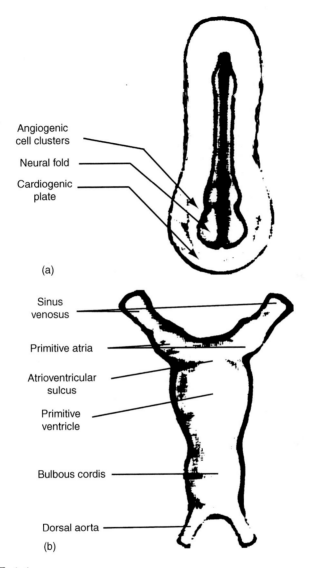

Angiogenic cell clusters

Neural fold

Cardiogenic plate

(a)

Sinus venosus

Primitive atria

Atrioventricular sulcus

Primitive ventricle

Bulbous cordis

Dorsal aorta

(b)

FIGURE 4.4
From top: (a) Sketch of embryo at 16/17 days showing the position of the cardiogenic plate. (b) Diagram of heart tube showing original regions of the heart during early fetal development. (Modified from McNulty, J. and Wellman, C., Heart Development. Original image at http://www.meddean.luc)

of the atria will occur. The sinus venosus, which looks like a paired horn structure, then forms several structures. First, the blood vessels supplying blood to the sinus venosus develop to form the superior and inferior vena cava. The right sinus horn is later incorporated into the right atrium and grows into the

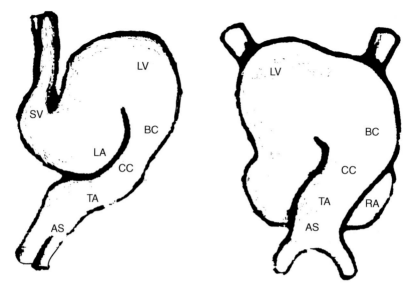

FIGURE 4.5

Mid-phase development of the fetal heart showing the developed atria and left ventricle. AS = aortic sac, BC = bulbus cordis, CC = conus cordis, LA = left atrium, LV = left ventricle, RA = right atrium, SV = sinus venosus, TA = truncus arteriosus. (Modified from McNulty, J. and Wellman, C., Heart Development. Original image at http://www.meddean.luc)

right auricle, whereas the left sinus horn forms the veins of the left ventricle; at the same time, the left atrium develops and pulmonary veins start to grow.

At the end of the seventh week, the human heart has reached its final stage of development. Because the fetus does not use its lungs, most of the blood is diverted to the systemic circulation, which is accomplished by a right to left shunting of blood between the two atria. At this stage, the bulbous cordis forms the left ventricle, whereas the proximal bulbous cordis forms the right ventricle, finally, the ventricles separate completely and the heart is almost fully developed. The final morphological change to the heart is the partitioning of the outflow tract. At this stage, the truncus arteriosus morphs into the aorta and the conus cordis becomes the pulmonary trunk. This is done by the development of a septum, which forms in the outflow tract, and the emergence of the two major blood vessels.

In the fetal heart's developmental stages, the heart resembles several other animal hearts. The initial tubelike heart is reminiscent of a fish heart, whereas the two-chambered heart resembles a frog heart. In the third phase, the three-chambered structure is similar to that of a snake or turtle. Only the final four-chambered heart structure uniquely distinguishes the human heart. The heart's rate of pumping oxygen-rich blood is fastest in infancy, that is, about 120 beats/min. As the child grows, the heart rate slows. A 7-year-old child's heart beats about 90 times/min, which stabilizes to about 70 beats/min by 18 years and remains in this region throughout adulthood.

4.2.4 The Heart Conduction System

The heart possesses a specialized nerve conduction system that allows it to beat autonomously even if devoid of control from the main nervous system. If the heart is removed from the body, it will continue beating for several hours if supplied with the appropriate nutrients.

There are four main components to the heart's conduction system: the sinoatrial (SA) node, internodal fiber bundles, AV node, and the AV bundle. The pacemaker of the heart responsible for controlling the heart rate is the SA node composed of a small group of specialized cardiac muscle situated in the superior (upper) part of the right atrium. It lies along the anterolateral margin of this chamber between the opening of the superior vena cava and the right auricle and has the special property of being able to produce self-excitation, which initiates a heartbeat. The fibers of the SA node fuse with the surrounding atrial muscle fibers allowing action potentials generated in the nodal tissue to spread to both atria causing atrial contraction. The speed of contraction is approximately 0.3 m/s. This action potential is carried at a higher speed by several internodal fiber bundles, which lie among the atrial muscle fibers. These fibers lead to the AV node, which is located in the right atrium near the lower part of the interatrial septurn.

The AV node (also known as the bundle of His) induces a short delay (approximately 0.1 s) in transmission of the action potential to the ventricles. This delay is important because it allows the atria to complete their contraction and pump all the blood into the ventricles before contraction occurs. The delay occurs not only within the fibers of the AV node itself but also at special junctional fibers that connect the node with ordinary atrial fibers.

The AV bundle then descends a short distance into the interventricular septurn, which divides into the right- and left-bundle branches. Each of these branches descends along its respective side of the interventricular septum immediately beneath the endocardium and spreads out into progressively smaller branches. These specialized muscle fibers are called Purkinje fibers (see Figure 4.1), which penetrate approximately one-third of the distance into the myocardium. The primary purpose of the Purkinje fibers is to conduct the action potential at approximately six times the velocity of an ordinary cardiac muscle (1.5–4.0 m/s) allowing the Purkinje fibers to channel a very rapid and simultaneous distribution of the action potential throughout the muscular walls of both ventricles. The fibers themselves terminate at the cardiac muscle within the ventricles, providing a pathway for action potentials to enter the ventricular muscle. When the impulse enters the ventricle muscle, it causes ventricular contraction, which proceeds upward from the apex of the heart toward its base.

A heartbeat starts in the right atrium when the SA node dispatches an impulse throughout the atria causing both atriums to contract. The impulse pauses in the AV node bundle allowing the atria to empty all their blood into the ventricles. The impulse then spreads through the ventricular muscles through the His–Purkinje system causing ventricular contraction. The impulse must follow this exact route to ensure proper heart function. Although the SA cells create the impulse that causes the heart to beat, other nerves function to change the rate at which the SA cells fire. These nerves are part of the autonomic nervous system, which is divided into the sympathetic nervous system and the parasympathetic nervous system. The sympathetic nerves increase the heart rate and the force of contraction, whereas the parasympathetic nerves inhibit or slow the rate of contraction.

4.2.5   The Cardiac Cycle

The heart pumps oxygen-rich blood to body organs and tissues and deoxygenated blood back to the lungs to be reoxygenated. This circulation system is vital to the sustainment of bodily functions and is referred to as the *cardiac cycle*. The major blood vessels that carry either oxygen-rich blood or deoxygenated blood can be divided into arteries, veins, and capillaries. Arteries carry oxygenated blood with the exception of the pulmonary artery, whereas veins carry deoxygenated blood with the exception of the pulmonary vein. At the top of the heart, two of the major blood vessels emerge, the aorta and the pulmonary artery (see Figure 4.6). The superior vena cava and inferior vena cava enter the right-hand side of the heart, whereas the pulmonary vein enters the left-hand side. It is sometimes easier to visualize the circulation system as being composed of two separate systems as seen in Figure 4.7.

The first system consisting of the right atrium and right ventricle is responsible for receiving deoxygenated blood from the body and sending it to the lungs where carbon dioxide can be exchanged for oxygen. Deoxygenated blood is carried by the body in veins, which cumulate in the superior and inferior vena cava. The superior vena cava carries the sum of deoxygenated blood from the upper half of the body, whereas the inferior vena cava carries deoxygenated blood from the lower half. These two veins join to form a single vena cava, and this blood vessel empties into the right atrium, which contracts and pushes the blood into the right ventricle. The time of atrium contraction is known as the *diastole phase*, and it is during this phase that the tricuspid valve opens to allow passage of blood into the ventricle. The term "diastole" is used to refer to heart relaxation during which the ventricle chambers are filled with blood. When blood is pumped from the atrium into the ventricle, the ventricular pressure increases with respect to the atrium pressure. To avoid a backflow of blood into the atrium, the AV valve (tricuspid in this case) is closed and prevented from opening. When it is full, the right ventricle contracts and pumps the deoxygenated blood into the pulmonary artery, which leads to the lungs. The phase of ventricle contraction is known as *systole* and the tricuspid valve is also closed to prevent blood flowing into the empty atrium.

The second system responsible for distribution of oxygenated blood to the body is composed of the left atrium and left ventricle. The newly oxygenated

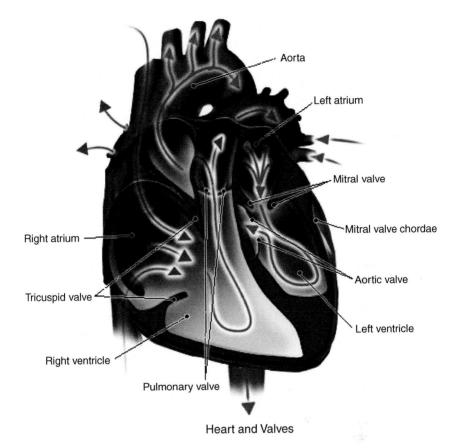

Heart and Valves

FIGURE 4.6 (See color figure following page 204.)
The major arteries and veins in the heart. (Courtesy of MedicineNet, www.medicinenet.com.)

blood in the lungs leaves through the pulmonary vein, which connects to the left atrium. The atrium pumps this blood into the left ventricle during the diastole phase. When the left ventricle contracts, blood is pumped out through the aorta and the bicuspid (mitral) valve prevents oxygenated blood from returning to the lungs. The aorta is the largest and thickest artery, which branches out to several arteries and smaller capillaries that end in cell walls of the various body tissues. Oxygen in the blood is passed to the cells by diffusion between capillary and cell walls where it is then used primarily for the generation of energy or other metabolic activities.

Despite the apparent segregation of the two systems, it should be noted that both atria contract together followed by the contraction of the ventricles. When the atria are in systole, the ventricles are relaxed (in diastole), whereas during ventricular systole the atria relax and remain in this phase even during a portion of ventricular diastole. This relaxation ensures that the amount

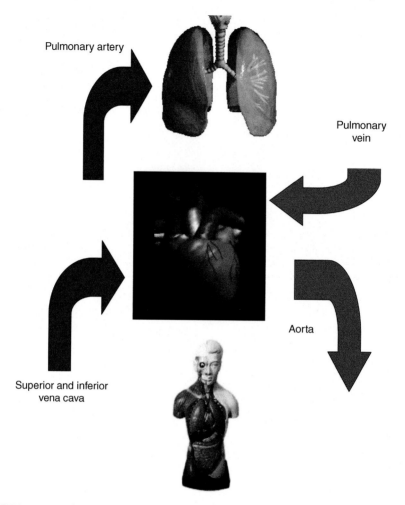

FIGURE 4.7 (See color figure following page 204.)
The blood circulation system depicting the flow of blood from heart to the lungs and the rest of the body. The red arrows show the flow of oxygenated blood, whereas the blue arrows depict the flow of deoxygenated blood.

of blood pushed out to the body is maximized. Being a fluid, blood flows from high-pressure to low-pressure areas, a property utilized effectively in the cardiac cycle. As each chamber of the heart fills, the interior pressure increases until it is sufficient to open the various one-way valves guarding the entry to the other chambers. The characteristic "lub-dup" sound of the heartbeat heard by the stethoscope is due to the opening and closing of these valves. For example, atrial pressure increases as blood from the veins enter and continue increasing during atrial systole. As the ventricles contract, the blood is forced in the reverse direction against the AV valves causing them to bulge into the atria and increasing the atrial pressure. This pressure increase acts naturally against the blood trying to enter the atria and keeps the AV valves shut, preventing blood regurgitation into the atria. During the diastole phase, both the left and right atria pump blood into empty ventricles against minimal resistance and, hence, have thinner walls.

Ventricles function briefly as closed chambers after they are completely filled, however, the pressure within them soon exceeds that of the aorta and pulmonary trunk until it is capable of forcing open the aortic and pulmonic semilunar valves. Blood then rushes out of the ventricles into the two large vessels. When the semilunar valves open, a rapid decline in intraventricular pressure occurs, which continues until the pressure within the ventricles is less than that of the atria. When this pressure differential is reached, blood within the atria pushes the AV valves open and begins to fill the ventricles once again. The right ventricle, which pumps blood to the lungs, has a moderately thick muscle layer composed of fibers intermediate in diameter between atrial and left ventricular muscle cells. The left ventricle, the strongest part of the heart, pumps blood through the high-pressure systemic arterial system and, therefore, possesses the thickest myocardium layer with the largest diameter muscle fibers. Consequently, we find the left-hand side of the heart significantly larger than the right-hand side.

The heart being a living organ also requires blood to sustain it. This circulation is provided by coronary arteries and veins. The coronary sinus returns blood from the heart wall itself and empties its contents into the right atrium. Although most of the blood from the heart wall is returned to the right atrium by the coronary sinus, there are additional microscopic veins called *venae cordis minimae*, which channel blood into all four chambers of the heart.

4.3   The ECG Waveform

The QRS complex is an electrical ventricular system and the most prominent waveform reflecting electrical activity within the heart. It serves as the foundation for automatic detection of heart rate and also as an entry point for classification schemes and ECG data-compression algorithms (Kohler et al., 2002). The QRS complex morphology depicts the mechanical action of the heart giving insight into how each chamber is operating. The waves of depolarization that spread through the heart during each cardiac cycle generate electrical impulses. These impulses propagate through various body fluids such as blood, up to the body's surface where they can be recorded using surface electrodes. These signals are then transmitted to an electrocardiograph, which amplifies and records the electrical activity resulting in a series of waveforms known as the ECG (or EKG in some cases).

The observed ECG originates from the spread of depolarization through the heart as blood is pumped through the various heart chambers. The fundamental QRS wave is composed of individual wave segments as seen in Figure 4.8.

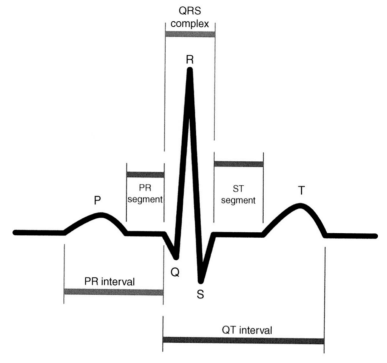

FIGURE 4.8
QRS complex showing the sinus rhythm regions.

The main features of the QRS wave that depict important information concerning cardiac health are as follows:

    a. P wave

    b. QRS complex

    c. T wave

    d. QRS intervals

4.3.1 P Wave

The initial P wave is caused by electrical activity originating from atrial contraction (systole). In CVDs, the P wave can become distorted and appear abnormal as seen, for example, in Figure 4.9. An *inverted* P wave caused by positive voltage changing to negative indicates that polarization direction of the atria is abnormal (Julian et al., 2005). This means that the origin of the pacemaker signal is not in the SA node, but could come from elsewhere, such as the atrium or the AV node. If the P wave has a characteristic *broadened* or *notched* look, this indicates a delay in the depolarization of the left atrium,

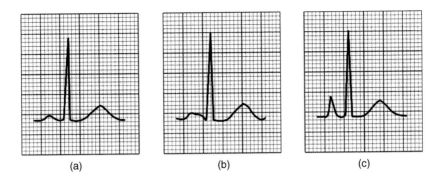

(a)   (b)   (c)

FIGURE 4.9
Lead II recordings of various P-wave measurements. (a) Normal, (b) broadened or notch, and (c) tall and peaked. (Reprinted with permission from Julian, D.G., Campbell-Cowan, J., and McLenachan, M.J., *Cardiology*, Saunders, Edinburgh, New York, 2005.)

which could mean problems in the conduction system. Right atrial enlargement (P pulmonale) can lead to tall P waves usually exceeding 3 mm on the ECG trace. In some cases, the P wave cannot be seen due to a junctional rhythm or SA block, whereas in other cases it may be replaced by small oscillations or fibrillation waves (atrial flutter).

4.3.2 QRS Complex

The QRS complex signifying ventricular contraction (systole) is composed of the Q wave, which is an initial negative deflection followed by the R wave, a positive (upward) deflection. Any negative deflection following immediately after the R portion is termed the S wave. Figure 4.10 shows some examples of abnormal QRS complexes where two or more R waves could be present and cases where the R portion is absent. The morphology of this wave depends on the ECG electrode/lead recording it, for example, if viewing from lead V1 (right-hand side of the heart), an R wave is seen due to septal depolarization and a large S wave is observed due to left ventricular forces acting away from the electrode. Abnormally large Q waves could be indicative of MI (described later), in contrast to a healthy Q wave, which does not normally exceed 2 mm in amplitude or 0.03 s in width. The QRS complex is usually not longer than 0.1 s and on average is of 0.06–0.08 s duration (Julian et al., 2005).

4.3.3 T Wave

After ventricular contraction, the ventricles relax (diastole) generating the T wave. Unlike the previous two wave segments, the T wave is a repolarization wave, which normally occurs for 0.25–0.35 s following ventricular depolarization. During this phase, the lower heart chambers are relaxing electrically

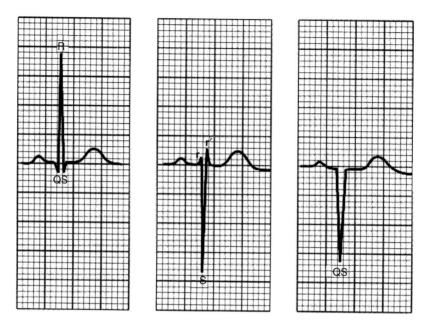

FIGURE 4.10
QRS variations in the ECG. (Reprinted with permission from Julian, D.G.,
Campbell-Cowan, J., and McLenachan, M.J., *Cardiology*, Saunders,
Edinburgh, New York, 2005.)

and preparing for their next muscle contraction. Atria repolarization is also
present, but the signal amplitude is usually masked by the larger QRS complex
due to ventricular contraction and is therefore difficult to observe. The T wave
points in the same direction as the QRS component because repolarization
occurs from epicardium to endocardium (see Section 4.2.2), in the direction
opposite to depolarization, which proceeds from endocardium to epicardium.

T waves are usually not taller than 5 mm in the standard leads. Abnormally
tall T waves may be indicative of MI, whereas flattened T waves can show
myxoedema as well as hypokalaemia. Hyperventilation and smoking could
cause slight T-wave inversion, whereas major T-wave inversions can be due to
myocardial ischaemia, infarctions, ventricular hypertrophy, and bundle branch
block.

4.3.4 QRS Segment Intervals

In addition to the shape of the waves, the time intervals are crucial in the evaluation of cardiac health. Between the beginning of the P wave and the onset of
the QRS complex is the PQ interval, since the Q wave is often absent this interval is also termed the PR interval and represents the time between the onset
of atrial contraction and the onset of ventricular contraction (normally about

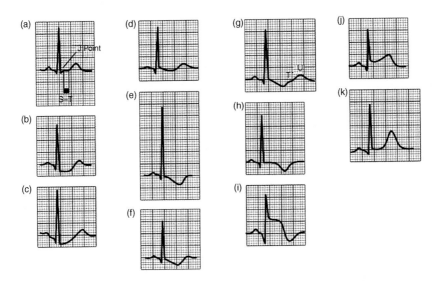

FIGURE 4.11

ST wave segments in the ECG. (Reprinted with permission from Julian, D.G., Campbell-Cowan, J., and McLenachan, M.J., *Cardiology*, Saunders, Edinburgh, New York, 2005.)

0.16 s). In patients with heart disease, especially with scarred or inflamed heart tissue, a longer PR interval may be observed as more time is required for the depolarization wave to spread through the atrial myocardium and the AV node. Shortened PR intervals could mean that the impulse originates in the junctional tissue or could be due to the Wolff–Parkinson–White syndrome (Julian et al., 2005).

The ST segment is a further interval important for detecting a variety of CVDs (see Figure 4.11). It usually appears as a levelled straight line between the QRS complex and the T wave. Elevated or depressed ST segments (depending on the ECG lead being observed) is a sign of MI as the heart muscle is damaged or does not receive sufficient blood, causing a disturbance in ventricular repolarization. *Pericarditis* can be detected by observing ST segments that are concave upward over many cardiac cycles. In *digitalis therapy* it has been found that ST segments are depressed with a gentle sagging, whereas the T wave remains normal or flattened. ST depressions can also indicate ventricular hypertrophy, acute myocardial ischemia, and sinus tachycardia, but the shape of the depressions are characteristic of the pathology.

The QT interval is approximately 0.35 s and provides important information about the state of ventricular contractions. Its duration shortens as heart rate increases and usually the QT interval does not exceed half the time between the previous RR interval for rates between 60 and 90 beats/min. This interval is sometimes difficult to measure because it cannot be clearly identified. Nevertheless, prolonged QT intervals can indicate the risk of ventricular tachycardia (VT) or the presence of certain drugs such as antidepressants.

4.4 The Electrocardiogram

The ECG is a representation of the electrical activity in the heart over time and is probably the single-most useful indicator of cardiac function. It is widely accepted that the ECG waveforms reflect most heart parameters closely related to the mechanical pumping of the heart and can be used to infer cardiac health. The ECG waveform is recorded from the body surface using surface electrodes and an ECG monitoring system. Any alteration in the transmission of impulses through the heart can result in abnormal electrical currents, which then appear as altered or distorted ECG waveforms. This phenomenon has been utilized successfully by cardiologists and clinical staff to diagnose most cardiovascular disorders. In the following, we will discuss the fundamental components of a standard ECG instrumentation.

4.4.1 ECG Signal Acquisition

All ECG instruments have horizontal and vertical calibration lines. Horizontal lines are utilized for voltage determination and are scaled such that 1 mV is equivalent to a deflection of 10 small divisions in either the upward (positive) or downward (negative) direction. Vertical lines on the ECG refering to time calibration for the period of the signal are generally calibrated so that 1 in. in the horizontal direction represents 1 s. Each inch is divided into five segments representing 0.2 s with smaller units further subdivided into 0.04 s segments. ECG traces are usually recorded on specialized ECG paper printed with thin (Julian et al., 2005) vertical lines 1 mm apart interspaced with thick vertical lines at 5 mm separation. This spacing is similar to that of the instrumentation itself and can be used to calculate the heart rate. Assuming a regular heartbeat, the heart rate can be computed by counting the number of small squares between two consecutive R waves; and dividing by 1500, alternatively, one can count the large squares and divide by 300.

4.4.2 ECG Electrodes and Leads

The ECG electrodes are sold for a variety of purposes ranging from temporary purposes to advanced high-quality designs for sophisticated quantitative measurements. ECG surface electrodes are manufactured by many biomedical companies, for example, Nikomed U.S.A. Inc. produces a variety of surface electrodes, some of which are depicted in Figure 4.12. Recessed electrodes have an interface paste between the skin and metal electrode. Reusable electrodes are usually made from silver, zinc, copper, and nickel alloys. The noise artifacts introduced in ECG due to power-line frequency and baseline instability can be reduced by using electrode paste (e.g., Ag/AgCl) to reduce impedance and intraelectrode offset potentials, which can saturate the amplifier. In most ECG recordings, disposable electrodes are preferable (Togawa et al., 1997) since they are smaller, more adhesive, have better conduction properties, and

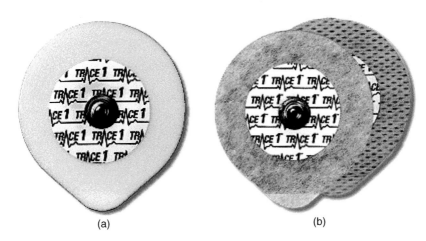

(a)   (b)

FIGURE 4.12

Surface ECG electrodes distributed by Nikomed U.S.A. Inc. (http://www.nikomedusa.com). The electrode (a) is made of foam and wet gel, whereas electrode (b) is made of solid gel and is porous allowing the skin to "breathe."

pose less risk of infection. Stainless-steel electrodes employ potassium citrate or potassium sulfate as the interface electrode paste, but their polarization considerably exceeds Ag/AgCl electrodes. Dry contactless or capacitive electrodes may be used in the case where amplifiers have high input impedance; these can be applied directly to the skin without gel paste.

Specialized electrodes have also been used according to the environment of operation. Utsuyama et al. (1988) used two electrodes (VL-00-S, Medicotest) for recording bipolar ECG in aquatic environments. These electrodes were sheathed in transparent plastic cases (6.5 cm diameter and 5 mm depth) and attached with adhesive tape to prevent electrode movement when the subjects dived into the water. Thin rubber skirts about 2 cm wide were also used to hold the electrodes in place.

The voltages obtained during a normal ECG depend upon the placement of the recording electrodes (leads) on the body's surface. There are several ECG-recording methods ranging from ambulatory recording, Holter tape, 12-lead recording, sleep ECG recording, and body surface potential mapping. The most widely used method is based on recording of ECG using 12-lead electrodes. Figure 4.13 shows a layout of the surface electrodes for the recording of 12-lead ECG. The term "lead" is used to distinguish recording a potential difference between a pair of electrodes as opposed to the voltage obtained by a single electrode. Table 4.1 summarizes the various leads that are described in the following.

Bipolar or standard leads are electrodes attached to the limbs and are usually termed leads I, II, and III. In lead I, the positive electrode is attached to the left arm, whereas the negative electrode is attached to the right arm. The potential difference is then obtained by subtracting the right-arm voltage

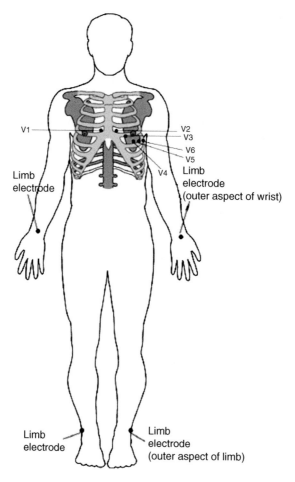

FIGURE 4.13
Diagram of 12-lead ECG positioning showing the lead positions. Details of lead positions are described in the text. (Reprinted from Prof. Peter Macfarlane, http://www.gla.ac.uk/CARE. With permission.)

from the left arm. Following this convention, lead II is the potential difference between the left leg and right arm, whereas lead III measures the difference between the left leg and left arm, from this we can deduce that the potential difference in lead II is the sum of leads I and III.

Unipolar leads measure the voltage at a single site by placing the positive electrode to the site and connecting the negative electrode to an indifferent potential (ground or small potential such as a Wilson terminal, which is composed of three limb electrodes connected through a $5000\,\Omega$ resistance). Additional leads such as V3R and V4R can be taken from the right-hand side of the chest and function similarly to V3 and V4. In some cases, the

TABLE 4.1

The 12-Lead ECG Leads and the Positions of the Positive and Negative Electrodes

Lead	Type	+ve Electrode	−ve Electrode
I	Bipolar	Left arm	Right arm
II	Bipolar	Left foot	Right arm
III	Bipolar	Left foot	Left arm
V1	Unipolar	4th intercoastal space to the right of the sternum	Ground
V2	Unipolar	4th intercoastal space to the left of the sternum	Ground
V3	Unipolar	Midway between V2 and V4	Ground
V4	Unipolar	5th intercoastal space	Ground
V5	Unipolar	Left anterior axillary line on the same latitude as V4	Ground
V6	Unipolar	Left midaxillary line on the same latitude as V4	Ground
aVR	Unipolar	Right arm	Ground
aVL	Unipolar	Left arm	Ground
aVF	Unipolar	Left foot	Ground

leads can be placed on higher levels such as the second, third, or fourth intercoastal spaces; examples of the ECG waveforms recorded from these leads are depicted in Figure 4.14.

4.4.3 Electrocardiogram Signal Preprocessing

While recording the ECG, various sources can potentially add noise to the recorded signal and corrupt the output signal. The presence of noise in signal systems can severely affect interpretation of the data and lead to an incorrect diagnosis. Significant sources of noise (Friesen et al., 1990) include the following:

a. Electrical interference from power lines adding 50 or 60 Hz power-line frequency. These include fundamental signals as well as their harmonics with an amplitude of up to a maximum of 50% peak-to-peak ECG amplitude.

b. Muscle contraction and muscle activity can generate high-frequency electromyography (EMG) noise.

c. Motion artifacts such as movement of the electrode over the skin surface.

d. Impedance changes at the skin/electrode interface due to temporary loss of contact or loose electrodes.

e. Baseline drift due to respiration.

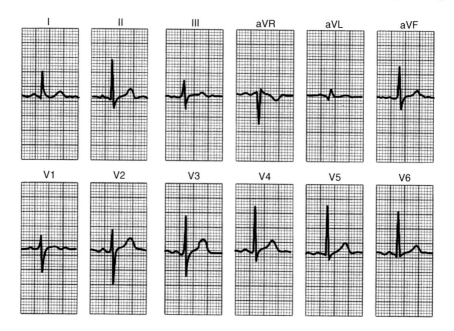

FIGURE 4.14

Example of QRS signals obtained from each of the leads in 12-lead ECG monitoring. (Reprinted with permission from Julian, D.G., Campbell-Cowan, J., and McLenachan M.J., *Cardiology*, Saunders, Edinburgh, NY, 2005.)

 f. Noise introduced due to instrumentation or electronic devices.

 g. Electrosurgical noise.

Preprocessing of the ECG signals is, therefore, necessary to reduce the various noise components and increase the signal-to-noise ratio (SNR) substantially. The ECG signal is filtered with a single bandpass filter with a 10–25 Hz passband. Zigel et al. (2000) investigated more advanced filtering by applying two separate infinite impulse response filters to different parts of the ECG waveform. In their scheme the P and T wave segments were filtered with a 0.1–50 Hz passband while the QRS complex was filtered using a 0.1–100 Hz passband.

4.4.4 Electrocardiogram Signal-Monitoring and Recording Instrumentation

Many biomedical companies specialize in manufacturing and marketing of health-monitoring devices, the most widely used being the ECG or EKG monitor. With considerable improvements, the ECG is no longer a large and clumsy machine confined to hospital bedsides, but is now available as a portable unit. A brief survey of medical companies listed on the Internet revealed several models, which we describe briefly.

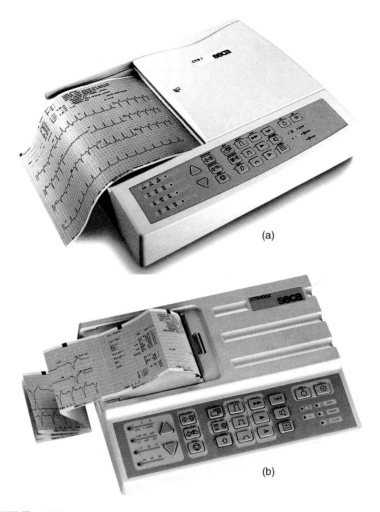

FIGURE 4.15

Portable Seca ECG recorders distributed by companies such as PMS Instruments Ltd., United Kingdom. CT6Pi Interpretive model, (a) is a 12 lead ECG recorder while the CT3000i Interpretive model (b) is a 3-lead ECG recorder. (www.seca.com)

Seca* ECG monitoring devices have long been used in hospitals and are generally trolley-based devices with large monitor screens. The company, however, has recently started producing portable lightweight ECG devices, such as seen in Figures 4.15 and 4.16, which are currently distributed by several companies, for example, PMS instruments Ltd. (United Kingdom). The standard 3-lead and 12-lead models can produce ECG traces on rhythm strips, make continuous measurements, and have basic software for interpretation of

---

*www.seca.com

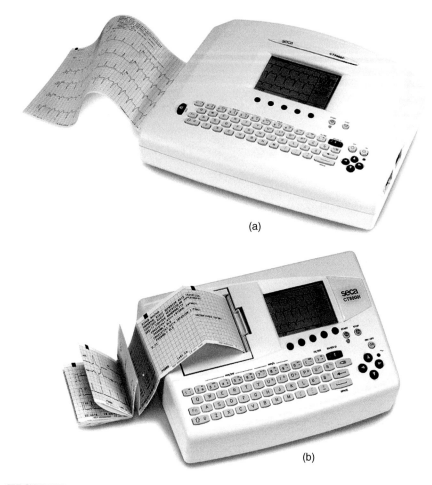

(a)

(b)

FIGURE 4.16
The more recent portables Seca CT8000P shown on the top and the CT8000i
on the bottom.

the data. Furthermore, the printouts can be obtained in real time, whereas
the batteries in portable mode give up to approximately 4 hours of usage
time. The portable devices range from 2.9 to 5 kg depending on the model.
The larger 12-lead ECG devices can also print ECG traces on A4 paper and
have lead-off warnings to detect loose or detached leads during measurements.
Prices range from £1000 to £2000 and are now more affordable for patients
confined to their homes.

The new advances in sensor technology, PDAs, and wireless communications
favor the development of a new type of monitoring system, which can provide patients with assistance anywhere and at any time. Of particular interest
are the monitoring systems designed for people with heart arrhythmias due
to the increasing prevalence of CVDs. PDAs and mobile phones can play a

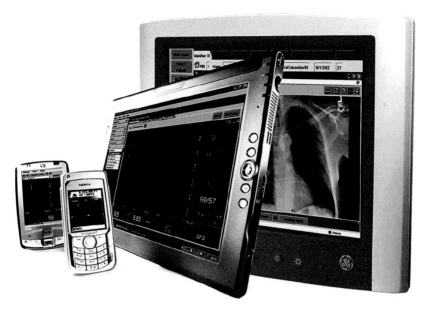

FIGURE 4.17
Webviewer and pocket ECG viewers manufactured by GE Healthcare are aimed at providing instant vital signs information. (http://www. gehealthcare.com. With permission.)

very important role in these kinds of systems because they are ubiquitous, portable, and can execute increasingly complex tasks. The Pocketview ECG interpretive device is the first device using PDA technology, which can store and download ECG traces from larger trolley-based ECG or portable ECG devices such as the EMIS, TOREX & VISION systems. This device is manufactured by MicroMedical Industries in Australia and was approved for use in the United States by the U.S. Food and Drug Administration (FDA) in 2002. Other portable ECG devices (Figure 4.17) now have wireless connectivity to computers and other display devices. In addition to this, portable handheld ECG devices (Figure 4.18) are fast gaining prominence. These devices are simpler having only a single lead, which can be used to detect atrial fibrillations or abnormal heartbeats.

ECG monitoring for patients and the bedridden elderly has long been possible with trolley-based ECG monitors (Figure 4.19), but these devices have been bulky and expensive. The newer ECG monitoring devices described earlier have started to shrink in size and are now more affordable. This technology is also set to replace the trolley-based ECG systems used to record uterine contractions during childbirth allowing continuous electronic monitoring of the fetal heart in the home. Such devices will enable expecting parents to monitor the growth stages of their baby and nurture stronger bonds with their child, an ability that was not possible until recently.

FIGURE 4.18
The FP180 portable easy ECG monitor is a signal-lead ECG for detecting atrial fibrillation and arrhythmias. (http://www.favoriteplus.com/handheld-ecg-ekg-monitor.php.)

4.5 Cardiovascular Diseases

In the general population, individuals with normal hearts have either very rapid or very slow heart rates, but extreme variation may be indicative of very serious cardiac disorders. In the following section, we will describe heart diseases that have been detected using ECG.

4.5.1 Arrhythmia

Cardiac arrhythmias are disturbances to the normal cardiac rhythm, which can vary in severity. According to the AHA,* arrhythmias occur when the

---

*http://www.americanheart.org

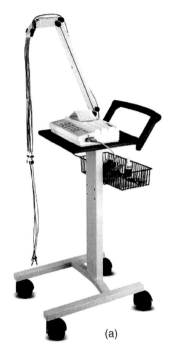
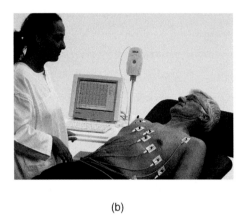

(a)   (b)

FIGURE 4.19
A Seca ECG trolley shown on the left and in use for patient ECG monitoring
on the right.

heart's natural pacemaker develops an abnormal rate or rhythm, when the
normal impulse conduction pathway is interrupted, or when another part of
the heart tissue takes over as the pacemaker. Many arrhythmias have no
known cause, whereas others (Coast et al., 1990) can develop from problems
in the cardiac pacemaker, in the ectopic pacemaker sites, or through abnormal propagation of pacing impulses in the cardiac conduction system. Several
risk factors have been identified, including coronary artery disease, high blood
pressure, diabetes, smoking, excessive use of alcohol or caffeine, drug abuse,
and stress. Some over-the-counter and prescription medications, dietary sup-
plements, and herbal remedies are also known to cause arrhythmias. Patients
with this disorder show disruptions in the synchronized contraction sequences
of the heart, which result in a reduced pumping efficiency and giving rise to
various types of arrhythmia disorders are depicted in Figure 4.20. Analysis
for arrhythmias can be undertaken using a 12-lead ECG where onset can be
detected, for example, by an early P wave in the ST segment.

The most common forms of arrhythmia are ectopic beats of a ventricular or
atrial origin, where a common characteristic of the disease is the occurrence
of premature beats (Silipo and Marchesi, 1998). Ventricular ectopic beats
(VEBs) exhibit a different waveshape from the normal QRS sinus rhythm

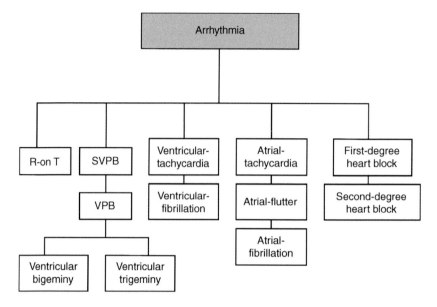

FIGURE 4.20
The various subgroups of cardiac arrhythmias. (Adapted from Kundu, M.,
Nasipuri, M., and Basw D., *IEEE Eng. Med. Biol. Mag.*, 28(2), 237–243,
1998. ©IEEE.)

and are more easily detected (Figure 4.21). Atrial ectopic beats, however, are
more difficult to detect since the waveshape is relatively similar to the normal
ECG. In severe cases, failure to produce synchronized heartbeats can lead to
life-threatening arrhythmias.

Impulses can arise in regions of the heart other than the SA node. Areas
of ectopic foci arise from irregular generation of impulses during the normal
cardiac rhythm and potentially initiate premature or ectopic beats. As a result
of the refractory period that follows the premature ectopic beat, the onset of
the next normal beat is delayed. When the next beat does occur, it may be
unusually strong and startling, giving rise to a pounding sensation within the
chest. Although the premature ectopic beats are not serious by themselves,
they may be indicative of underlying heart damage. Conditions that may lead
to the production of ectopic foci include the following:

1. Development of calcified regions within the heart that press upon and
   irritate cardiac muscle fibers

2. Focal areas of ischemia

3. Stress

4. Abnormal AV node stimulation resulting from the toxic effects of caffeine, nicotine, and certain drugs

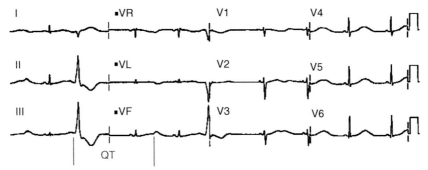

Ventricular premature beats

Long QT intervals and long distance between normal beats

FIGURE 4.21

ECG of ventricular premature beats that is evidenced by long QT intervals and long periods between two normal beat cycles. (Modified from Dean Jenkins and Stephen Gerred, http://www.ecglibrary.com/ecghome.html. With permission.)

The normal healthy heart beats between 60–80 beats/min and variations in this can reflect several disorders. Conditions where the heart beats slower than 60 beats/min are known as *bradycardia* (from brady, meaning slow). This is seldom a problem unless there is prolonged or repeated symptoms such as fatigue, dizziness, lightheadedness, or fainting. Physically fit individuals have resting heart rates slower than 60 beats/min. Bradycardia becomes a problem when there is insufficient blood flow to the brain. This condition occurs more in elderly people and can be corrected by electronic pacemakers to increase the heart rhythm. It may also occur in patients with atherosclerotic lesions of the carotid sinus region in the carotid artery.

Tachycardia (from tachy, meaning fast, and cardia, meaning heart) or *tachyarrhythmia* is a condition in which the heart beats too rapidly, for example, more than 100 beats/min. The beats may be regular or irregular and can produce palpitations, rapid heart action, chest pain, dizziness, lightheadedness, or near fainting because the blood is circulated too quickly to effectively service the tissues with oxygen. VT occurs when rapid beating starts in the ventricles and reduces the heart's ability to pump adequate blood to the brain and other vital organs. This is a dangerous arrhythmia that can change without warning into *ventricular fibrillation*, which is one of the most serious arrhythmias. When ventricular fibrillation occurs, the lower chambers quiver, and the heart is unable to pump any blood, the patient soon collapses and sudden cardiac death follows unless medical assistance is provided. If treated in time, VT and ventricular fibrillation can be changed to normal rhythms with proper electrical shocks. The characteristic rapid beating can also be controlled with medications or by locating and treating the focus of rhythm disturbances. One way to correct this disorder is by using an implantable cardioverter–defibrillator (ICDs).

In *atrial fibrillation*, the atria quiver, as a result, blood is not pumped completely into the ventricles and, thus, clots. This disorder is found in about 2.2 million Americans and leads to *stroke*. Stroke occurs when the blood clot in the atria leaves the heart and gets lodged in the brain. Approximately, 15% of strokes occur in people with atrial fibrillation.

4.5.2 Myocardial Infarction

Acute myocardial infarction (AMI or MI) is commonly known as heart attack. The onset of MI can be sudden but is often characterized by varying degrees of chest pain or discomfort, weakness, sweating, nausea, vomiting, and arrhythmias, sometimes resulting in loss of consciousness. These are often life threatening, demanding immediate medical services. The etiology of this disease is the interruption of blood supply to a part of the heart causing death or scarring of that heart tissue. When oxygen is insufficient, the heart muscle cells in the myocardium die and can no longer contract or pass impulses onward. This can occur when the coronary arteries that nourish the heart muscles become blocked and blood supply is insufficient to meet their metabolic needs. The process of slow muscle death due to lack of oxygen is called *infarction*, and Figures 4.22 and 4.23 depict the state of infarcted cardiac muscles. Since the affected area can vary in size, the severity of heart attacks is also modulated.

There are several types of pathology, for example, 90% of transmural AMIs occur from a thrombus overlying an ulcerated or fissured stenotic plaque. This platelet aggregation together with coronary atherosclerosis causes vasospasm

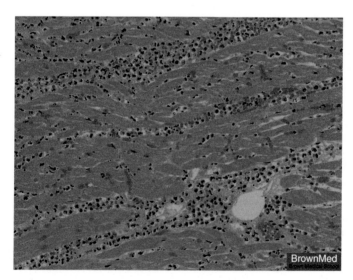

FIGURE 4.22 (See color figure following page 204.)
View of myocardial fibers after approximately 48 hours post myocardial infarction. Fibers depict maximum acute inflammation and loss of nuclei striations. Irreversible cell death occurs after 15–20 minutes of onset. (Courtesy of A/Prof. Calvin E. Oyer, M.D., Brown Medical School.)

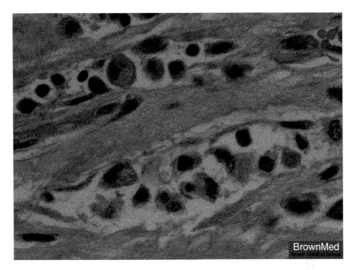

FIGURE 4.23 (See color figure following page 204.)
View of dead myocardial fibers after 7–9 days of post myocardial infarction. Cell macrophages, fibroblasts, and debris from dead myocytes can still be seen. (Courtesy of A/Prof. Calvin E. Oyer, M.D., Brown Medical School.)

or blockages. Other infarction patterns include transmural infarction, which involves the whole left ventricular wall from endocardium to epicardium. Isolated infarcts of the right ventricle and right atrium can also occur but are extremely rare. Subendocardial infarction occurs when multifocal areas of necrosis (muscle death) are confined to the inner one-third to half of the left ventricular wall and are different from those of transmural infarction. In these cases diagnosis is reached through a combination of medical history, ECG findings, and blood tests for cardiac enzymes.

The most important treatment for MI is to restore blood flow to the heart muscle usually by dissolving the artery clot (thrombolysis) or opening the artery with a surgical balloon (angioplasty). ECG recordings can be used to confirm the diagnosis; findings suggestive of MI are elevations of the ST segment and changes in the T wave. After MI, changes in the Q segment representing scarred heart tissue can be observed and ST-segment elevation can be used to distinguish STEMI ("ST-elevation myocardial infarction"). A normal ECG does not rule out an MI and so individual ECG leads need to be inspected for abnormalities and to localize any potential problems. For example, leads V1–V4 can localize problems in the heart anterior wall, leads II and III for problems in the inferior wall, and leads V1 and V2 in the posterior wall. Figure 4.24 depicts an example of ECG morphologies related to the onset of acute inferior MI.

4.5.3 Myocardial Ischemia

If blood supply is restricted but not completely blocked, partial damage may occur to the cardiac muscle, this restriction is known as *ischemia*, which

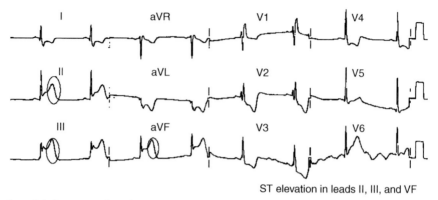

ST elevation in leads II, III, and VF

Acute inferior myocardial infarction

FIGURE 4.24

ECG of acute inferior myocardial infarction depicting ST elevation in the inferior leads II, III, and aVF and a reciprocal ST depression in the anterior leads. (Modified from Dean Jenkins and Stephen Gerred, http://www.ecglibrary.com/ecghome.html. With permission.)

occurs due to factors such as atherosclerosis in the blood vessels. A major symptom is angina pectoris, which is characterized by paroxysmal attacks of chest pain usually caused by ischemia, which does not induce infarction. There are several types where the typical angina is stable angina characterized by paroxysms of pain related to exertion and relieved by rest. Prinzmetal's angina occurs at rest and is caused by reversible spasm in normal to severely atherosclerotic coronary arteries. In this case, the ST segment of the ECG is either elevated or depressed during the attacks. Unstable angina is associated with prolonged pain, pain at rest, or worsening of pain similar to stable angina patients but the symptoms vary or are unstable. In extreme cases, sudden cardiac death can result usually within 1 hour of a cardiac event or in the absence of symptoms.

4.5.4 Blood Vessel and Valve Diseases

Several diseases affect the AV valves, aortic valve, and pulmonic valves resulting in narrowing or leaking, which reduces their efficiency. The majority of valvular heart disease cases are due to congenital abnormalities or from inflammation due to diseases such as rheumatic fever or syphilis. In the case of rheumatic fever, the mitral and aortic valve are most frequently affected.

Thickening of the mitral valve results in narrowing of the orifice, *mitral stenosis*, which blocks the flow of blood from the left atrium to the left ventricle causing increased pressure in the left atrium due to incomplete emptying. As a result, cardiac output is reduced because there is inadequate filling of the left ventricle. Mitral stenosis also causes characteristically abnormal heart sounds due to turbulence in blood flow and opening and closing of the narrowed valve.

Valvular insufficiency occurs when a valve cusp is shortened and prevented from closing completely due to some unknown mechanism. Blood then regurgitates into the preceding chamber where it may clot leading eventually to more serious disorders such as stroke. In the case of mitral valve insufficiency, blood flows backward into the left atrium when the left ventricular pressure exceeds that of the left atrium. As a result, the left atrial pressure increases and cardiac output declines. In serious cases, an increase in pressure of the pulmonary circulation is observed, which causes pulmonary oedema (swelling). In valvular disorders, when the atria fills some blood leaks into the ventricles and forces them to accommodate a larger than normal volume. This overload can result in dilation and increase in size (hypertrophy) of the ventricle, which can ultimately lead to high blood pressure.

4.5.5 Congenital Heart Diseases

Congenital malformations of the heart are disorders due to defects during the growth of a fetus in the womb. These heart diseases typically include openings in the interatrial or the interventricular septa, respectively as atrial septal defects (ASD) and ventricular septal defects (VSD). The most common ASD is the result of a patent foramen ovale, that is, the failure of the fetal opening between the two atria to close at birth. In some individuals, a very small opening may persist throughout life without resulting in clinical symptoms (asymptomatic). Abnormal embryonic development of the interatrial wall may result in much larger openings and in rare cases, the interatrial septum may be completely absent causing a condition, which is clinically termed "common atrium."

The most frequently occurring congenital heart defect is VSD and is generally associated with the upper or membranous portion of the interventricular septum. This disorder occurs in approximately 1 out of every 10,000 live births. Single or multiple septal defects can also occur in the muscular portion of the ventricular septum although these are less common. Multiple VSDs in the muscular portion give rise to a condition known as the "swiss cheese" form of VSD. Although some VSDs are diagnosed at birth, others may go undetected for 4 or 5 years after birth, for example, cases in which there is complete absence of the interventricular septum known as cortriloculare biatriatum or a three-chambered heart.

4.5.6 Other Diseases

There are many other CVDs that can be detected using ECG readings, for example, left ventricular hypertrophy can be detected by an increase in the amplitude of the R waves from left-chest leads or increase in S-wave amplitudes from the right-chest leads. In right ventricular hypertrophy, the R waves are dominant in V1, V3R, and V4R. In disorders where the left bundle branch is blocked (see Figure 4.25), the interventricular septum is activated from the right-hand side resulting in abnormal ventricle depolarization. If the right bundle branch is blocked, the septum is activated normally, but abnormalities

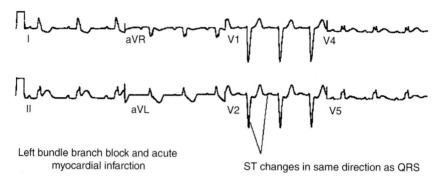

Left bundle branch block and acute
myocardial infarction

ST changes in same direction as QRS

FIGURE 4.25

ECG of left bundle block leading to acute myocardial infarction. It can be seen that ST changes occur in the same direction as the QRS. ST elevation however is higher than would have been expected from left bundle block alone and the Q waves in two consecutive lateral leads indicate the presence of myocardial infarction. (Modified from Dean Jenkins and Stephen Gerred, http://www.ecglibrary.com/ecghome.html).

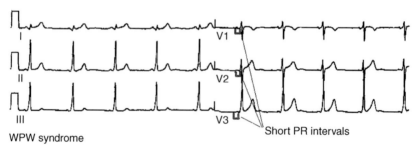

WPW syndrome

Short PR intervals

FIGURE 4.26

ECG depicts Wolf-Parkinson-White syndrome showing short PR intervals and slurred upstroke to the QRS wave and broadening of the QRS wave. (Modified from Dean Jenkins and Stephen Gerred, http://www.ecglibrary.com/ecghome.html).

are observed in the R wave (Julian et al., 2005). If different conduction pathways are present, it may cause the impulse to arrive too early at the ventricles, a condition known as Wolff–Parkinson–White syndrome (Figure 4.26).

4.6   ECG Feature Extraction Methods

Once the ECG waveform has been amplified and filtered, it is usually processed further to extract more information. ECG feature extraction is designed to derive differential characteristics from the waveforms to aid detection or diagnosis. Feature extraction is important in the postprocessing of ECG readings because a good set of features will have sufficient discriminative power to facilitate the diagnosis or investigation. In this section, we first review the mature field of QRS detection algorithms followed by an outline of feature-extraction methods, which have proved to be successful.

4.6.1 QRS Detection Algorithms

Characteristic features of the QRS complex such as physical morphology, its time, and frequency domain characteristics, provide much information about cardiac health. There has been a considerable amount of work devoted to the design of QRS detection algorithms over the last 30 years. The review by Friesen et al. (1990) provides several references to QRS detection and provides a comparison of nine QRS detection algorithms categorized according to the technique used to find the QRS complex. Their preliminary test results revealed that none of the algorithms were able to detect the QRS complexes without false positives for all the noise types at the highest noise levels. It was found that algorithms based on amplitude and slope showed the best performance for EMG-corrupted ECG. An algorithm using a digital filter had the best performance for the composite-noise-corrupted data. In a more recent review article, Kohler et al. (2002) outline nonlinear transform algorithms and computational techniques such as ANNs, genetic algorithms, wavelet transforms, filter banks, and heuristic methods for QRS detection. We now proceed to describe briefly the various methods used in QRS-detection algorithms.

4.6.1.1 Amplitude and Derivative Algorithms

The most fundamental algorithm is based on amplitude and first derivatives of the ECG signal (Morizet-Mahoudeaux et al., 1981). Let $X(n) = X(0)$, $X(1), \ldots, X(N)$ represent the discrete samples of an ECG signal. An amplitude threshold, θ, is first computed as a fraction of the maximum ECG signal amplitude as follows:

$$\theta = \gamma \max[X(n)] \tag{4.1}$$

where $\gamma = 0.3, 0.4$ have been popularly used. The first derivatives, $y'(n)$, of the ECG waveform are taken for parts of the wave with amplitudes exceeding the amplitude threshold. Several methods have been used such as

$$y'(n) = X(n+1) - X(n) \tag{4.2a}$$

$$y'(n) = X(n+1) - X(n-1) \tag{4.2b}$$

$$y'(n) = 2X(n+2) - X(n+1) - X(n-1) - 2X(n-2) \tag{4.2c}$$

where the values of $X(n)$ in this case can either be restricted by the amplitude threshold or not. If restricted by the threshold θ, the first derivative equations

(Equation 4.2) can be used subject to the following constraint:

$$X(n) = \begin{cases} |X(n)| & \text{if } |X(n)| > \theta \\ \theta & \text{if } |X(n)| \leq \theta \end{cases} \tag{4.3}$$

The second derivative (Ahlstrom and Thompkins, 1983) can also be computed as

$$y''(n) = X(n+2) - 2X(n) - X(n-2) \tag{4.4}$$

Some algorithms also use linear combinations of the first and second derivatives (Balda, 1977), for example,

$$z(n) = 1.3|y'(n)| + 1.1|y''(n)| \tag{4.5}$$

4.6.1.2 Digital Filter Algorithms

Some QRS-detection algorithms are based on enhancing certain portions of the waveform through bandpass filtering (Okada, 1979; Dokur et al., 1997; Fancott and Wong, 1980; Suppappola and Sun, 1994; Sun and Suppappola, 1992). A simple method (Okada, 1979) is to bandpass filter the ECG signal and process it using a nonlinear function

$$y_2(n) = y_1(n) \left[\sum_{k=-m}^{k=m} y_1^2(n+k) \right]^2 \tag{4.6}$$

where $y_1(n)$ is the filtered ECG signal. Multiplication of backward difference (MOBD) algorithms (Suppappola and Sun, 1994; Sun and Suppappola, 1992) use products of adjacent sample derivatives to obtain the QRS feature

$$y_1(n) = \prod_{k=1}^{N} |X(n-k) - X(n-k-1)| \tag{4.7}$$

where N defines the order of the model. Additional constraints can be imposed on the signal to avoid the effects of noise. Other algorithms (Hamilton and Thompkins, 1986) have utilized threshold values on the bandpass filtered signal. QRS peaks are detected by comparing amplitudes with a variable v, where v is the most recent maximum sample value. If, for example, the following ECG samples fall below $v/2$, a peak is detected. It is also possible to use recursive and nonrecursive median filters (Yu et al., 1985) based on the assumption that the QRS peaks occur within the passband of the filters. These filters have the following form:

$$y(n) = \text{median}[y(n-m), \ldots, y(n-1), X(n), X(n+1), \ldots, X(n+m)] \tag{4.8a}$$

$$y(n) = \text{median}[X(n-m), \ldots, X(n-1), X(n), X(n+1), \ldots, X(n+m)] \tag{4.8b}$$

The median is obtained by arranging the elements in the vector according to their values and taking the midpoint as the filter output to approximate where the QRS waveform is occurring, on the assumption that the wave is symmetrical.

Adaptive filters (Hamilton and Thompkins, 1988) have also been employed where the objective is to model the ECG signal as a superposition based on past signal values

$$X(n) = \sum_{k=1}^{P} a_k(n)X(n-k) \tag{4.9}$$

where a_k are time variant coefficient, which adapts according to input signal statistics. A midprediction filter (Dandapat and Ray, 1997) has also been used with the following:

$$X(n) = \sum_{k=-P}^{P} a_k(n)X(n-k) \tag{4.10}$$

The standard adaptation rules such as the LMS algorithm are used to obtain the coefficients. In addition, some authors (Kyrkos et al., 1987) have suggested using the differences between the coefficients to capture the statistical changes in the signal. This is modeled as

$$D(n) = \sum_{k=1}^{P} |a_k(n) - a_k(n-1)|^2 \tag{4.11}$$

and also as a combination of the difference between two short-time energies of the residual error in the consecutive segments

$$D_e(n) = \sum_{k=n}^{n+m} e^2(k) - \sum_{k=n}^{n-m} e^2(k) \tag{4.12}$$

Another type of filtering technique is match filtering where a matched filter is applied after digital filtering to improve the SNR:

$$y(n) = \sum_{k=1}^{N} h_k(n)X(n-k) \tag{4.13}$$

where $h_k(n)$ is the impulse response of the time-reversed template of the QRS waveform to be detected. The impulse response is obtained from the first few cardiac cycles but must be manually determined. Filter banks have been used to generate downsampled subband signals under the assumption that the QRS complex is characterized by simultaneous occurrence of ECG frequency components in the subbands.

4.6.1.3   Transform and Classification Algorithms

Signal transforms have also been used to detect peaks in the ECG. The wavelet transform is a popular method that can be used to capture time and frequency characteristics of the signal and is closely related to the filter bank method. Most of the wavelet methods applied to ECG peak detection are derived from the Mallat and Hwang (Mallat, 1999) algorithm for singularity detection and classification. Peak detection is accomplished by using local maxima of the wavelet coefficients and computing the singularity degree, that is, the local Lipschitz regularity, α, estimated from

$$\alpha_j = \ln|Wf(2^{j+1}, n^{j+1})| - \ln|Wf(2^j, n^j)| \qquad (4.14)$$

Heart rate variability (HRV) algorithms examine the intervals between heartbeats to detect potential pathologies and have, in the past, used Poincare plots for two-dimensional visualization of the classes. Brennan et al. (2001a) later showed that these Poincare plots extracted only linear information from the intervals and neglected nonlinear information, thus, motivating the use of nonlinear CI techniques. They also proposed a network oscillator interpretation for the information depicted by the Poincare plots (Brennan et al., 2002) as a transform method to extract HRV information. The authors later investigated signal-processing algorithms, specifically, the integral pulse frequency modulation (IPFM) model (Brennan et al., 2001b) to modulate the heart intervals into digital pulses and a spectrum count method to interpret the results.

ANNs have also been used to detect the QRS complex, the more common networks being the MLP, RBF networks, and the learning vector quantization (LVQ) networks. In QRS detection, NNs have been applied as nonlinear predictors (Hu et al., 1993; Xue et al., 1992), where the objective is to predict the value of the $X(n + 1)$ sample based on previous samples. The reversed logic is used where the network is trained to recognize the nonexistence of the QRS waveform well. Portions of the waveform that contain the QRS will cause large network output errors, $e(n)$, and can be used to infer the location of the QRS. Suzuki (1995) developed a self-organizing QRS-wave-recognition system for ECGs using an ART2 (adaptive resonance theory) network. This network was a self-organizing NN system consisting of a preprocessor, an ART2 network and a recognizer. The preprocessor was used to detect R points in the ECG and segment it into cardiac cycles. The input to the ART2 network was one cardiac cycle from which it was able to approximately locate one or both the Q and S points. The recognizer established search regions for the Q and S points where the QRS wave was thought to exist. The average recognition error of this system was reported to be less than 1 ms for Q and S points.

Genetic algorithms have also been used to design optimal polynomial filters for the processing of ECG waveforms (Poli et al., 1995). Three types of filters have been applied: quasilinear filters with consecutive samples, quasilinear

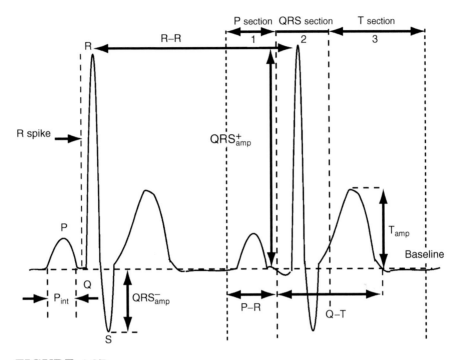

FIGURE 4.27
Extraction of physical ECG features for use with a weighted diagnostic distortion measure and beat segmentation. (Adapted from Zigel, Y., Cohen, A., and Keitz, A., *IEEE Trans. Biomed. Eng.*, 47(10), 1308–1316, 2000. ©IEEE.)

filters with selected samples, and quadratic filters with selected samples. The weakness of genetic algorithms is that it can become slow due to the many possible recombinations and exhaustive search for the best genes.

4.6.2 Morphological Features

Once the QRS complexes have been located, it is possible to extract further information from the physical features such as maximum amplitude, length of QT segment, and duration of RT segment. Figure 4.27 shows several physical features that have been employed for discriminating CVDs.

4.6.3 Statistical Features

Higher-order statistics (HOS) refers to cumulants with orders three and higher or computed quantities that are linear combinations of lower-order moments and lower-order cumulants. The moments and cumulants are calculated and then employed as features under the assumption that the ECG samples are random variables drawn from an unknown distribution, $P(x)$.

The moment about zero, $\mu_n(0)$ (raw moment), of a continuous distribution, $P(x)$, is defined mathematically as

$$\mu_n(0) = \int x^n P(x) \tag{4.15}$$

In the case of digital signals, we assume a discrete distribution of $P(x)$ and define the moment as

$$\mu_n(0) = \sum x^n P(x) \tag{4.16}$$

where summation is taken over all available discrete samples. Moments can be taken about a point a in which case Equation 4.16 is computed as

$$\mu_n(a) = \sum (x - a)^n P(x) \tag{4.17}$$

The first moment, μ_1, is the mean, the second moment, μ_2, the variance, and the third moment, μ_3, the skewness of the distribution, which measures the degree of asymmetry. The fourth-order moment is kurtosis, which measures the peakedness in the distribution. Central moments are moments taken about the mean of the distribution.

The cumulants, κ_n, of a distribution are derived from the Fourier transform of the probability density function, $P(x)$, with Fourier coefficients, $a = b = 1$. The higher-order cumulants are found to be linear combinations of the moments. Assuming raw moments, the first four cumulants can be written explicitly as

$$\kappa_1 = \mu_1$$
$$\kappa_2 = \mu_2 - \mu_1^2$$
$$\kappa_3 = 2\mu_1^3 - 3\mu_1\mu_2 + \mu_3$$
$$\kappa_4 = -6\mu_1^4 + 12\mu_1^2\mu_2 - 3\mu_2^2 - 4\mu_1\mu_3 + \mu_4 \tag{4.18}$$

HOS methods for feature extraction have been used on QRS complexes to reduce the relative spread of characteristics belonging to a similar type of heart rhythm (Osowski et al., 2004; Osowski and Tran, 2001). Tu et al. (2004) applied signal-processing techniques to detect the P wave when atrial fibrillation occurred using statistical features derived from a histogram of the signal and a genetic algorithm for selecting the best features.

4.6.4   Transform Features

A more robust feature-extraction method is the wavelet transform (al Fahoum and Howitt, 1999), which has been successfully employed to extract informa-tion for arrhythmia classification. Their features included six energy descrip-tors derived from the wavelet coefficients over a single-beat interval from the ECG signal and nine different continuous and discrete wavelet transforms were used as attributes of the feature vector. It was found that a Daubechies wavelet transform could provide an overall correct classification of 97.5% with 100% correct classification for both ventricular fibrillation and VT.

In addition to detecting CVDs, wavelet transforms can also be used alone to explore ECG waveform and its characteristics. Multiscale features of the wavelet transform have been extracted from the QRS complex (Li et al., 1995) and used to successfully distinguish between high P or T waves, noise, baseline drift, and motion artifacts. It was shown that the detection rate for QRS waveforms was above 99.8% for the Massachusetts Institute of Technology–Beth Israel Hospital (MIT–BIH) database and the P and T waves could be correctly detected even in the presence of serious baseline drift and noise. Sahambi et al. (1997) believed that the timing difference between various intervals of the QRS wave were more important than the subjective assessment of ECG morphology. In these cases, the wavelet transformation was employed to detect the onset and offsets of P and T waves in the QRS complex and the system was developed to continually apply the wavelet transform to ECG waveforms.

It is also desirable to compress the ECG waveform and only transmit the compressed information between ECG coders and decoders. In this aspect, compression was achieved by transmitting just the wavelet coefficients (Anant et al., 1995). Anant et al. (1995) proposed an improved wavelet compression algorithm for ECG signals by applying vector quantization to the wavelet coefficients. Vector quantization was applied for long ECG duration and low dynamic range that allowed the feature integrity of the ECG to be retained using a very low bit-per-sample rate. Other compression algorithms have used the discrete symmetric wavelet transform (Djohan et al., 1995) to achieve compression ratios of 8:1 with percent root mean square difference (PRD) of 3.9%, compared to earlier algorithms such as the AZTEC, which has a compression ratio of 6.8:1, with PRD = 10.0% and the fan algorithm compression ratio of 7.4:1, with PRD = 8.1%. Linear predictors with wavelet coefficients have also been used in ECG coding schemes to prevent maximal reconstruction error in the QRS region (Ramakrishnan and Saha, 1997). The authors normalize the periods of heartbeats by multirate processing and applying the DWT to each beat after amplitude normalization, thereby, increasing the usefulness of the wavelet correlations computed. The orthonormal wavelet transform has also been used for ECG compression where improved compression ratios of 13.5:1–22.9:1 have been reported using an optimum bit-allocation scheme (Chen and Itoh, 1998; Chen et al., 2001). More advanced applications such as mobile telecardiology models have used specialized wavelet-based ECG data compressions such as the optimal zonal wavelet-based ECG data compression (Istepanian and Petrosian, 2000). In this system, the optimal wavelet algorithm achieved a maximum compression ratio of 18:1 with low PRD ratios. Furthermore, the mobile telemedical simulation results demonstrated that ECG compression with bit error rates (BER) of less than 10–15 were possible and provided a 73% reduction in total mobile transmission time with clinically acceptable reconstruction of received signals. This particular approach is advantageous because it provides a framework for the design and functionality of GSM-based wireless telemedicine systems and integration for the next-generation mobile telecardiology systems.

4.6.5   Feature Selection Methods

It is usual to find that a subset of the feature set would produce better classification rather than using all the features. This is because inclusion of redundant features with poor discriminative power may confuse the classifier. Feature selection is, therefore, of utmost importance in developing a more accurate model for biomedical systems. There are several available methods by which to select an optimal set of features to represent the problem. Some of the applied ones are listed as follows:

a. *Principal component analysis (PCA)*. In large data sets with many features, it may be more efficient to find a smaller and more compact feature representation using a feature transformation. One method is to use PCA, which applies a projection to the features to find a reduced representation. Suppose that the training set has n training examples, that is, $\boldsymbol{X} = \{\boldsymbol{x}_1, \boldsymbol{x}_2, \ldots, \boldsymbol{x}_j\}$, then the algorithm constructs principal components, P_k, which are linear combinations of the original features $\boldsymbol{X}$. This can be written as

$$P_k = a_{k1}\boldsymbol{x}_1 + a_{k2}\boldsymbol{x}_2 + \cdots + a_n\boldsymbol{x}_{kn} \tag{4.19}$$

where $\sum_i a_{ki}^2 = 1$. The principal component vectors are constructed so that they are orthogonal to one another and hence have maximum variance (principal components). Generally, the training data is first normalized to zero mean and unit variance before application of the PCA algorithm. PCA has been used in many ECG applications, some of which include filtering of ECG beats (Kotas, 2004), analysis of features extracted from the QRS complex (Costa and Moraes, 2000), computer simulations of the cardiovascular system (Emdin et al., 1993), individual beat recognition (Biel et al., 2001), and diagnosis of CVDs from ECG (Stamkopoulos et al., 1998; Wenyu et al., 2003; Zhang and Zhang, 2005).

b. *Genetic and evolutionary methods*. This is an unsupervised method that uses an evolutionary approach to prune the feature set. A clustering method is used to determine separability of the features (Mitra et al., 2002) and iteratively retain only the most separable features. It can be used in conjunction with classifiers such as SVM (Xu and Yuan, 2003) to select the features that minimize generalization errors.

c. *Hill climbing*. Assuming p number of features, this method begins by selecting one feature and building a classifier based on it. The feature giving the highest accuracy is retained and the next from the remaining $p-1$ features is selected and combined with it. This is repeated until all features have been combined and the set that gives the highest accuracy is the optimal feature set. This technique and the following two methods have been used in applications such as gait analysis (see Chapter 7).

d. *Hill descent.* This is the reverse of the hill-climbing method, where all p features are used first and one feature is removed and the remainder used to train a classifier.

e. *Receiver operating characteristics area.* A simple thresholding method can be used to compute the receiver operating characteristics (ROC) area of single features. Area values closer to unity indicate higher-feature separability and are likely to contain more discriminative information.

4.7 Computational Intelligence for Diagnosis of Cardiovascular Diseases

One important use of CI techniques in cardiology is the diagnosis of heart disease from ECG recordings. A classifier is usually trained on preprocessed ECG signals to detect abnormalities in the ECG waveform and have been widely applied to detection of arrhythmias, ectopic beats, MIs, and myocardial ischemias. Most current research focuses on publicly available databases (Silipo and Marchesi, 1998; Kohler et al., 2002), which are briefly described as follows:

1. The MIT–BIH database provided by MIT and Boston's BIH is perhaps the most extensively used due to the large volume of data and ease of interpretation. This database contains a collection of 10 different databases, which can be used for various test purposes. These include the arrhythmia database, noise stress test database, ventricular tachyrhythmia database, ST change database, malignant ventricular arrhythmia database, atrial fibrillation/flutter database, ECG compression test database, supraventricular arrhythmia database, long-term database, and normal sinus rhythm database. The arrhythmias database is generally used for evaluation of ECG detection classifiers and contains 48 recordings of 30-min ECG waveforms. The analog waveforms were sampled at 360 Hz using a 11-bit resolution over a 10 mV range and contain a total of 116137 QRS complexes for detection. A key characteristic of this database is that the individual records are corrupted by noise such as abnormal waveshapes, noise, and artifacts (e.g., record 108 and 207) making classification challenging.

2. The chronic alterations database developed from a common standards for ECG (CSE) project by the University of Leuven, Belgium,

and supported by the European Union. It contains 1000 multilead (12–15-lead ECG) recordings (Bemmel and Williams, 1990).

3. The AHA database (American Heart Association) contains 155 records of ventricular arrhythmia detectors of the AHA consisting of signals digitized at a sampling rate of 250 Hz and a resolution of 12 bits over a range of 20 mV. The individual records contain 2.5 hours of unannotated signal and 30 min of annotated ECG signals with eight groups of different ectopic excitations.

4. The Ann Arbor Electrogram Libraries Jenkins (Jenkins) contains 800 intercardiac electrograms (EGMs) and surface ECGs, now expanded to four volumes of databases. The ECGs were recorded using unipolar and bipolar leads. Experiments on this database are usually performed for evaluation of implantable cardiac devices such as pacemakers.

5. Other databases include the European ST-T Database (National Research Council [NRC]), QT Database (Laguna et al., 1997), MGH database (Massachusetts General Hospital), and IMPROVE data library (Technical Institute of Finland).

The QRS complex is usually the ECG region of most interest with the R spike easiest to detect due to its large amplitude. Design of an automated ECG diagnosis system (Figure 4.28) involves collecting the raw ECG signals, performing some signal processing, that is, digitization and filtering, feature selection, and training the selected classifier. We summarize in the following sections some progress in the design of automated ECG recognition using NNs, SVMs, fuzzy logic, genetic algorithms, and hybrid systems, which are combinations of the aforementioned techniques.

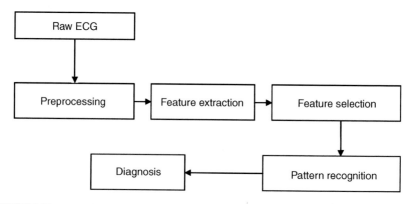

FIGURE 4.28
Diagram of ECG signal processing to detection using computational intelligence.

4.7.1 Artificial Neural Networks

NNs are probably the most widely applied technique for ECG diagnosis as evidenced by the extensive literature (exceeding 1000 references) on various architecture networks used for ECG waveform-recognition problems. Design considerations typically include the number of nodes in the hidden layer, how many hidden layers are required, as well as the effect of variations in the learning constant and momentum term. Differences in the reported findings may depend on the feature-extraction methods employed, and given their variations it is not possible to exhaustively deal with them here. Prior to feature extraction using signal-processing methods, biochemical markers from blood samples were used as the main features for detecting AMI using ANNs (Lisboa, 2002).

One of the earliest application of NNs was beat detection in ECG readings due to the simplicity of MLP implementation and its widespread success in many other pattern-recognition problems. Abnormal heartbeats or arrhythmias were first classified using the MLP trained with the popular back-propagation (BP) algorithm. Watrous and Towell (1995) used an MLP for discrimination between normal and ventricular beats. Marques et al. (1994) have compared several NN ECG classifiers for detecting normal rhythms, ventricular hypertrophies, and myocardial infarcts demonstrating good classification accuracies of above 80%. An MLP for classifying normal, ventricular, and fusion rhythms using PCA features has been reported by Nadal and Bosan (1993), but the system is more complex due to the requirement of computing the various principal components. A comparison between MLP, LVQ, and Kohonen's SOM was done by Zhu et al. (1991) for recognition of normal beats against ventricular premature beats (VPBs) demonstrating a better overall performance using the MLP network. A system of three ANNs trained to detect VEBs in the AHA database was proposed by Chow et al. (1992). Two ANNs were trained for each record using a training set consisting of five normal beats obtained from the record and seven VEBs obtained from a larger database. These two ANNs worked as local pattern recognizers for each record, whereas the third ANN was trained to arbitrate between the first two ANNs and provide an overall global decision. This system was used to replace the beat-classification logic of an existing VEB detector and was evaluated using 69 records from the AHA database. Gross VEB sensitivity improved from 94.83% to 97.39% with the ANN system in place, whereas gross VEB positive prediction improved from 92.70% to 93.58%. Oien et al. (1996) considered a three-class problem where the NSR and two different abnormal rhythms were distinguished. Detection of such beats are practically very important for treat or no-treat decisions encountered, for example, during operation of a semi-automatic defibrillation device. In their work, AR parameters and samples of the signal's periodogram were combined into feature vectors and fed into an MLP. Training and testing were performed using signals from two different ECG databases and the best net sizes and feature vector dimensions were decided upon by means of empirical tests. Other beat-recognition systems using NNs have also been developed (Yeap et al., 1990; Simon and Eswaran,

TABLE 4.2

Performance of Various Neural Networks–Based ECG Beat Recognition
Systems

Classifier Type	Number of Beat Types	Accuracy (%)
MLP1 (Hu et al., 1993)	2	90.0
MLP1 (Hu et al., 1993)	13	84.5
MLP2 (Izeboudjen and Farah, 1998)	12	90.0
MLP-Fourier (Minami et al., 1999)	3	98.0
SOM-LVD (Yu et al., 1997)	4	92.2
MLP-LVQ (Oien et al., 1996)	2	96.8
Hybrid Fuzz NN (Osowski and Tran, 2001)	7	96.06

1997), and a summary of the performance of some successful beat recognition
system is provided in Table 4.2.

The success of NNs attracted further research in applications such as detection of MIs where in several reported studies, the ST-T segment of the ECG
waveform yielded substantially discriminative information for the recognition
of MI and ischemic episodes based on large amounts of ECG data. Edenbrandt
et al. (1993a,b) studied two thousand ST-T segments from a 12-lead ECG and
visually classified them into seven groups. These constituted labeled data,
which were then divided into a training set and a test set where computer-measured ST-T data for each element in the training set with the corresponding label formed inputs for training various configurations of NNs. It was found
that the networks correctly classified 90–95% of the individual ST-T segments
in the test set. However, it should be noted that the success of NNs depends
largely on the size and composition of the training set making it difficult to
design a generalized system. Edenbrandt and coworkers also note that careful
incorporation of NNs with a conventional ECG interpretation program could
prove useful for an automated ECG diagnosis system if the aforementioned
factors could be dealt with consistently.

In more specific work, Maglaveras et al. (1998) examined the European
ST-T database using an NN-based algorithm to detect ST-segment elevations
or depressions, which could signify ischemic episodes. The performance of
their system was measured in terms of beat-by-beat ischemia detection and
the number of ischemic episodes. The algorithm used to train the NN was an
adaptive BP algorithm and reported to be 10 times faster than the classical BP
algorithm. The resulting NN was capable of detecting ischemia independent
of the lead used and gave an average ischemia episode detection sensitivity
of 88.62%, whereas the ischemia duration sensitivity was 72.22%. ST-T segment information was also used by Yang et al. (1994) to detect inferior MI. A
total of 592 clinically validated subjects, including 208 with inferior MI, 300
normal subjects, and 84 left ventricular hypertrophy cases, were used in their
study. From this, a total of 200 ECGs (100 from patients with inferior MI

and 100 from normal subjects) were used to train 66 supervised feedforward NNs with a BP algorithm. A comparison between the best-performing network using QRS measurements only and the best using QRS and ST-T data was made by assessing a test set of 292 ECGs (108 from patients with inferior MI, 84 from patients with left ventricular hypertrophy, and 100 from normal subjects). These two networks were then implanted separately into a deterministic Glasgow program to derive more uniform classification thresholds with the inclusion of a small inferior Q criterion to improve the specificity of detecting inferior MI. This produced a small loss of specificity when compared to the use of the NN alone. However, the 20% gain in sensitivity as opposed to a 2% loss in overall specificity was considered tolerable.

In a different study, Heden et al. (1994) used combinations of QRS and ST-T measurements as features for inputs into an ANN to detect MIs. A total of 1107 ECGs from patients who had undergone cardiac catheterization were used to train and test the NNs for the diagnosis of MI. During the learning process, the networks were able to correctly diagnose anterior or inferior wall MI after seeing two-thirds of the ECG data. The remaining data were then employed as a test set and the performance of the trained NNs was compared with conventional electrocardiographic criteria. The sensitivity for the diagnosis of anterior MI was 81% for the best network and 68% for the conventional criteria ($p < 0.01$), both having a specificity of 97.5%. The corresponding sensitivities of the network and the criteria for the diagnosis of inferior MI were 78% and 65.5% ($p < 0.01$), respectively, compared at a specificity of 95%.

Data from 12-lead ECG have also been combined with patient history and clinical findings to serve as features for classification. Heden et al. (1997) applied these features to early diagnosis of AMI to construct a decision support system for automated interpretation of ECG, which could be employed by less-experienced physicians. The objective was to investigate whether ANNs could be used as reliable classifiers to improve several aspects of conventional rule-based interpretation programs. A total of 1120 computerized ECGs were studied from patients with AMI, and 10,452 control ECGs recorded at an emergency department. Measurements from the 12 ST-T segments of each ECG lead together with the correct diagnosis were used as training data. Following training, performance of the NNs was compared to a widely used ECG interpretation program and to the classification of an experienced cardiologist. It was interesting that the NNs demonstrated higher sensitivities and discriminant power than both the interpretation program and the cardiologist. The sensitivity of the NNs was reportedly 15.5% (95% confidence interval, 12.4–18.6) higher than that of the interpretation program compared at a specificity of 95.4% ($P < 0.00001$) and 10.5% (95% CI, 7.2–13.6) higher than the cardiologist at a specificity of 86.3% ($P < 0.00001$). Other large-scale studies have also been performed on a group of patients using 12-lead ECG over 5 years involving 1120 confirmed cases and 10,452 controls (Fricker, 1997). A 20 s trace was generated from ST-T measurements from each of the leads providing inputs to 72 units of an MLP with a single hidden layer. The cross-validation procedure showed a 15.5% improvement in sensitivity over

rule-based criteria compared to the 10% detection rate by AMI expert cardiologists restricted to reading the ECG without information on patient history
or clinical findings. These results further reinforce the potential of using ANNs
to improve automated ECG interpretation for AMI and in enhancing decision
support systems.

The use of 12-lead ECG data gave rise to the concern that errors due to
electrode placement could severely affect the results. Heden et al. (1995),
therefore, undertook a series of investigations into the effect of electrode
misplacement during the ECG recording. It was hypothesized that this error
could cause incorrect interpretation, misdiagnosis, and subsequent lack of
proper treatment for the patient. The specific objective of their study was
to develop ANNs that yielded peak sensitivities for the recognition of right-
or left-arm lead reversal at a very high specificity. Comparisons were made
against two other widely used rule-based interpretation programs in a study
based on 11,009 ECGs recorded from emergency department patients. Each
of the ECGs was used to computationally generate an ECG with right- or
left-arm lead reversal and different NNs and rule-based criteria were used
depending on the presence or absence of P waves. The networks and the criteria all showed a very high specificity (99.87–100%). NNs outperformed the
rule-based criteria both when P waves were present (sensitivity 99.1%) or
absent (sensitivity 94.5%) compared to 93.9% and 39.3%, respectively, for the
rule-based criteria. This increase in performance translates to an additional
100,000–400,000 right- or left-arm lead reversals being detected by NNs for
the estimated 300 million ECGs recorded annually worldwide.

NNs have also been applied to the recognition of a wider range of CVDs
better known as multiclass problems. Bortolan and coworkers (Bortolan and
Willems, 1993; Bortolan et al., 1996), for example, performed a study on diagnostic classification of resting 12-lead ECGs using NNs. A large electrocardiographic library (the CORDA database established at the University of Leuven,
Belgium) was utilized where the diagnosis and classification had been validated by electrocardiographic-independent clinical data. From this database,
a subset of 3253 ECG signals with single diseases was selected and seven diagnostic classes were considered including normal, left, right, and biventricu-
lar hypertrophy, and anterior, inferior, and combined MI. Classification was
performed using a feedforward NN and the BP algorithm for the training
and comparisons made against linear discriminant analysis and logistic discriminant analysis methods. Several network combinations were trained by
varying the parameters of the architecture and considering subsets or clusters of the original learning set. The results demonstrated better performance
using the NN approach as compared to classical statistical analysis. Further
study (Bortolan et al., 1996) was conducted on two architectures of NNs in
detail concentrating on the normalization process, pruning techniques, and
fuzzy preprocessing by the use of RBFs. Work by Silipo and Marchesi (1998)
revolved around the detection of arrhythmia, chronic myocardial disease, and
ischemia by NNs (see Figure 4.29) using information in the ECG signal.
This was also extended to ambulatory ECG arrhythmic event classification

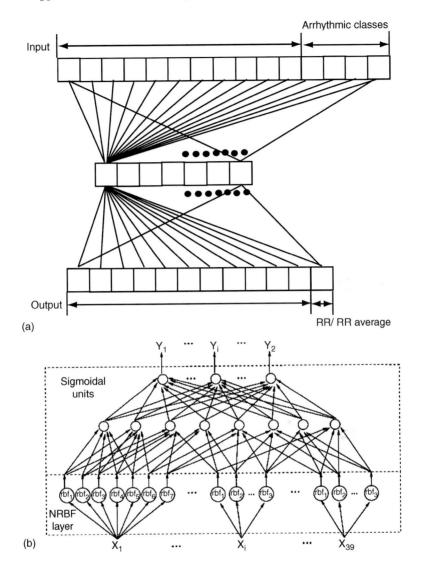

FIGURE 4.29
An autoassociator network and an RBF network used for arrhythmia detection. (From Silipo, R., and Marchesi, C., *IEEE Trans. Signal process.* 46(5), 1417–1425, 1998. ©IEEE.)

(Silipo et al., 1995) and compared against several traditional classifiers including the median method (heuristic algorithm), which was set as a quality reference. Another BP-based classifier was designed as an autoassociator for its peculiar capability of rejecting unknown patterns. The first test was aimed at discriminating normal beats from ventricular beats and the second was to distinguish among three classes of arrhythmic event. The ANN approach was

TABLE 4.3

Performance Comparison of Neural Networks Against Traditional Classifiers for Specific Beat Disorders

Beat Type	Neural Network (%)	Traditional Classifiers (%)
Normal beats	98	99
Ventricular ectopic beats	98	96
Supraventricular ectopic beats	96	59
Aberrated atrial premature beats	83	86

Source: From Silipo, R., Gori, M., Taddai, A., Varanini, M., and Marchesi C., *Comp. Biomed. Res.*, 28(4), 305–318, 1995.

more reliable than the traditional classifiers in discriminating several classes of arrhythmic event and a comparison of these results is shown in Table 4.3. The authors also examined the management of classifier uncertainty using two concurrent uncertainty criteria. The criteria allowed improved interpretation of classifier output and reduced the classification error of the unknown ventricular and supraventricular arrhythmic beats. It was found that the error in the ventricular beats case was maintained close to 0% on average and for supraventricular beats was approximately 35%. This procedure improved the usefulness of the ANN approach with the advantage of minimizing uncertainty and reducing global error.

In addition to heart diseases, NNs have been used in other ECG applications. Reaz and Lee (2004), for example, proposed an NN approach to separate the fetal ECG from the combined ECG of both the maternal and fetal ECGs using the adaptive linear network (ADALINE). The ADALINE approach was similar to adaptive filtering techniques where the input was the maternal signal, whereas the target signal was the composite signal. The NN emulates the maternal signal as closely as possible and then discerns the maternal ECG component from the composite abdominal ECG. The network error is computed as the difference between the composite abdominal ECG and the maternal ECG, which is the required fetal ECG. This technique works well due to high correlation between maternal ECG signals and abdominal ECG signals of pregnant women. In addition, a graphics user interface (GUI) program written in Matlab was written to detect the changes in extracted fetal ECG by adjusting the different values of momentum, learning rate, and initial weights used by the NN. It was found that filtering performs best using a high learning rate, low momentum, and small initial weights.

Genetic algorithms have been applied as evolutionary methods to iteratively select the best ECG features and the best classifiers by mimicking the natural selection of biological systems. Olmez et al. (1997) have applied a genetic algorithm to train a restricted coulomb energy network (GARCE), fundamentally an MLP network, except that the second layer is selected using a

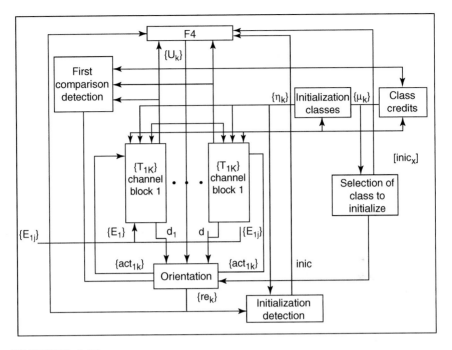

FIGURE 4.30

A multichannel adaptive resonance theory network (MART). (Proposed by Barro, S., Fernandez-Delgado, Vila-Sobrino, J., Regueiro, C., and Sanchez, E., *IEEE Eng. Med. Biol. Mag.*, 17, 45–55, 1998. ©IEEE.)

genetic algorithm a variable node layer during training. When Fourier transform coefficients of the QRS complex in the R-wave region were used as input features, comparisons with an MLP network showed improved training efficiency using GARCE. An adaptive NN as seen in Figure 4.30 was used by Barro et al. (1998) to deal with a larger range of patients. Their network was based on the ART model (Carpenter and Grossberg, 1987; Carpenter et al., 1991) with segment inputs of the ECG signal. Template matching was used to detect new QRS waveforms and a difference function was employed to calculate the global difference and classify the waveforms.

4.7.2 Support Vector Machines

SVMs are a recent classifier formulation (early 1990s) and are only beginning to find use in ECG classification. The advantage of using SVMs is that good performance can still be obtained using smaller datasets. A general SVM recognition system is similar to that of an NN system with the difference that the NN classifier is replaced by an SVM classifier as seen in Figure 4.31.

Rojo-Alvarez et al. (2002a,b) initially hypothesized that ventricular EGM onset could be used to discriminate supraventricular tachycardias and VTs.

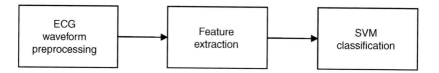

FIGURE 4.31

Classification of ECG waveforms using an SVM-based system.

This method of obtaining features was shown to exceed features derived from heart rate, QRS width, or coefficients of the correlation waveform analysis. The SVM classifier was first tuned using a bootstrap resampling procedure to select proper SVM parameters and subsequently managed to achieve 96% specificity and 87% sensitivity, but no comparisons were made against NNs. Incremental learning was then used to establish a trade-off between population data and patient-specific data. The authors argued that many machine-learning algorithms build a black box statistical model for the differential diagnosis without knowledge of the underlying problem. The SVM, however, can be used to extract information from this black box model via support vectors representing the critical samples for the classification task. Rojo-Alvarez et al. (2002b) proposed a SVM-oriented analyses for the design of two new differential diagnosis algorithms based on the ventricular EGM onset criterion.

Millet-Roig et al. (2002) applied SVMs on wavelet features extracted from a mixture of MIT–BIH, AHA, and data from the Hospital Clinico de Valencia in Spain. A holdout method was used where 70% of the dataset was applied to training the SVM, whereas 30% was reserved for testing. The SVM classifier was compared against an NN and logistic regression that statistically fitted an equation based on the probability of a sample belonging to a class Y. It was found that the SVM performance was comparable to the NN and outperformed it in one case, whereas it was significantly superior than the logistic regression method.

SVMs have also been used in combination with genetic algorithms for classification of unstable angina. Sepulveda-Sanchis et al. (2002) employed genetic algorithms to select proper features from ECG waveforms collected by four hospitals from 729 patients. These patients fulfilled several preselection criteria such as having no history of MI but suffering from unstable angina. It was found that the Gaussian kernel worked best for their data achieving a successful prediction rate of 79.12% with a low false-positive rate.

SVMs have also been used in the design of rate-based arrhythmia-recognition algorithms in implantable cardioverter-defibrillators. In the past, these algorithms were of limited reliability in clinical situations and achieved improved performance only with the inclusion of the morphological features of endocardial ECG. Strauss et al. (2001) applied a coupled signal-adapted wavelet together with an SVM arrhythmia detection scheme. During electrophysiological examination data, segments were first recorded during normal sinus rhythm (NSR) and ventricular tachycardia (VT), following which consecutive beats were selected as morphological activation patterns of NSR and VT. These patterns were represented by their multilevel decompositions derived from a signal-adapted and highly efficient lattice structure-based wavelet decomposition technique. This proposed wavelet technique was employed to maximize the class separability and takes into account the final classification of NSR and VT by SVMs with radial and compactly supported kernels. In an automated analysis of an independent test set, the hybrid scheme classified all test patterns correctly and was shown to be more robust than the individual techniques.

4.7.3 Hidden Markov Models

HMMs have also been used to model the ECG signal as a series of discrete samples drawn from a random process with unseen observations. This technique is basically a nondeterministic probabilistic finite state machine, which can be constructed inductively as new information becomes available. The HMM technique is widely used in the area of automatic speech recognition and deoxyribonucleic acid (DNA) modeling and, thus, applied to model a part or all of the ECG waveform.

It is known that ventricular arrhythmias can be classified accurately using just the QRS complex and R-R intervals (Feldman, 1983), but if the arrhythmia is supraventricular, the P wave of the ECG has proven to be a more useful feature since it corresponds to ventricular diastole or relaxation prior to the QRS wave. Since the P wave is smaller than the QRS and is sometimes attenuated, it is seldom extracted and used as a feature. Coast et al. (1990) demonstrated that one way to incorporate P-wave information is to model the entire ECG signal as a single parametric model using HMMs. In doing so, they reported at least 97% classification accuracies on data from the AHA database.

To model the entire ECG waveform using HMMs, one relies on a stationarity assumption of the ECG waveform. Thoraval et al. (1992, 1994) argued that this approach had potential to generate severe errors in practice if the model was inaccurate. They instead proposed a new class of HMMs called modified continuous variable duration HMMs coupled with a multiresolution front-end analysis of the ECG to account for the specific properties of the ECG signal and reported increased performances compared to classical HMMs. Further work using continuous probability density function HMMs (Koski, 1996) for the ECG signal analysis problem have also yielded improved results on segmented ECG signals, mainly because the assumption of stationarity holds more often there.

Andreao et al. (2006) presented analysis of an original HMM approach for online beat segmentation and classification of ECGs. The HMM framework was applied because of its ability to model beat detection and perform segmentation and classification of ECG signals. Their work extends to a large range of issues such as waveform modeling, multichannel beat segmentation and classification, and unsupervised adaptation to the patient's ECG. The performance of their HMM system was evaluated on the two-channel QT database in terms of waveform segmentation precision, beat detection, and classification. Their segmentation results compared favorably to other systems but increased beat detection with sensitivity of 99.79% and positive prediction of 99.96% was achieved using a test set of 59 recordings. Premature ventricular contraction (PVC) beats were also detected by using an original classification strategy.

4.7.4   Fuzzy Methods

Fuzzy methods are frequently applied to the design of ECG-recognition systems based on rules. Features are treated as linguistic variables and given membership values that allow a coarser classification or decision rule. A general fuzzy system is shown in Figure 4.32 where the features are assumed to have been preselected. Fuzzy systems differ in the fuzzy features, fuzzy reasoning, and fuzzy rules used to obtain the diagnosis.

Xie et al. (1997) applied a fuzzy neural network (FNN) model because their ECG features possessed numerical ranges that overlapped for both normal and disease types. The features in their experiments consisted of physical waveform measurements such as ST horizontal depression, ST pseudoischemic depression, and T-wave depression. The model operated on seven distinct fuzzy rules used in combination with a standard NN output to obtain the classifier decision. Tests were conducted on a total of 42 ECG cases, of which 14 cases were classified as normal, whereas the remaining cases were ischemia instances. Their FNN model had a better detection rate for normal as compared to

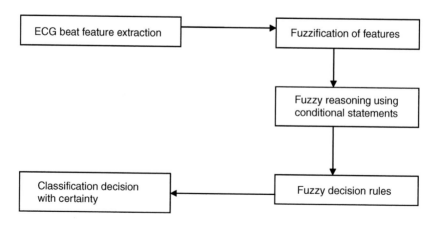

FIGURE 4.32

General fuzzy system for detection and diagnosis of cardiovascular diseases using ECG.

the diseased, which is generally not desirable. Later, Zong and Jiang (1998) showed that a better selection of fuzzy linguistic variables and decision rules could be used to classify five different ECG beat types over 20 arrhythmias obtained from the popular MIT–BIH database. The fuzzy variables included premature beats, QRS widths, and beat times with the classification decision achieved by selecting a maximum likelihood method with a threshold level of 0.5. Good prediction accuracies of up to 99.77% were achieved for beat types such as supraventricular premature beats but results for VEBs (63.01%) were low.

Silipo et al. (1999) investigated the relevance of ECG features in their design of an arrhythmia classifier and argued that if the number of features is high, the interpretability or significance of the corresponding fuzzy rules may be overshadowed by the number of rules required. To overcome this, the feature set should be studied *a posteriori* to select the best feature set containing the most useful features. They then applied the information gain measure for a fuzzy model on 14 different physical ECG features obtained from ECG samples in the MIT–BIH database. It was discovered that the QRS width was the most useful in the discrimination of a three-class arrhythmia problem as compared to ST-segment levels or the negative of the QRS area.

Ifeachor et al. (2001) described fuzzy methods for handling imprecision and uncertainty in computer-based analysis of fetal heart rate patterns and ECG waveshapes during childbirth. CI models, based on fuzzy logic techniques, that explicitly handle the imprecision and uncertainty inherent in the data obtained during childbirth were proposed. The ability to handle imprecision and uncertainty in clinical data is critical to removing a key obstacle in electronic fetal monitoring.

The use of fuzzy logic has also been reported in the design of a knowledge-based approach to the classification of arrhythmias. Kundu et al. (1998) have put forward a knowledge-based system to recognize arrhythmia based on the QRS complex, P waveshape, and several other physical features. Several fuzzy rules based on fuzzification of these features were implemented, for example,

> If average heart rate of the ECG cycle is *high*, then diagnosis is *sinus tachycardia*.

It was argued that the fuzzy classification increased the granularity of the diagnosis as opposed to the hard classification methods. In addition, it was argued that this method could deal with incomplete knowledge of the input data, however, experiments on 18 datasets demonstrated only limited improvements in diagnostic accuracy. Beat recognition was also pursued by Osowski and Tran (2001), who applied an FNN model on statistical ECG features shown in Figure 4.33. In their design, the second- to fourth-order cumulants were extracted from the ECG signal because these features were less sensitive of minor variations in waveform morphology. These cumulants were calculated for the QRS segment of the ECG signal. They reported a decreased error rate of 3.94% over the classification of seven ECG beat types.

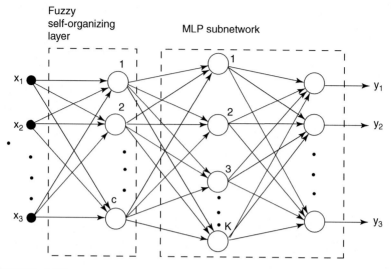

FIGURE 4.33

The neurofuzzy system for beat recognition. (Proposed by Osowski, S. and Tran, H., *IEEE Trans. Biomed. Eng.*, 42(11), 1137–1141. ©IEEE.)

4.7.5   Expert Systems: A Combined Hybrid System

In previous classification systems, a single classifier was shown to work exceptionally well on the set of patients but was inconsistent for other patients due to large variations in ECG morphology. Even if the dataset size was increased it would still be difficult to cover every patient, hence, a combined adaptable system or expert system was required to incrementally adjust to new patients. Expert systems have primarily been designed as beat-recognition systems for diagnosis of cardiovascular systems employing several different feature extraction methods and a variety of classifier types. Gardner and Lundsgaarde (1994) had earlier conducted a survey to measure the acceptance of physicians and nurses of an expert system called the health evaluation through logical processing (HELP) clinical information system, which was rule-based and contained a large medical database. The questionnaire survey was completed by 360 attending physicians and 960 staff nurses practicing at the hospital, and consisted of fixed-choice questions with a Likert-type scale supplemented by free-text comments. The questions covered topics such as computer experience, general attitudes about impact of the system on practice, ranking of available functions, and desired future capabilities. The participant comments supported future development and deployment of medical expert systems in hospitals, quite contrary to the fear that physicians would refuse to accept recommendations of a medical expert system.

Yu et al. (1997) developed a patient-adaptable automatic ECG beat classifier using a mixture of experts (MOE). Their system uses statistical histogram

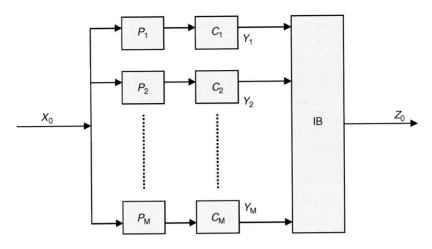

FIGURE 4.34

General expert-based system using voting for classification of ECG beats. P_i blocks represent different processing and feature-extraction methods, whereas C_i blocks represent the classifiers and their respective outputs y_i. The integrating block, IB combines the decisions of the classifiers for a single expert system decision.

features in which the final decision was obtained as a linear estimate of the various experts. The system incorporated a global expert trained on a large patient database and several local experts specific to particular patients. This allowed incremental data to be added to the overall system by the addition of local experts. A gating function was then employed to scale the outputs of the global and local experts to derive a single decision value.

A similar system using SVMs as the base classifiers was developed by Osowski et al. (2004), who extracted higher-order statistical features and Hermite basis function expansions from the ECG waveforms in the MIT–BIH database. This system was applied to classify 13 beat types using single SVM classifiers combined through a voting method as seen in Figure 4.34. The integration of the individual classifier decisions was achieved by training the entire system and using a weighted matrix to minimize the training error. This matrix was adjusted by least squares computed from the pseudoinverse and singular value decomposition techniques. The output was a vector with elements representing the membership of the training or test example to a particular class. The position of the highest value was used to assign the class to a test pattern. The average error found over the 13 beat classes was 6.28% solely using HOS feature inputs and 5.43% using Hermite coefficients as features. The combination of these features, however, produced poorer results. The authors concluded that an expert system combining classifiers with average performance may provide better overall recognition.

Ham and Han (1996) investigated the use of fuzzy ART mapping (ARTMAP) to classify cardiac arrhythmias based on the QRS complex. The system was tested on the data from the MIT–BIH database where cardiac beats for normal and abnormal QRS complexes were extracted, scaled, and Hamming windowed. After bandpass filtering, a sequence of 100 samples for each QRS segment was obtained and two conditions analyzed; normal and abnormal PVC. From each of these sequences, two linear predictive coding (LPC) coefficients were generated using Burg's maximum entropy method. These two LPC coefficients were utilized with the mean-square value of the QRS complex segment as features for each condition to train and test a fuzzy ARTMAP NN. The test results indicated that the fuzzy ARTMAP NN could classify cardiac arrhythmias with greater than 99% specificity and 97% sensitivity.

Expert systems have also been used to achieve knowledge integration such as in the case of coronary-care monitoring. The clinical impact of these expert systems has not been remarkable due to the difficulties associated with the design and maintenance of a complete knowledge base. Model-based systems, however, represent an alternative to these problems because they allow efficient integration of deeper knowledge regarding the underlying physiological phenomena being monitored. These systems are specifically designed for cardiac rhythm interpretation. The cardiac arrhythmia recognition by model-based ECG matching (CARMEM) has the ability to provide online parameter adaptation to simulate complex rhythms and to match observed ECG signals (Hernndez et al., 2000). It is hoped that this model could be useful for explanation of the origin of cardiac arrhythmias and provide a contribution toward their robust characterization of coronary-care units.

4.8   Concluding Remarks

This chapter examined applications of CI techniques for the diagnoses of CVDs by processing the ECG signal. This field has matured to the extent where most of the algorithms available have been tried and tested on ECG recordings. The trend now is toward producing smaller and more portable ECG devices incorporating PDA technology and other technologies such as wireless communication. Motivation in commercialization of ECG devices stems from several evolving factors, which include improvements in the health care system and response times, increase in awareness of one's personal well-being, and the need for personal health-monitoring devices. From this perspective, there is an emerging requirement for automated heart-monitoring systems, which require better intelligence to be incorporated as opposed to the era where heart monitoring was reserved for cardiologists. Nevertheless, the fundamental challenges in the design of automated systems remain, the most prominent being the need to develop more accurate CI techniques to aid in correct diagnosis. It is this key requirement that we believe will continue to motivate further research in this field using CI techniques.

References

Ahlstrom, M. and W. Thompkins (1983). Automated high-speed analysis of holter tapes in microcomputers. *IEEE Transactions on Biomedical Engineering 30*, 651–657.

al Fahoum, A. and I. Howitt (1999). Combined wavelet transformation and radial basis neural networks for classifying life-threatening cardiac arrhythmias. *Medical and Biological Engineering and Computing 37(5)*, 566–573.

American Heart Association. AHA Database. ECRI, PA.

Anant, K., F. Dowla, and G. Rodrigue (1995). Vector quantization of ECG wavelet coefficients. *IEEE Signal Processing Letters 2(7)*, 129–131.

Andreao, R., B. Dorizzi, and J. Boudy (2006). ECG signal analysis through hidden Markov models. *IEEE Transactions on Biomedical Engineering 53(8)*, 1541–1549.

Balda, R. (1977). The HP ECG analysis program. In *Trends in Computer Processed Electrocardiograms,* J. H. van Bemnel and J. L. Williams (Eds.), pp. 197–205, Amsterdam: North Holland.

Barro, S., M. Fernandez-Delgado, J. Vila-Sobrino, C. Regueiro, and E. Sanchez (1998). Classifying multichannel ECG patterns with an adaptive neural network. *IEEE Engineering in Medicine and Biology Magazine 17*, 45–55.

Bemmel, J. and J. Williams (1990). Standardization and validation of medical decision support systems: the CSE project. *Methods of Information in Medicine 29*, 261–262.

Biel, L., O. Pettersson, L. Philipson, and P. Wlde (2001). ECG analysis: a new approach in human identification. *IEEE Transactions on Instrumentation and Measurement 50(3)*, 808–812.

Bortolan, G., C. Brohet, and S. Fusaro (1996). Possibilities of using neural networks for ECG classification. *Journal of Electrocardiolgy 29*, 10–16.

Bortolan, G. and J. Willems (1993). Diagnostic ECG classification based on neural networks. *Journal of Electrocardiolgy 26*, 75–79.

Brennan, M., M. Palaniswami, and P. Kamen (2001a). Do existing measures of Poincare plot geometry reflect nonlinear features of heart rate variability? *IEEE Transactions on Biomedical Engineering 48(11)*, 1342–1347.

Brennan, M., M. Palaniswami, and P. Kamen (2001b). Distortion properties of the interval spectrum of IPFM generated heartbeats for heart rate variability analysis. *IEEE Transactions on Biomedical Engineering 48(11)*, 1251–1264.

Brennan, M., M. Palaniswami, and P. Kamen (2002). Poincare plot interpretation using a physiological model of HRV based on a network of oscillators.

American Journal of Physiology: Heart and Circulatory Physiology 283(5), 873–886.

Carpenter, G. and S. Grossberg (1987). ART2: self-organizing of stable category recognition codes for analog input patterns. *Applied Optics 26(23)*, 4919–4930.

Carpenter, G., S. Grossberg, and D. Rosen (1991). ART 2-A: an adaptive resonance algorithm for rapid category learning and recognition. *International Joint Conference on Neural Networks 1*, 151–156.

Chen, J. and S. Itoh (1998). A wavelet transform-based ECG compression method guaranteeing desired signal quality. *IEEE Transactions on Biomedical Engineering 45(12)*, 1414–1419.

Chen, J., S. Itoh, and T. Hashimoto (2001). ECG data compression by using wavelet transform. *IEICE Transactions on Information and Systems E76-D(12)*, 1454–1461.

Chow, H., G. Moody, and R. Mark (1992). Detection of ventricular ectopic beats using neural networks. *Computers in Cardiology*, 659–662.

Coast, D., R. Stern, G. Cano, and S. Briller (1990). An approach to cardiac arrhythmia analysis using hidden Markov models. *IEEE Transactions on Biomedical Engineering 37(9)*, 826–836.

Costa, E. and J. Moraes (2000). QRS feature discrimination capability: quantitative and qualitative analysis. *Computers in Cardiology*, 399–402.

Dandapat, S. and G. Ray (1997). Spike detection in biomedical signals using midprediction filter. *Medical and Biological Engineering and Computing 35(4)*, 354–360.

Department of Health and Welfare (2002). Australias health 2002. Canberra: AIHW. Available at http://www.heartfoundation.com.au/

Djohan, A., T. Nguyen, and W. Tompkins (1995). ECG compression using discrete symmetric wavelet transform. *IEEE 17th Annual Conference Engineering in Medicine and Biology Society 1*, 167–168.

Dokur, Z., T. Olmez, E. Yazgan, and O. Ersoy (1997). Detection of ECG waveforms by using neural networks. *Medical Engineering and Physics 198*, 738–741.

Edenbrandt, L., B. Devine, and P. Macfarlane (1993a). Neural networks for classification of ECG ST-T segments. *Journal of Electrocardiology 26(3)*, 239–240.

Edenbrandt, L., B. Heden, and O. Pahlm (1993b). Neural networks for analysis of ECG complexes. *Journal of Electrocardiology 26*, 74.

Emdin, M., A. Taddei, M. Varanini, J. Marin Neto, C. Carpeggiani, and L'Abbate (1993). Compact representation of autonomic stimulation on cardiorespiratory signals by principal component analysis. *Proceedings of Computers in Cardiology 1*, 157–160.

Fancott, T. and D. Wong (1980). A minicomputer system for direct high speed analysis of cardiac arrhythmia in 24 h ambulatory ECG tape recordings. *IEEE Transactions on Biomedical Engineering 27*, 685–693.

Feldman, C. (1983). Computer detection of cardiac arrhythmias: historical review. *American Review of Diagnostics 2(5)*, 138–145.

Fricker, J. (1997). Artificial neural networks improve diagnosis of acute myocardial infarction. *Lancet 350*, 935.

Friesen, G., T. Jannett, M. Jadallah, S. Yates, S. L. Quint, and H. Nagle (1990). A comparison of the noise sensitivity of nine QRS detection algorithms. *IEEE Transactions on Biomedical Engineering 37(1)*, 85–98.

Gardner, R. and H. Lundsgaarde (1994). Evaluation of user acceptance of a clinical expert system. *Journal of the American Medical Informatics Association 1*, 428–438.

Ham, F. and S. Han (1996). Classification of cardiac arrhythmias using fuzzy ARTMAP. *IEEE Transactions on Biomedical Engineering 43(4)*, 425–429.

Hamilton, P. and W. Thompkins (1986). Quantitative investigation of QRS detection rules using MIT/BIH arrhythmia database. *IEEE Transactions on Biomedical Engineering 33*, 1157–1165.

Hamilton, P. and W. Thompkins (1988). Adaptive matched filtering for QRS detection. *Proceedings of the 10th Annual International Conference of IEEE Engineering in Medicine and Biology Society Conference 1*, 147–148.

Heden, B., L. Edenbrandt, W. Haisty, and O. Pahlm (1994). Artificial neural networks for the electrocardiographic diagnosis of healed myocardial infarction. *American Journal of Cardiology 75(1)*, 5–8.

Heden, B., M. Ohlsson, L. Edenbrandt, R. Rittner, O. Pahlm, and C. Peterson (1995). Artificial neural networks for recognition of electrocardiographic lead reversal. *American Journal of Cardiology 75(14)*, 929–933.

Heden, B., H. hlin, R. Rittner, and L. Edenbrandt (1997). Acute myocardial infarction detected in the 12-lead ECG by artificial neural networks. *Circulation 96*, 1798–1802.

Hernández, A., G. Carrault, F. Mora, and A. Bardou (2000). Overview of carmem: a new dynamic quantitative cardiac model for ECG monitoring and its adaptation to observed signals. *Acta Biotheoretica 48(3–4)*, 303–322.

Hu, Y., W. Tompkins, J. Urrusti, and V. Afonso (1993). Applications of artificial neural networks for ECG signal detection and classification. *Journal of Electrocardiology 26*, 66–73.

Ifeachor, E., J. Curnow, N. Outram, and J. Skinner (2001). Models for handling uncertainty in fetal heart rate and ECG analysis. *Proceedings of the 23rd Annual International Conference of the IEEE Engineering in Medicine and Biology Society 2*, 1661–1667.

Istepanian, R. and A. Petrosian (2000). Optimal zonal wavelet-based ECG data compression for a mobiletelecardiology system. *IEEE Engineering in Medicine and Biology Magazine 4(3)*, 200–211.

Izeboudjen, N. and A. Farah (1998). A new neural network system for arrhythmia's classification. *Proceedings of the Neural Network Conference 1*, 208–216.

Jenkins, J. Ann Arbor Electrogram Libraries. Available at http:// electrogram.com.

Julian, D. G., J. Campbell-Cowan, and M. J. McLenachan (2005). *Cardiology* (8th ed.) Edinburgh, NY: Saunders.

Kohler, B., C. Hennig, and R. Orglmeister (2002). The principles of software QRS detection. *IEEE Engineering in Medicine and Biology Magazine 21(1)*, 42–57.

Koski, A. (1996). Modelling ECG signals with hidden Markov models. *Artificial Intelligence in Medicine 8(5)*, 453–471.

Kotas, M. (2004). Projective filtering of time-aligned ECG beats. *IEEE Transactions on Biomedical Engineering 51(7)*, 1129–1139.

Kundu, M., M. Nasipuri, and D. Basu (1998). A knowledge-based approach to ECG interpretation using fuzzy logic. *IEEE Engineering in Medicine and Biology Magazine 28(2)*, 237–243.

Kyrkos, A., E. Giakoumakis, and G. Carayannis (1987). Time recursion predictive techniques on QRS detection problems. *Proceedings of the 9th Annual International Conference of IEEE Engineering in Medicine and Biology Society 1*, 1885–1886.

Laguna, P., R. Mark, A. Goldberger, and G. Moody (1997). A database for evaluation of algorithms for measurement of QT and other waveform intervals in the ECG. *Computational Cardiology 24*, 673–676.

Li, C., C. Zheng, and C. Tai (1995). Detection of ECG characteristic points using wavelet transforms. *IEEE Transactions on Biomedical Engineering 42(1)*, 21–28.

Lisboa, P. (2002). A review of evidence of health benefit from artificial neural networks in medical intervention. *Neural Networks 15*, 11–39.

Mackay, J. and G. A. Mensah (2005). *The Atlas of Heart Disease and Stroke*, World Health Organization (WHO), www.who.int/cardiovasculardiseases/resources/atlas/en.

Maglaveras, N., T. Stamkopoulos, C. Pappas, and M. Gerassimos Strintzis (1998). An adaptive backpropagation neural network for real-time ischemia episodes detection: development and performance analysis using the european ST-T database. *IEEE Transactions on Biomedical Engineering 45(7)*, 805–813.

Mallat, S. (1999). *A Wavelet Tour of Signal Processing* (2nd ed.) New York: Academic Press.

Marques, J., A. Goncalves, F. Ferreira, and C. Abreu-Lima (1994). Comparison of artificial neural network based ECG classifiers using different feature types. *Computers in Cardiology 1*, 545–547.

Martini, F. (2006). *Fundamentals of anatomy and physiology* (7th ed.) San Francisco, CA: Pearson Benjamin Cummings.

Massachusetts General Hospital. Massachusetts General Hospital/Marquette Foundation Waveform Database. Dr. J. Cooper. MGH Anesthesia Bioengineering Unit, Boston, MA.

Massachusetts Institute of Technology. MIT–BIH ECG Database. Available at http://ECG.mit.edu.

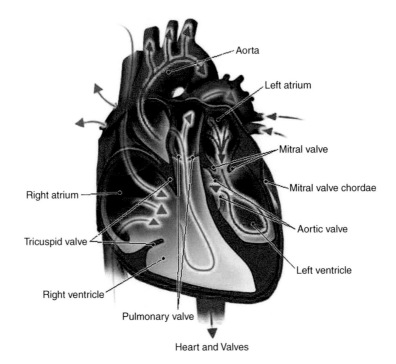

Aorta

Left atrium

Mitral valve

Mitral valve chordae

Right atrium

Aortic valve

Tricuspid valve

Left ventricle

Right ventricle

Pulmonary valve

Heart and Valves

COLOR FIGURE 4.6
The major arteries and veins in the heart. (Courtesy of MedicineNet, www.medicinenet.com.)

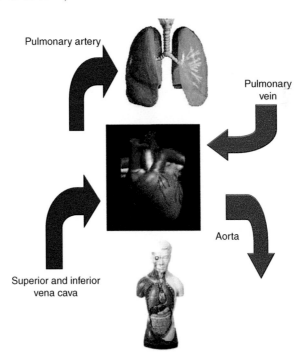

Pulmonary artery

Pulmonary vein

Aorta

Superior and inferior vena cava

COLOR FIGURE 4.7
The blood circulation system depicting the flow of blood from heart to the lungs and the rest of the body. The red arrows show the flow of oxygenated blood, whereas the blue arrows depict the flow of deoxygenated blood.

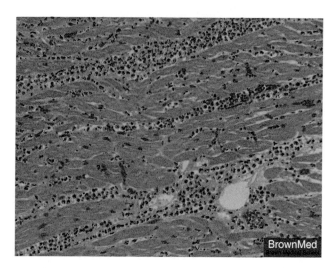

COLOR FIGURE 4.22
View of myocardial fibers after approximately 48 hours post myocardial infarction. Fibers depict maximum acute inflammation and loss of nuclei striations. Irreversible cell death occurs after 15–20 minutes of onset. (Courtesy of A/Prof. Calvin E. Oyer, M.D., Brown Medical School.)

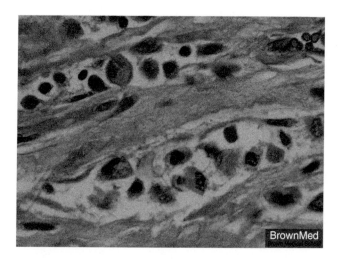

COLOR FIGURE 4.23
View of dead myocardial fibers after 7–9 days of post myocardial infarction. Cell macrophages, fibroblasts, and debris from dead myocytes can still be seen. (Courtesy of A/Prof. Calvin E. Oyer, M.D., Brown Medical School.)

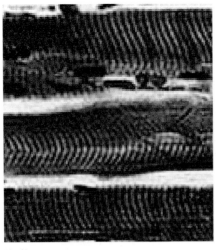
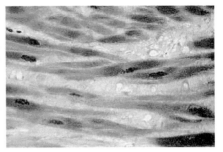

(Left)   (Right)

COLOR FIGURE 5.1

Left: Micrograph of striated muscle showing muscle fibers in skeletal muscle. The striated appearance comes from the hundreds of thousands of sarcomeres in a layered formation. *Right*: Micrograph of smooth muscle fiber, which do not have the striated characteristics of skeletal muscle fibers. (Courtesy of http://people.eku.edu/ritchisong/301notes3.htm.)

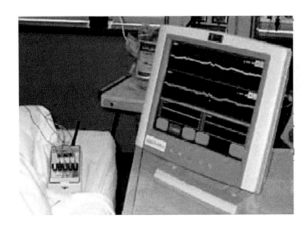

COLOR FIGURE 6.2

An EEG-monitoring system setup showing wireless EEG electrode connections to the patient's bedside and a display monitor showing EEG signals. (From Thakor, N. V., Hyun-Chool, S., Shanbao, T., et al., *IEEE Eng. Med. Biol. Mag.*, 25, 20–25, 2006. ©IEEE.)

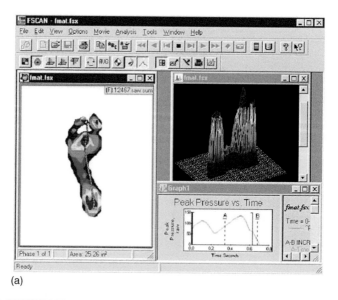

(a)

COLOR FIGURE 7.3

(a) Examples of foot pressure distribution using the F-scan system (www.tekscan.com).

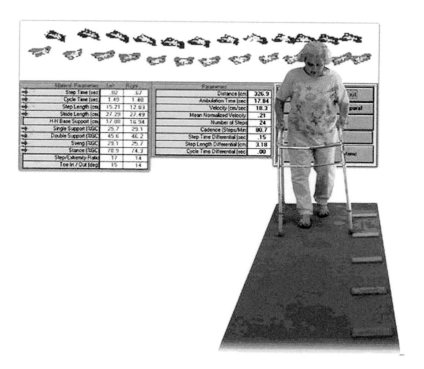

COLOR FIGURE 7.4

GAITRite walking analysis system showing portable 14 in. carpet with embedded sensors for recording temporospatial parameters and an output from GAITRite system (electronic footprint and time/distance data) (http://www.gaitrite.com/).

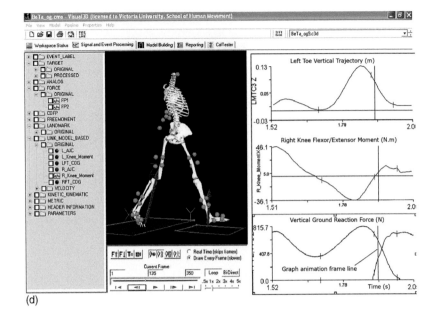

COLOR FIGURE 7.5

(d) Typical output from the Optotrak system and C-Motion Visual3D software package (www.c-motion.com. showing skeleton model, plots of knee joint angle, joint moment and vertical ground reaction forces from VU Biomechanics Lab).

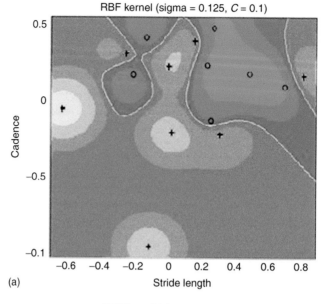

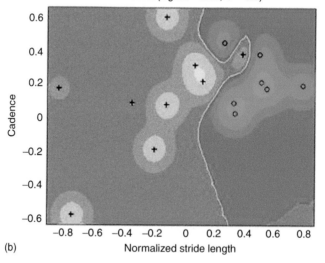

COLOR FIGURE 7.16

Scatter plots showing the test data and surfaces the hyperplane using raw and normalized gait data by the SVM classifier with RBF kernel. (a) With raw stride length and cadence and (b) with normalized stride length and cadence. (From Kamruzzaman, J. and Begg. R. K., *IEEE Trans. Biomed. Eng.*, 53, 2479–2490, 2006. ©IEEE.)

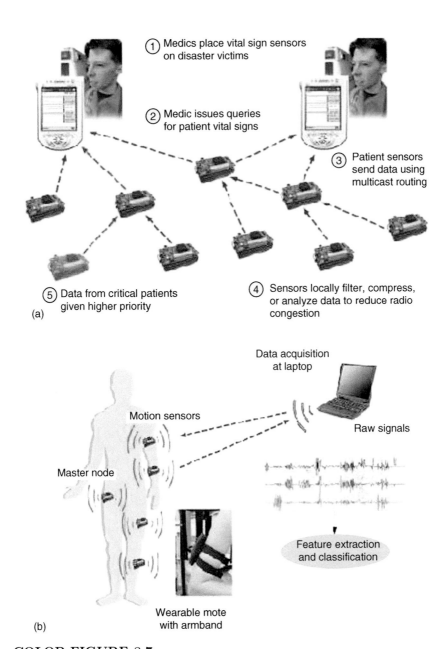

COLOR FIGURE 8.7

(a) The Codeblue architecture consisting of Telos or Pluto motes used for emergency response. (b) Limb motion monitoring for rehabilitation in stroke patient. (Photo courtesy of Codeblue homepage; http://www.eecs. harvard.edu/~mdw/proj/codeblue/, accessed 15/4/2007.)

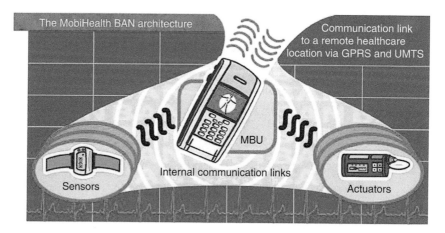

COLOR FIGURE 8.8

Mobihealth BAN architecture using GPRS/UMTS wireless protocol. (http://www.mobihealth.org/.)

McNulty, J. and C. Wellman (1996). Heart Development. Available at http://www.meddean.luc.edu/lumen/MedEd/GrossAnatomy/thorax0/heartdev/.

Millet-Roig, J., R. Ventura-Galiano, F. Chorro-Gasco, and A. Cebrian (2000). Support vector machine for arrhythmia discrimination with wavelet transform-based feature selection. *Computers in Cardiology*, 407–410.

Minami, K., H. Nakajima, and T. Toyoshima (1999). Real-time discrimination of ventricular tachyarrhythmia with Fourier-transform neural network. *IEEE Transactions on Biomedical Engineering 46*, 179–185.

Mitra, P., C. Murthy, and S. Pal (2002). Unsupervised feature selection using feature similarity. *IEEE Transactions on Pattern Analysis and Machine Intelligence 24(3)*, 301–312.

Morizet-Mahoudeaux, P., C. Moreau, D. Moreau, and J. Quarante (1981). Simple microprocessor-based system for on-line ECG arrhythmia analysis. *Medical and Biological Engineering and Computing 19(4)*, 497–500.

Nadal, J. and Bosan, M. (1993). Classification of cardiac arrhythmias based on principal component analysis and feedforward neural networks. *Computers in Cardiology*, 341–344.

National Research Council (NRC). European ST-T Database. Institute of Clinical Physiology, Department of Bioengineering and Medical Informatics. Pisa, Italy. Available at http://www.ifc.pi.cnr.it.

Oien, G., N. Bertelsen, T. Eftestol, and J. Husoy (1996). ECG rhythm classification using artificial neural networks. *IEEE Digital Signal Processing Workshop 1*, 514–517.

Okada, M. (1979). A digital filter for the QRS complex detection. *IEEE Transactions on Biomedical Engineering 26*, 700–703.

Olmez, T., Z. Dokur, and E. Yazgan (1997). Classification of ECG waveforms by using genetic algorithms. *Proceedings of the 19th Annual International Conference of the IEEE Engineering in Medicine and Biology Society 1*, 92–94.

O'Rourke, R. A. (Ed.) (2005). *Hurst's the Heart: Manual of Cardiology* (11th ed.) New York: McGraw-Hill, Medical Pub. Division.

Osowski, S., L. Hoai, and T. Markiewicz (2004). Support vector machine-based expert system for reliable heartbeat recognition. *IEEE Transactions on Biomedical Engineering 51(4)*, 582–589.

Osowski, S. and H. Tran (2001). ECG beat recognition using fuzzy hybrid neural network. *IEEE Transactions on Biomedical Engineering 48(11)*, 1265–1271.

Poli, R., S. Cagnoni, and G. Valli (1995). Genetic design of optimum linear and nonlinear QRS detectors. *IEEE Transactions on Biomedical Engineering 42(11)*, 1137–1141.

Ramakrishnan, A. and S. Saha (1997). ECG coding by wavelet-based linear prediction. *IEEE Transactions on Biomedical Engineering 44(12)*, 1253–1261.

Reaz, M. and S. Lee (2004). An approach of neural network based fetal ECG extraction. *Proceedings of the 6th International Workshop on Enterprise Networking and Computing in Healthcare Industry 1*, 57–60.

Rojo-Alvarez, J., A. Arenal-Maiz, and A. Artes-Rodriguez (2002a). Discriminating between supraventricular and ventricular tachycardias from EGM onset analysis. *IEEE Engineering in Medicine and Biology Magazine 21(1)*, 16–26.

Rojo-Alvarez, J., A. Arenal-Maiz, and A. Artes-Rodriguez (2002b). Support vector black-box interpretation in ventricular arrhythmia discrimination. *IEEE Engineering in Medicine and Biology Magazine 21(1)*, 27–35.

Sahambi, J., S. Tandon, and R. Bhatt (1997). Using wavelet transforms for ECG characterization. An online digital signal processing system. *IEEE Engineering in Medicine and Biology Magazine 16(1)*, 77–83.

Schlant, R. C., R. W. Alexander, R. A. O'Rourke, R. Roberts, and E. H. Sonnenblick (Eds) (1994). *The Heart, Arteries and Veins* (8th ed.) New York: McGraw-Hill, Health Professions Division.

Sepulveda-Sanchis, J., G. Camps-Valls, E. Soria-Olivas, S. Salcedo-Sanz, C. Bousono-Calzon, G. Sanz-Romero, and J. Marrugat-Iglesia (2002). Support vector machines and genetic algorithms for detecting unstable angina. *Computers in Cardiology 1*, 413–416.

Silipo, R., M. Gori, A. Taddei, M. Varanini, and C. Marchesi (1995). Classification of arrhythmic events in ambulatory electrocardiogram, using artificial neural networks. *Computers and Biomedical Research 28(4)*, 305–318.

Silipo, R. and C. Marchesi (1998). ECG signal compression using analysis by synthesis coding. *IEEE Transactions on Signal Processing 46(5)*, 1417–1425.

Silipo, R., W. Zong, and M. Berthold (1999). ECG feature relevance in a fuzzy arrhythmia classifier. *Computers in Cardiology 1*, 679–682.

Simon, B. and C. Eswaran (1997). An ECG classifier designed using modified decision based neural networks. *Computers and Biomedical Research 30(4)*, 257–272.

Stamkopoulos, T., K. Diamantaras, N. Maglaveras, and M. Strintzis (1998). ECG analysis using nonlinear PCA neural networks for ischemia detection. *IEEE Transactions on Signal Processing 46(11)*, 3058–3067.

Strauss, D., G. Steidl, and J. Jung (2001). Arrhythmia detection using signal-adapted wavelet preprocessing for support vector machines. *Computers in Cardiology*, 497–500.

Sun, Y. and S. Suppappola (1992). Microcontroller based real time QRS detection. *Biomedical Instrumentation and Technology 26(6)*, 477–484.

Suppappola, S. and Y. Sun (1994). Nonlinear transforms of ECG signals for digital QRS detection: a quantitative analysis. *IEEE Transactions on Biomedical Engineering 41*, 397–400.

Suzuki, Y. (1995). Self-organizing QRS-wave recognition in ECG using neural networks. *IEEE Transactions on Neural Networks 6(6)*, 1469–1477.

Technical Institute of Finland IMPROVE data library. Available at http://www.vtt.fi/tte/samba/projects/improve.

Thoraval, L., G. Carrault, and J. Bellanger (1994). Heart signal recognition by hidden Markov models: the ECG case. *Methods of Information in Medicine 33(1)*, 10–14.

Thoraval, L., G. Carrault, and F. Mora (1992). Continuously variable duration hidden Markov models for ECG segmentation. *Proceedings of the 14th Annual International Conference of the IEEE Engineering in Medicine and Biology Society 2*, 529–530.

Togawa, T., T. Tamura, and P. A. Oberg (1997). *Biomedical transducers and instruments*. Boca Raton, FL: CRC Press.

Tu, C., Y. Zeng, X. Ren, S. Wu, and X. Yang (2004). Hybrid processing and time-frequency analysis of ECG signal. *Proceedings of the 26th Annual International Conference of the Engineering in Medicine and Biology Society 1*, 361–364.

Utsuyama, N., H. Yamaguchi, S. Obara, H. Tanaka, J. Fukuta, S. Nakahira, S. Tanabe, E. Bando, and H. Miyamoto (1988). Telemetry of human electrocardiograms in aerial and aquatic environments. *IEEE Transactions on Biomedical Engineering 35(10)*, 881–884.

Watrous, R. and G. Towell (1995). A patient adaptive neural network ECG patient monitoring algorithm. *Computers in Cardiology 1*, 229–232.

Wenyu, Y., L. Gang, L. Ling, and Y. Qilian (2003). ECG analysis based on PCA and SOM. *Proceedings of the 2003 International Conference on Neural Networks and Signal Processing 1*, 37–40.

Xie, Z. X., H. Z. Xie, and X. B. Ning (1997). Application of fuzzy neural network to ECG diagnosis. *International Conference on Neural Networks 1*, 62–66.

Xu, J. and Z. Yuan (2003). A feature selection method based on minimizing generalization bounds of SVM via GA. *IEEE International Symposium on Intelligent Control 1*, 996–999.

Xue, Q., Y. Hu, and W. Tompkins (1992). Neural newtork based adaptive matched filtering for QRS detection. *IEEE Transactions on Biomedical Engineering 39*, 317–329.

Yang, T., B. Devine, and P. Macfarlane (1994). Use of artificial neural networks within deterministic logic for the computer ECG diagnosis of inferior myocardial infarction. *Journal of Electrocardiolgy 27*, 188–193.

Yeap, T., F. Johnson, and M. Rachniowski (1990). ECG beat classification by a neural network. *Proceedings of the 12th Annual International Conference of the IEEE Engineering in Medicine and Biology Society 1*, 1457–1458.

Yu, B., S. Liu, M. Lee, C. Chen, and B. Chiang (1985). A nonlinear digital filter for cardiac QRS complex detection. *Journal of Clinical Engineering 10*, 193–201.

Yu, H. H., S. Palreddy, and W. Tompkins (1997). A patient-adaptable ECG beat classifier using a mixture of experts approach. *IEEE Transactions on Biomedical Engineering 44(9)*, 891–900.

Zhang, H. and L.Q. Zhang (2005). ECG analysis based on PCA and support vector machines. *Proceedings of the International Conference on Neural Networks and Brain 2*, 743–747.

Zhu, K., P. Noakes, and A. Green (1991). ECG monitoring with artificial neural networks. *Proceedings of the 2nd International Conference on Artificial Neural Networks*, 205–209.

Zigel, Y., A. Cohen, and A. Katz (2000). ECG signal compression using analysis by synthesis coding. *IEEE Transactions on Biomedical Engineering 47(10)*, 1308–1316.

Zong, W. and D. Jiang (1998). Automated ECG rhythm analysis using fuzzy reasoning. *Computers in Cardiology 1*, 69–72.

5

Computational Intelligence in Analysis of Electromyography Signals

---

5.1 Introduction

The human skeletal muscular system is primarily responsible for providing the forces required to perform various actions. This system is composed of two subsystems; namely, the nervous system and the muscular system, which together form the neuromuscular system. The CNS provides control via nerve signals and innervation of the muscles. Nerves can be thought of as wires conducting electrical currents, where the nerve heads (nuclei) originate in the spinal column and their long axonal bodies extend far and deep, innervating individual motor units in various muscles. The skeletal–muscular system consists of muscle groups attached to bones via tendons and movement is produced when nerve signals cause muscle contractions and relaxations that either pull or release the bone. As with the previous cardiovascular system and any other physiological system of the human body, the neuromuscular system is also susceptible to diseases. Neuromuscular diseases include disorders originating in the nervous system (e.g., spinal column conduction), in the neuromuscular junctions, and in the muscle fibers (e.g., motor units). These disorders have different degrees of severity ranging from minor loss of strength to amputation due to neuron or muscle death. In more severe disorders such as amyotrophic lateral sclerosis (ALS), death is almost certain.

As in the previous chapter, early and accurate diagnosis is crucial for improved patient prognosis and increasing opportunity for full recovery. In most situations, clinical examination is inadequate to diagnose and localize the disorder (Preston and Shapiro, 2005) because many different disorders could cause a particular symptom. Unlike CVDs in which the heart is the sole organ involved, neuromuscular diseases can be caused by several different muscle fibers or nerves, sometimes far removed from the symptoms. Accurate localization of the disorder is, therefore, of paramount importance so that more focused treatment can be administered, for example, a patient could present with acute pain in the knee joints or muscles due to problems with a spinal disc. Currently, electrodiagnostic (EDX) studies consisting of nerve conduction studies (NCS) and electromyography (EMG) are used for evaluating and diagnosing patients with neuromuscular disorders. EMG was introduced as a method of studying neuromuscular states based on cell action potentials during muscle activity. Features of the EMG waveform (0.01–10 mV and 10–2000 Hz on average) may be indicative of the location, etiology, and type of pathology. For example, EMG pulse duration indicates the location and net metabolic state of the muscle fiber (Emly et al., 1992), whereas sporadic spikes may indicate myopathy. Note, however, that EDX techniques assist doctors in their diagnosis but are rarely useful for confirming the diagnosis and in difficult cases more intrusive methods such as muscle biopsies or more sophisticated imaging techniques such as ultrasound or MRI are required.

The interpretation of EMG readings is usually performed by trained and skilled neurologists who in addition to examining EMG waveforms also use techniques such as needle conduction studies and muscle acoustics. Problems arise when there are too few experts to meet the demand of patients and, therefore, it is becoming increasingly important to develop automated diagnostic systems based on EMG readings. This need provides scope for the application of CI techniques for the detection and classification of neuromuscular disorders based on EMG processing. These intelligent systems will assist medical specialists in detecting anomalies in the neuromuscular system. Other applications of EMG signal processing such as motion control of prostheses, ergonomics, and intention detection will also benefit from improvements in CI technology.

The aim of intelligent diagnostic and artificially controlled neuromuscular systems is to first preprocess the raw EMG signal and then extract distinctive information or features. Features that can be extracted include time and frequency domain information, Fourier coefficients, autoregressive coefficients, wavelet coefficients, and a wide range of quantities derived from other signal-processing techniques. This information can then be used as input data for classifiers such as NNs and SVMs, which can classify neuromuscular diseases or detect the intention to move. Designing a truly practical and accurate automated diagnostic system, however, remains challenging, for example, EMG readings tend to be patient-dependent as their amplitude and duration can vary across patient age. Young children especially have ever changing EMG waveform morphologies due to their rapid growth making it difficult to design a general automated detection system based on readings from one age group. The challenge in this area is to devise signal-processing techniques that preserve or capture important discriminatory information so as to provide a good set of features for classification.

In this chapter, we will first describe the mechanics of muscle and nerve connection involved in actuating a limb. This is essential for understanding the origins of the MUAP, which is the primary biosignal measured in EMG studies. The following sections elaborate several neuromuscular disorders in an effort to introduce the many pathologies that can be detected using EMG. We then examine the methods of processing the recorded EMG signals to extract useful information. Indeed, the optimal signal-processing method to apply is dependent on the application of the EMG signal, such as detecting a particular neuromuscular disease or using the EMG as a control signal for an orthotic device. In the final sections, we review CI techniques employed thus far to detect disorders and for research in areas such as rehabilitation therapy.

5.2   The Human Muscle Physiology

5.2.1   Anatomy of the Muscle

There are three types of human muscles, skeletal muscle, smooth muscle, and cardiac muscle, which have specific functions such as motion and maintaining posture, providing support, and heat generation. Muscles are excited by neurons that cause them to contract. Extensibility allows muscles to stretch when pulled, whereas contractility enables muscles to shorten. In addition, muscles are elastic, which means that they return to their original length after a period of lengthening or shortening, a characteristic critical for enabling us to perform the necessary motions required to negotiate our everyday environments. As muscles can be thought of as conduits that translate our thoughts into action, let us now examine them more closely to better understand their structure and function.

Skeletal muscles move bones and provide us with mechanical forces. Figure 5.1 gives a microscopic view of these skeletal muscles highlighted with

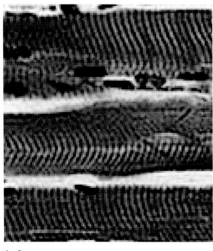
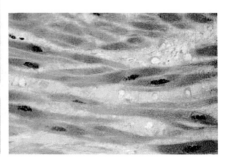

Left   Right

FIGURE 5.1 (See color figure following page 204.)

Left: Micrograph of striated muscle showing muscle fibers in skeletal muscle. The striated appearance comes from the hundreds of thousands of sarcomeres in a layered formation. *Right*: Micrograph of smooth muscle fiber, which do not have the striated characteristics of skeletal muscle fibers. (Courtesy of http://people.eku.edu/ritchisong/301notes3.htm.)

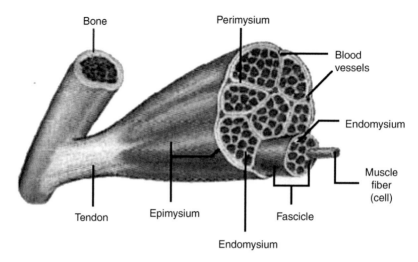

Bone   Perimysium   Blood vessels   Endomysium   Muscle fiber (cell)   Tendon   Epimysium   Fascicle   Endomysium

FIGURE 5.2

Cross-section of skeletal muscle and connection to the bone via tendons. (Courtesy of http://people.eku.edu/ritchisong/301notes3.htm.)

iodine. On closer inspection, we find that skeletal muscles have a striated or striped appearance, which is a unique characteristic of this type of muscle. The appearance is caused by numerous sarcomere filaments. Skeletal muscles are attached to bones by tendons, connective tissue ensheathed in a layer known as the *epimysium*. The muscles groups are composed of *fascicles*, which are fiber bundles surrounded by a layer called the *perimysium* as seen in Figure 5.2.

The cell membrane of each muscle fiber is known as the *sarcolemma*, which acts as a conductor for impulses, much like neurons do for nerve impulses. The muscle cells (see Figure 5.3) are composed of myofibrils that are bundles of sarcomeres. Sarcomeres are multiprotein complexes composed of three types of myofilaments called actin, myosin, and titin. Actin is a protein monomer that forms thin filaments in the sarcomere whereas myosin forms thick filaments. Each thick myofilament is surrounded by six thin myofilaments. The elastic system of the muscle is composed of the large protein structure titin (or connectin). The sarcomere is defined as the segment in between two Z-lines that are actually formed by overlapping myofilaments and appear as dark and distinct bands in electron micrographs. The region around the Z-lines is known as the isotropic band (I-band) and the region in between the Z-lines is the anisotropic band (A-band). Within the A-band itself is a paler H-band composed of M-lines. The I-band is mainly composed of actin filaments whereas the A-band contains myosin filaments. The interaction between actin and myosin filaments together with calcium ions provides the contraction properties of muscle. The interaction between these components is commonly referred to as the sliding cross-bridge system, which will be described in greater detail later.

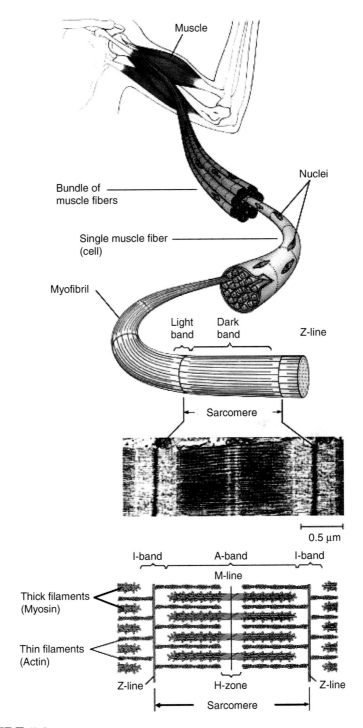

FIGURE 5.3

Diagram depicts the physiology of the human skeletal muscle, from the muscle fiber to the position of the sarcomeres. (Courtesy of http://people.eku.edu/ritchisong/301notes3.htm.)

Smooth muscles are found in the linings of blood vessels, intestines, and hollow structures of organs. These are involuntary muscles, which have spindle-shaped muscle fibers arranged in a sheet as seen in the electron micrograph in Figure 5.1. There are two forms of smooth muscle. The first is visceral muscle found in blood vessel linings and hollow parts of the organs such as the reproductive organs. The second form is the multiunit muscle usually seen in large blood vessels (valves), eyes (adjusting the lenses or the iris), and the base of the hair follicle. These groups of muscles do not possess T-tubes, have minimal sarcoplasmic reticulum, and hence have little mobility. The muscle fibers are composed of thick and thin myofilaments but do not form sarcomeres. The myofilaments do not contain troponin as in the skeletal muscles and rather than the troponin–tropomyosin interaction activating the cross-bridge system, a calcium–calmodulin complex system is in place.

Cardiac muscle, which was briefly discussed in Chapter 4 is the third class of muscle. This muscle is special because it is an involuntary, mononucleated, striated muscle found only in the heart. The cardiac muscle is myogenic in that it does not require electrical impulses from the nervous system for contraction. The rate of contraction, however, is controlled by electrical nerve signals originating from the SA node. When one cardiac muscle fiber is placed next to another, contraction of one muscle will stimulate contraction in the adjacent muscle fiber. This results in systematic propagation of the contraction action throughout the heart. The T-tubes in cardiac muscles run along the Z-bands and are shorter whereas the sarcoplasmic reticulum lacks terminal cisternae.

5.2.2 Muscle Contraction: Sliding Cross-Bridge Action

The sarcolemma or cell membrane of muscle fibers contains several hollow passages known as *transverse tubes* or "T-tubes" for short that wrap around the myofibril and connect to other locations in the sarcolemma. They do not open into the interior of the muscle because their sole function is to conduct action potentials travelling on the surface of the sarcolemma deep into the muscle core where the sarcoplasmic reticulum resides. The sarcoplasmic reticulum is a hollow reservoir for storing Calcium ions (Ca^{2+}), which is a key chemical component for sustaining contraction. The membrane of the sarcoplasmic reticulum as seen in Figure 5.4 is equipped with "pumps," which work using active transport in which energy is used to move ions across the membrane. The membrane also has "gates," which allow the ions through. Calcium is continuously pumped into the sarcoplasmic reticulum from the cytoplasm of the muscle fibers (sarcoplasm). In the relaxed state, the sarcoplasmic reticulum has a high concentration of calcium ions compared to the muscle fibers or myofibrils. The ionic gates are closed and calcium ions cannot flow back into the muscle fibers creating a large ion diffusion gradient.

The motion of muscle fibers is due to the interaction between thick and thin myofilaments. The myosin protein molecule or thick myofilament has a headlike structure known as a cross-bridge, which forms connections with actin protein molecules (Figure 5.5). This cross-bridge contains adenosine

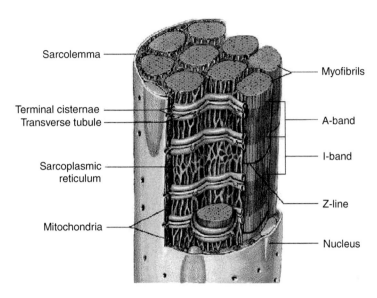

FIGURE 5.4
Cross-section of the sarcoplasmic reticulum showing the T-tubes, myofibrils, and sarcolemma. (Courtesy of http://people.eku.edu/ritchisong/ 301notes3.htm.)

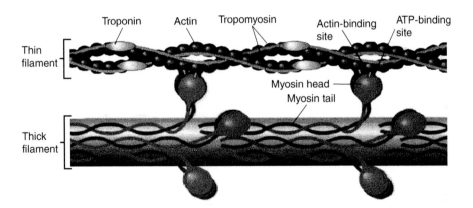

FIGURE 5.5
Diagram depicts the thick myofilaments composed of myosin protein structures. (Courtesy of http://people.eku.edu/ritchisong/301notes3.htm.)

triphosphate (ATP) molecules, which represent stored energy to be released and also molecule-binding sites for actin molecules. Thin myofilaments are composed of actin, troponin, and tropomyosin protein molecules. Actin molecules form the long chain of the myofilament as seen in Figure 5.6 whereas

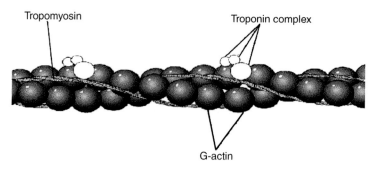

FIGURE 5.6
Diagram depicts the thin myofilaments composed of actin, troponin, and tropomyosin protein structures. (Courtesy of http://people.eku.edu/ritchisong/301notes3.htm.)

tropomyosin are single individual chains, which wrap around the actin chain. At the end of the tropomyosin chain lies a troponin molecule. In the muscle's relaxed state, the tropomyosin chain lies in contact with the myosin heads and although this configuration is maintained, the muscle remains relaxed. Troponin molecules have binding sites for calcium ions, which cause the molecule structure to change shape in the presence of calcium ions.

At the onset of contraction, a nerve impulse travels down the sarcolemma into the sarcoplasmic reticulum. As the electrical impulse crosses the surface of the sarcoplasmic reticulum, the ionic gates open and calcium ions diffuse into the myofibril filaments along the concentration gradient. In the presence of calcium ions, troponin molecules change shape and drag the tropomyosin chains along causing the myosin head to come into contact with an actin molecule. When contact is achieved, the myosin head binds with an actin molecule in the presence of ATP. This chemical combination transforms ATP into adenosine diphosphate (ADP), which is accompanied by a release of energy. The chemical equation for this transformation is represented by

$$A + M + ATP \Rightarrow A + M + ADP + P + Energy \qquad (5.1)$$

where P is inorganic phosphate. This sudden release of energy causes the myosin head to swivel or move, which pulls the actin chain or thin myofilament along. A combination of these movements in the many hundreds of thousands of thin myofilaments is observed as a single united movement of the muscle fiber.

When this motion is complete, ATP fills the binding sites on the myosin heads and disconnects the bonds between the myosin and actin molecules causing the myosin heads to swivel in reverse and the chemical reaction (Equation 5.1) again results in the myosin head binding with another actin molecule further along the chain. The process is repeated and continues as

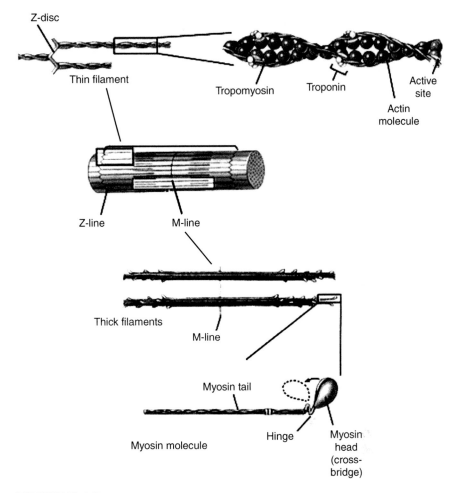

FIGURE 5.7
Structure of thin and thick myofilaments in cross-bridge motion. (Courtesy of http://people.eku.edu/ritchisong/301notes3.htm.)

long as there are sufficient calcium ions present. In this manner, the thick myofilaments are observed as pulling themselves along the thin myofilaments or sliding and the muscle contracts. Figure 5.7 shows the position of the thin and thick myofilaments, which allow motion via the cross-bridge mechanism.

It is possible for the supply of ATP molecules to diminish as motion is prolonged, causing a reduction in the ability of myosin heads to attach to new actin molecules. The range of motion in this case is reduced even in the presence of sufficient calcium ions. This condition occurs naturally and is more commonly known as *muscle fatigue*. When the nerve impulse stops, the ionic gates in the sarcoplasmic reticulum close and the calcium ions are pumped back into the reservoir via active transport as before. Without calcium ions, the troponin molecules cease motion and the tropomyosin filaments lie in contact with the myosin heads as in the initial stage. The muscle then returns to its equilibrium or "relaxed state."

There are two basic types of contractions; the first is *isotonic* contraction where muscle length decreases and more force is generated than is required by the load. The second contraction is *isometric* contraction where the muscle does not shorten and the load is greater than the force produced. *Twitching* refers to sudden movements of the muscle due to nerve impulses, an action that can also be subdivided into isotonic twitches and isometric twitches. Other contractions include passive lengthening and eccentric contractions or active lengthening.

5.2.3   The Motor Unit Action Potential

The nerve impulse, which travels into the sarcoplasmic reticulum and initiates motion is generated by a collective set of action potentials (Preston and Shapiro, 2005). The primary nerve connected to each muscle is the motor neuron (see Figure 5.8). The motor neuron is the "live wire," which channels the electrical impulse into the muscle. There are two major types of motor neurons, the somatic motor neuron and the autonomic motor neuron. The former is of particular interest here because it is the motor neuron that connects the nervous system to the skeletal muscles. The autonomic motor neurons include sympathetic and parasympathetic nerves, which innervate the cardiac muscle

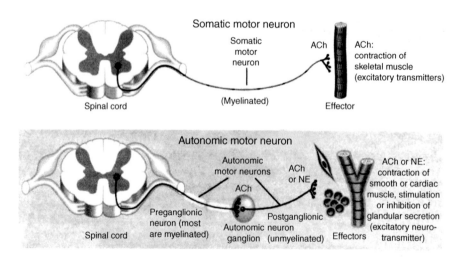

FIGURE 5.8

Motor neuron connection showing the neuron cell body in the spinal cord with axon extending toward the effector. Both somatic and autonomic motor neuron connections are shown.

and the peripheral nervous system giving us primary sensory information such as touch and temperature. A *motor unit* consists of one alpha motoneurone, a motor axon, and skeletal muscle fibers innervated by the terminals of the axon (Preedy and Peters, 2002).

The motor neuron is an example of an *efferent* nerve in the body's nervous system. Efferent nerves carry impulses away from the central nervous system such as the spinal cord to effector units such as muscles and glands. In contrast, *afferent* nerves convey impulses to the CNS such as nerves from the skin and other sensory organs. The motor neuron has its cell body in the CNS with a long axon extending out toward the effectors; in this case the muscles. The axon body has several branching dendrites and ends in a neuromuscular junction (NMJ) with the muscle. A diagram of this nerve connection is depicted in Figure 5.8 and an electron micrograph of the NMJ is shown in Figure 5.9. The axon of the motor neuron is usually sheathed in a myelin insulation, which is typically composed of concentric spirals of Schwann cells. The myelin sheath extends along the whole axon except at the nodes of Ranvier (gaps between adjacent Schwann cells) and near the NMJ (Figure 5.10).

The main cell head of the motor neuron is also known as the anterior horn cell and lies in the ventral gray matter of the spinal cord. The axonal body, which extends out has an electrically active cell membrane, which is semipermeable to anions and cations and is always permeable to large negative ions.

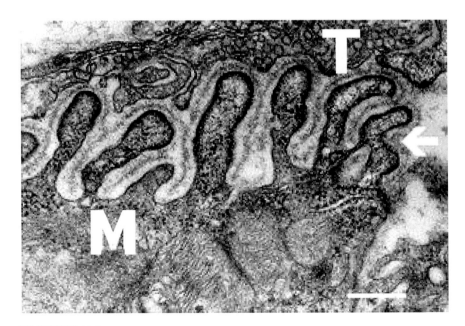

FIGURE 5.9
An electron micrograph of the NMJ is shown. Label T indicates the axon terminal and M indicates the muscle fiber. (Courtesy of http://www.nimh. nih.gov/Neuroinformatics/shins04.cfm.)

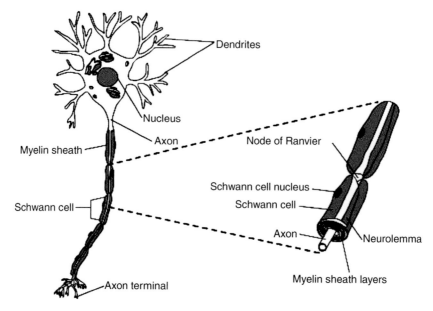

FIGURE 5.10

The composition of the myelin insulation for the motor neuron axon is shown here. The gaps between Schwann cells for the uninsulated parts of the axon are known as the nodes of Ranvier. (Courtesy of http://www.apptechnc.net/w̃indelspecht/nervous/pages/motor_neuron_jpg.htm.)

At rest, this membrane is relatively impermeable to sodium ions (Na^+). There is also a sodium/potassium (Na^+/K^+) pump, which actively pumps sodium outside the axonal body in exchange for potassium ions resulting in a concentration gradient across the membrane. At equilibrium, the resting membrane potential is in the range of -70 to -90 mV with the axonal body more negatively charged.

The cell membrane has sodium gates composed of molecular pores that open and close due to changes in the cell-membrane potential. When current is injected into the axon, it causes depolarization, which opens the gates allowing sodium ions to rush into the axon along the concentration gradient. If the depolarization exceeds 10–30 mV above the resting potential, it generates an action potential. A positive feedback system then comes into play whereby a single depolarization causes successive depolarizations to occur and progressively more sodium gates open. Action potentials are all or no events and propagate in both directions from the point of origin along the axon. This means that if an action potential begins, it will continue to travel along the axon until it reaches both ends of the nerve. The opening of the sodium gate is limited to about 1–2 ms before a secondary inactivation gate causes the cell membrane to become impermeable to sodium ions and the sodium flow stops. This flow of sodium ions into the axonal body has reversed the concentration gradient and the concentration of potassium ions inside the axonal body is lower than the exterior. During this time, the potassium channels open and potassium ions flow back into the axonal body along the concentration gradient returning the voltage to negative and restoring the cell membrane to its resting equilibrium potential.

The series of depolarizations continue along the axonal body until it reaches the NMJ at which the electrical impulse crosses using an electrical–chemical–electrical interface from nerve to muscle. The interface is formed by two special membranes, one on the nerve end and one on the muscle, which are separated by a synaptic cleft (Figure 5.9). As the action potential reaches the nerve membrane, voltage-gated-calcium channels activate causing calcium ions to enter. The consequent increase in calcium ion concentration activates the release of the neurotransmitter acetylcholine, which diffuses across the synaptic cleft and binds with receptors on the muscle membrane. The binding causes sodium gates on the muscle side to open and initiates a depolarization followed by propagation of an action potential in the muscle. This impulse is the initiator of the muscle fiber motion described earlier.

A MUAP occurs when depolarization in a motor neuron causes action potentials in all muscle fibers of the motor unit. Conduction velocity is dependent on the diameter of the axonal body, with higher speeds in larger axonal bodies. If the axon is unmyelinated, the conduction speed rarely exceeds 3–5 m/s as opposed to myelinated axons like the motor neuron where speeds of 35–75 m/s can be achieved. Depolarization only occurs at each node of Ranvier and the current jumps between the myelin gaps to cause depolarizations in adjacent nodes, known as *saltatory conduction*. The nervous system controls the level of muscle contraction by altering the firing rate of the motor neurons and varying the number of motor units recruited for a particular movement.

5.3   Neuromuscular Disorders

Neuromuscular disorders are generally diseases of the peripheral nervous system, for example, nerve disorders and NMJ disorders. These disorders can be categorized depending on the location or point of origin. Neuropathies are disorders of the nerves themselves whereas myopathies are complications with the muscle usually caused by muscle degradation or muscle death. In this section, we introduce the various disorders that affect the neuromuscular system by highlighting their clinical manifestations, etiologies, symptoms, and where appropriate their prognosis. Table 5.1 summarizes some of the common disorders, whereas a more extensive set of muscular diseases can be found, for example, at the Muscular Dystrophy Association (MDA)* homepage.

---

*http://www.mdausa.org/disease/

TABLE 5.1

Table Shows Categorization of Neuromuscular Disorders and some Specific
Diseases Associated with Each Group

Neuropathy (Mono and Poly)	**Sensory neuronopathy**
Entrapment	Paraneoplastic
Demyelinating	Autoimmune
Axonal	Infections
Mononeuritis multiplex	Toxins
Radiculopathy	**Plexopathy**
Disk herniation	Radiation induced
Spondylosis	Entrapment
Neoplastic	Neoplastic
Infarction	Diabetic
Infections	Hemorrhagic
Inflammatory	Inflammatory
Motor neuron disease	**Myopathy**
Amyotrophic lateral sclerosis	Muscular dystrophy
Polio	Congenital
Spinal muscular atrophy	Metabolic
Monomelic amyotrophy	Inflammatory
Neuromuscular junction disorders	Toxic
Myasthenia gravis	Endocrine
Lambert–Eaton myasthenic syndrome	
Congenital Myasthenic Syndrome	
Botulism	

Source: Adapted from Preston, D. and Shapiro, B. in *Electromyography and Neuromuscular Disorders: Clinical-Electrophysiologic Correlations*, Elsevier, Philadelphia, PA, 2005.

5.3.1   Neuropathy Disorders: Mononeuropathies and Polyneuropathies

Neuropathic disorders can be categorized into mononeuropathies and polyneuropathies depending on the number of nerves involved. In mononeuropathies, a single nerve is involved causing pain and some disability, whereas in polyneuropathies a group or all of the nerves can be affected. In both cases, the nerve is affected either through disease or some form of injury.

The most common form of mononeuropathy in the upper extremities of the body (arms and hands) is *median nerve entrapment* in the wrist. Entrapment means that the nerve is compressed or placed in an abnormal position that disrupts its function. The disorder usually occurs in the carpal tunnel of the wrist where the nerve is compressed causing pain and disabling the hand. EMG is generally used to differentiate this problem from lesions of the proximal median nerve, brachial plexus, and cervical nerve roots. The median nerve entrapment is often mistaken with *proximal median neuropathy* during clinical examination, which makes it difficult to diagnose, particularly if the proximal median neuropathy is mild. EMG studies have been shown to alleviate this diagnostic problem by providing more accurate localization of the trapped nerve. The next most common neuropathy is *ulnar neuropathy*, which is difficult to localize using either EMG or other EDX methods. Although it is commonly due to compression of the ulnar nerve at the elbow, it can be caused by entrapments of the nerve at the wrist making it difficult to pinpoint. Ulnar neuropathy at the wrist is rare and not easily detected during the early stages. Nerve lesions due to this include palmar motor lesions, which affect muscles supporting the palmar motor branch (located in the palm of the hand) and sensory nerve lesions, which affect only the sensory nerve branches. *Radial neuropathy* is less frequently studied than median and ulnar neuropathies. The radial nerve is a major nerve that extends from the brachial plexus to the wrist and is also susceptible to entrapments. There are three general lesions: the spinal grove, axilla lesions, and lesions of the posterior interosseous and superficial radial sensory nerves. Figure 5.11 shows the anatomy of the radial nerve and its connections to the brachial plexus whereas Figure 5.12 illustrates radial nerve connections to various muscle structures in the arm.

The most common disorder in the lower extremities is *peroneal neuropathy*, commonly in the fibular neck of the calf where the peroneal nerve runs. The peroneal nerve (Figure 5.13) is derived from nerve roots in the spinal column and passes down through the lumbosacral plexus to the sciatic nerve. It branches into the common peroneal nerve and the tibial nerve before forming secondary nerve branches such as the deep peroneal nerve and the superficial peroneal nerve (Figure 5.14). Patients with peroneal neuropathy experience sensory disturbances of the calf and foot area usually reporting numbness and weakness. Another disorder is *femoral neuropathy*, which is generally not observed in EMG laboratories because the more common lumbar plexus lesions in peroneal neuropathy have a similar appearance. The femoral nerve runs parallel to the femoral artery and innervates the quadriceps muscles. Patients with femoral neuropathy usually report buckling knees, difficulty in raising the thigh, dragging of feet, and some numbness and weakness around the thigh and calf. If the ankle and soles of the feet are in pain, tarsal tunnel syndrome is suspected. This is caused by distal tibia nerve entrapment in the ankle. Although similar to the carpal tunnel syndrome in the wrist, the tarsal tunnel syndrome is extremely rare and more often observed in older patients.

Mononeuropathies are not restricted to limbs but may also occur in the face and around the cranium (head), which are innervated by several major nerves depicted in Figure 5.15. Although EMG studies are most suited to the study of the limbs, they may also be employed in the investigation of cranial nerves. Mononeuropathies afflicting the facial and trigeminal nerves are the most common lesions found. The facial nerve (also known as the cranial nerve VII) is composed of several different nerve fiber bundles, including nerves to the facial muscles, parasympathetic motor fibers to the salivary and lacrimal glands, taste fibers to a major segment of the tongue, somatic sensory fibers serving the ear surface and a small part of the external auditory meatus. The most common facial mononeuropathy is facial nerve palsy, which generally manifests as the idiopathic Bell's palsy. This facial nerve dysfunction is also

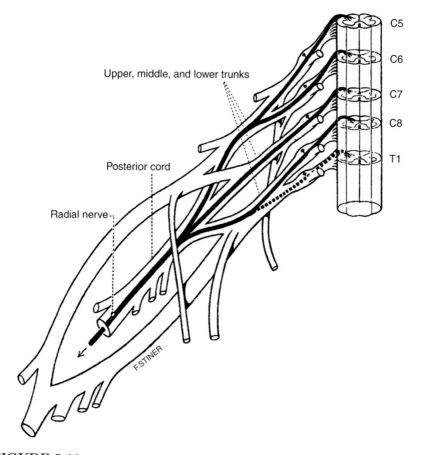

Upper, middle, and lower trunks

C5

C6

C7

C8

T1

Posterior cord

Radial nerve

F.STINER

FIGURE 5.11

The radial nerve is derived from the posterior cord of the brachial plexus. It receives innervation from all three trunks of the brachial plexus and contributions from the C5–T1 nerve roots. (Reprinted with permission from Preston, D., and Shapiro, B., *Electromyography and Neuromuscular Disorders: Clinical-Electrophysiologic Correlations*, Elsevier, Philadelphia, PA, 2005.)

associated with several disorders such as diabetes, herpes zoster, lymphoma, leprosy, and stroke. The symptoms depend on the location, pathophysiology, and severity of the lesion, for example, a central lesion presents as weakness in the lower facial region. Compression and injury to the facial nerve can also cause hemifacial spasm, which is characterized by involuntary contractions and spasms in the face. Trigeminal neuropathy is another disorder affecting the trigeminal nerve (cranial nerve V), which is responsible for carrying nerve impulses to the face and motor fibers. Damage or dysfunction in this nerve causes numbness over the ipsilateral face and difficulty in opening the mouth or chewing. Trigeminal neuralgia is a less common disorder where severe pain

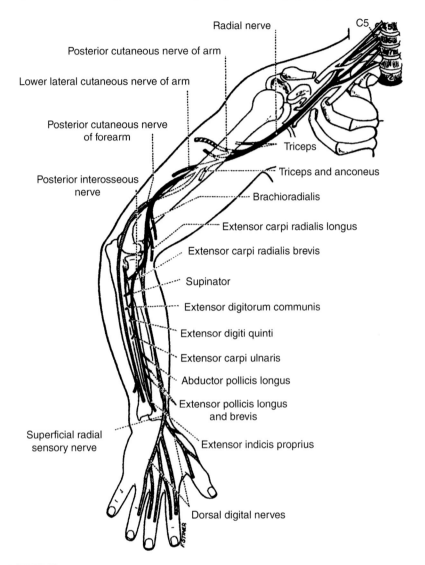

FIGURE 5.12

In the upper arm, the radial nerve first derives several cutaneous nerves of the arm followed by muscular branches to the triceps brachii. It then wraps around the humerus before descending down the elbow and deriving several smaller nerves, which branch into the hand and the fingers. (Reprinted with permission from Preston, D., and Shapiro, B., *Electromyography and Neuromuscular Disorders: Clinical-Electrophysiologic Correlations*, Elsevier, Philadelphia, PA, 2005.)

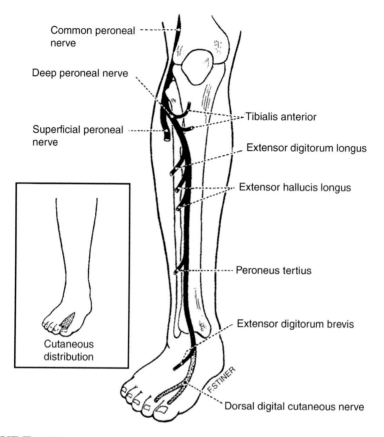

Common peroneal nerve

Deep peroneal nerve

Superficial peroneal nerve

Tibialis anterior

Extensor digitorum longus

Extensor hallucis longus

Peroneus tertius

Extensor digitorum brevis

Cutaneous distribution

F.STINER

Dorsal digital cutaneous nerve

FIGURE 5.13

The peroneal nerve anatomy showing the nerve branches into the calf, feet, and toes. (Reprinted with permission from Preston, D. and Shapiro, B., *Electromyography and Neuromuscular Disorders: Clinical-Electrophysiologic Correlations*, Elsevier, Philadelphia, PA, 2005.)

is present in one or more branches of the trigeminal nerve and can be triggered by a light touch or brushing of the facial skin causing acute pain.

Although mononeuropathies are the result of injury to a single nerve, polyneuropathies arise from disorders in a nerve bundle or a whole group of peripheral nerves. Despite several etiological symptoms, polyneuropathies are similar due to the limited manner in which sensory nerves can react to malfunction. There are several distinct disorders that include variants of chronic inflammatory demyelinating polyneuropathy and acute inflammatory demyelinating polyneuropathy (AIDP) characterized by sudden onset. Examination of the nerve fibers involved and the fiber size can help pinpoint the etiology; however, EMG is of limited usefulness because patients with pure small fiber polyneuropathy have normal electrophysiological characteristics.

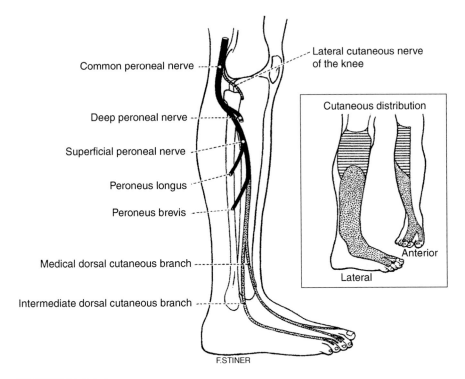

Common peroneal nerve

Lateral cutaneous nerve of the knee

Deep peroneal nerve

Superficial peroneal nerve

Peroneus longus

Peroneus brevis

Medical dorsal cutaneous branch

Intermediate dorsal cutaneous branch

Cutaneous distribution

Anterior

Lateral

F.STINER

FIGURE 5.14
The common, deep, and superficial peroneal nerves are shown. (Reprinted with permission from Preston, D. and Shapiro, B., *Electromyography and Neuromuscular Disorders: Clinical-Electrophysiologic Correlations*, Elsevier, Philadelphia, PA, 2005.)

Pathophysiological characteristics of nerve injury arise from two major sources, axonal loss and fiber demyelination.

5.3.2 Motor Neuron Disease

Motor neuron disease (MND) designates serious diseases characterized by degeneration of the motor neurons, leading to muscle atrophy and in more severe cases death. EMG readings in these cases show MUAPs with either longer than normal durations, larger amplitudes (Preston and Shapiro, 2005), or irregular firing patterns (Hassoun et al., 1994). These observations are a result of temporal dispersion of action potentials in the area either due to slowed conduction in the nerve fibers or increased conduction in the end plate zone.

Although recognized much earlier, the most common form of MND was first described in 1869 by French neurologist Jean-Mertin Charcot. This disease is known as amyotrophic lateral sclerosis derived from the Greek word *amyotrophic* or "no nourishment to the muscle" and lateral sclerosis referring

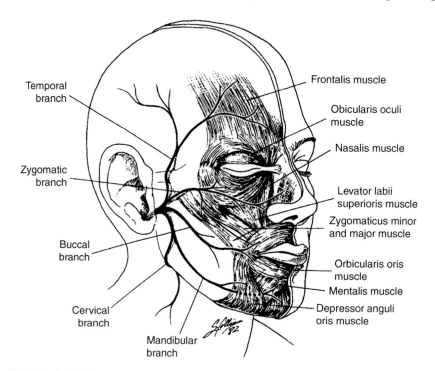

FIGURE 5.15

The facial nerve and its major branches are depicted innervating some of the major face muscles. (Reprinted with permission from Preston, D. and Shapiro, B., *Electromyography and Neuromuscular Disorders: Clinical-Electrophysiologic Correlations*, Elsevier, Philadelphia, PA, 2005.)

to scarring of the nerves in the lateral spinal column due to nerve death. ALS unfortunately possesses a uniformly poor prognosis, being fatal in all cases, with death on average of 3 years after onset with approximately 10% of patients surviving longer. The diseases usually affect those above 55 years of age but is sometimes also found in younger children. Several ALS variants have been identified including progressive bulbar palsy, progressive muscular atrophy, and primary lateral sclerosis. Both upper and lower motor neurons are affected whereas the sensory and autonomic functions are spared. Symptoms include muscle atrophy, cramps, spreading weakness, slowness in movements, and spasticity. Death usually results from respiratory failure as the nerves controlling the respiratory process die and respiratory control is lost whereas other complications include pulmonary embolus, sepsis, and pneumonia. At this point in time, the exact etiology for ALS is still unknown but several unproven speculations ranging from genetics, viral infections to toxic poisoning have been suggested.

Another group of MNDs, less common and less fatal is the *atypical* MNDs. These diseases have similar symptoms to ALS but can be differentiated using

EMG studies. Infectious atypical MNDs can be secondary to other diseases such as paralytic poliomyelitis, West Nile encephalitis (caused by a single ribonucleic acid (RNA) strand virus carried in mosquitoes), or retroviral disorders (HIV). Rarer disorders include monomelic amyotrophy usually found in younger males, motor neuron injury due to electrical shocks, or radiation and paraneoplastic injury due to certain cancers. There are also inherited MNDs including disorders such as familial ALS, spinal muscular atrophy, Kennedy's disease, and spastic paraplegia.

5.3.3   Radiculopathy

A group of disorders that have been successfully referred to EMG study is radioculopathy because imaging studies are inadequate in detecting the etiology. Radiculopathy is caused by a degenerative bone and disc disease usually around the cervical and lower lumbosacral segments of the spinal column. Disorders in this group are characterized by pain and paresthesis radiating from the nerve root, sometimes accompanied by sensory loss and paraspinal muscle spasm. Sensory loss are minor and rarely develop into frequent sensory disturbances. From an etiological viewpoint, the disease is generally caused by structural lesions in the spinal column, herniated spinal discs, impingement of bone in the spinal column, and metastatic tumors of the spine. A very common disorder from this is spondylosis where a patient suffers acute pain in the neck and has difficulty sitting up, kneeling, or walking. Other causes of radiculopathies include tumor infiltration, nerve infarctions, and infections from other diseases such as Lyme's disease, herpes zoster, cytomegalovirus, and herpes simplex. Differential diagnosis includes a less common class of disorders with similar characteristics known as plexopathies, which include brachial plexopathy and lumbosacral plexopathy. Plexopathies generally result from traumatic injuries such as overstretching a muscle, automobile accidents, and gunshot or knife wounds.

5.3.4   Myopathy

Myopathy disorders are associated with injury to the muscle group itself, usually skeletal muscles. The most common form is muscular dystrophy, which results in degenerative muscle states, which also forms one of the major myopathy disorders and it can be either hereditary or acquired. EMG readings in these cases are characterized by MUAPs with shorter durations and reduced amplitudes, which are thought to be caused by a decrease in muscle fibers within the motor unit.

Congenital muscular dystrophy (CMD) is a disorder occuring before birth due to genetic mutations affecting the building of necessary proteins for the muscles. The disease can cause mental retardation and disability for life. Duchenne muscular dystrophy (DMD) or pseudohypertrophic disorder affects involuntary muscles such as the cardiac muscle. The disorder is genetic in origin and usually detected in children aged between 2 and 6 years, causing muscles to waste away due to the lack of a protein called dystrophin that functions to preserve the integrity of muscle cells. This disorder is severe and survival after the age of 30 is rare. Becker muscular dystrophy (BMD) disorder is similar to DMD except that onset is in the adolescence. It eventually spreads to the heart muscle potentially leading to heart failure. Emery–Dreifuss muscular dystrophy (EDMD) is a slowly progressing disease caused by gene mutations, which result in defects of emerin, lamin A, or lamin C proteins in the membrane surrounding the muscle cell nucleus. It occurs in children and results in weakness and wasting of the upper body. Myotonic muscular dystrophy (MMD) or Steinert's disease can occur in early adulthood but is not as severe as its congenital form. This is due to a gene defect associated with repeated sections of DNA on either chromosome 3 or chromosome 19. The disease progression is slow beginning with weakness in the facial muscles and spreading throughout the body. Distal muscular dystrophy (DD) is also due to gene mutation and affects young to middle-aged adults where muscles in the leg tend to degenerate but the disease is not life threatening. Another gene mutation disorder is limb–girdle muscular dystrophy (LGMD), which affects the waist and lower limbs. The disease is slowly progressive causing wasting of the limbs and some cardiopulmonary complications. Other similar disorders caused by genetic deficiencies or mutations include facioscapulohumeral muscular dystrophy and oculopharyngeal muscular dystrophy (OPMD), which affects the face and eyes.

5.3.5 Neuromuscular Junction Disorders

The NMJ acts as an interface for transferring electrical nerve impulse to impulses in the muscle itself. Diseases of the NMJ usually disrupt the receptor interaction and inhibit the function of the neurotransmitter acetylcholine. EMG readings show distinct instability of the MUAP waveform (jitter) where no two successive waveforms are exactly the same (Hassoun et al., 1994). This characteristic is due to instability in the damaged NMJ and can be best detected by single-fiber electromyography (SFEMG).

There are several variants of NMJ disorders, each with a different manifestation. Myasthenia gravis (MG) is an autoimmune disorder where the body's immune system attacks the NMJ area targeting the acetylcholine receptor or muscle-specific kinase, which helps organize acetylcholine receptors on the muscle cell membrane. The etiology of the disorder is unclear although some theories of bacteria or viruses as being likely causes exist. The disease seems to affect women more than men causing weakness or tiredness in voluntary muscles. A similar disorder is Lambert–Eaton myasthenic syndrome (LES) associated with weakness in the lower limbs with about 60% of the cases accompanied by cancer. Congenital myasthenic syndrome (CMS) presents prior to birth causing defects in the genes necessary to produce the required NMJ proteins. Symptoms and prognosis vary depending on the severity of the disorder because problems could occur with the presynaptic and postsynaptic part, which respectively affect nerve and muscle cells, and the synaptic parts between nerve and muscle cells.

5.4   The Electromyograph

EMG studies are more sensitive than clinical examinations in determining muscle fiber types involved in a disorder. EMG studies may demonstrate abnormal sensory nerve conduction although the patient still possesses normal motor-nerve conduction. This changes the diagnosis from peripheral neuropathy to sensory neuropathy and narrows the possible disorders afflicting the patient (differential diagnosis). It is also possible for EMG to reveal underlying pathophysiological problems due to axonal loss or demyelination. This will increase the clinician response time and allow some disorders to be treatable. Furthermore, EMG studies can assist a clinician in making a diagnosis without requiring a muscle biopsy. We now examine the components and processes involved in using EMG.

5.4.1   The Electromyograph and Instrumentation

The primary recorder of MUAPs is the electromyograph, an instrument used to record the action potential of the motor unit. There are two basic types of EMGs, needle or fine wire EMG and surface EMG (sEMG). EMG instrumentation includes electrodes, a signal acquisition system with audio and visual displays, biosignal amplifiers, and both low-pass and bandpass filters. A good EMG system produces minimum noise and artifacts. Most EMG instruments are available with standard settings for signal characteristics such as gain, common mode rejection ratio, input impedance, and filter bandwidth (Preedy and Peters, 2002).

5.4.2   EMG Electrodes

Needle electrodes or intramuscular wire electrodes are preferred over surface electrodes because they can access the individual motor units directly and measure the emanating action potentials more accurately. These electrodes are made of stainless steel and can be single or multistrand wires connected to a hypodermic needle. The most commonly used needle variation is the concentric ring needle with a monopolar single electrode as seen in Figure 5.16. Other needle electrodes include single fiber EMG (sFEMG) and macro EMG (Togawa et al., 1997). Unlike needle electrodes, wire electrodes are more easily flexed and suited for kinesiology studies. They are typically of small diameter and usually made of platinum, silver, nickel, or chromium insulated by either nylon, polyurethane, or Teflon. The electrodes are inserted into the muscle fibers by way of a cannula, which is later withdrawn. The wires are physically removed following the recording. The depth and location of insertion is important to obtain accurate readings. Typically, a

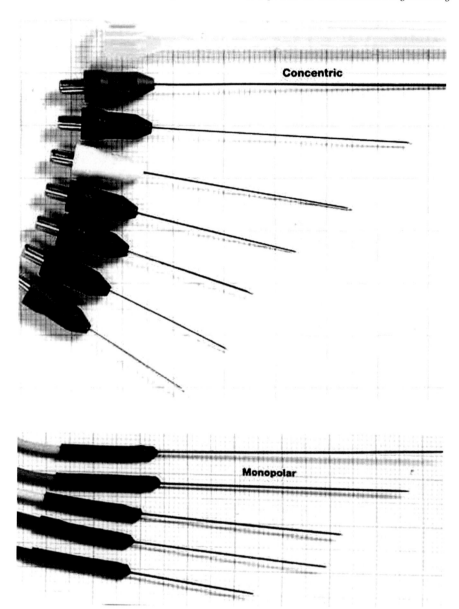

FIGURE 5.16

Examples of concentric and monopolar EMG needle electrodes sold by Med Products. (From http://www.medproducts.com. With permission.)

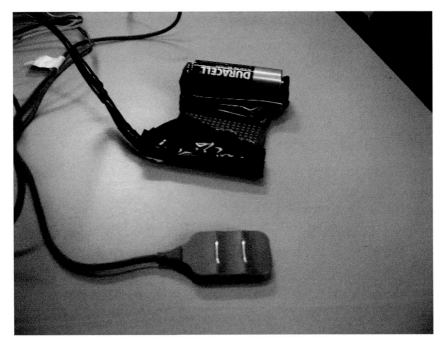

FIGURE 5.17
Surface electrodes used in the DelSys system by researchers of the biosignals lab in the Royal Melbourne Institute of Technology (RMIT), Australia.

ground electrode is used and measurements are made with respect to ground voltage. The ground electrode is generally placed in an area where there is no muscle activity, for example, over the elbow. The drawback of using these electrodes is the great discomfort caused to patients or test subjects.

Surface EMG electrodes are a more practical alternative to needle and wire methods because they are noninvasive. These electrodes are made of Ag/AgCl and come in the form of discs or bars. It is advisable to use gel or paste to obtain a good interface for signal acquisition as opposed to using dry or isolated electrodes, which possess inherent higher-noise corruption. Before attaching these electrodes, alcohol or medical napkins are first applied to cleanse the skin and the area is shaved to avoid motion artifacts. A surface electrode is shown in Figure 5.17 and a five channel sEMG experimental setup for measuring limb motion is depicted in Figure 5.18. More sophisticated surface electrodes such as the high-spatial resolution sEMG (HSR–EMG) electrode array have been developed by Reucher et al. (1987).

Recently, wireless electrodes (Ohyama et al., 1996) have been proposed as a means of reducing the level of crosstalk seen in standard wired electrodes. These electrodes have a Ag/AgCl interface but very high impedance, which means that it can measure surface potentials without requiring skin preparation or conductive paste. The authors demonstrated that signals

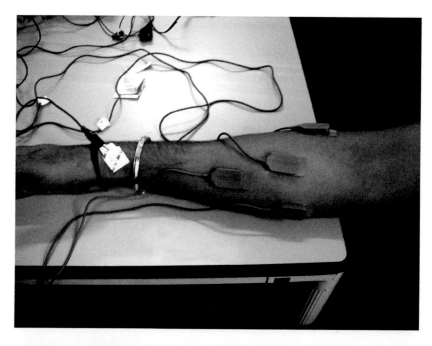

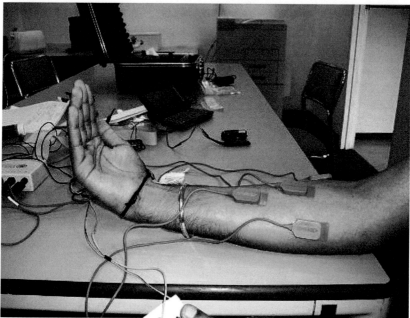

FIGURE 5.18

Experimental setup using five channel sEMG for detecting limb motion. Experiments performed in the biosignals lab of RMIT, Australia.

transmitted up to distances of 20 m suffered from minimal lag and distortion, making the wireless electrode a viable alternative for obtaining more accurate signals.

5.4.3   Signal Acquisition

EMG data are recorded in the hospital EMG lab by an electromyographer using either noninvasive surface EMG or needle EMG depending on the sensitivity of the apparatus available. The suspected area of disorder is identified for EMG recording, for example, the biceps brachii in the upper arm. The EMG is then triggered to record for a predetermined time after which the acquired signal is differentially amplified, bandpass filtered, and then digitized. Most modern EMG instrumentation provides these basic facilities such as the DelSys system.

All EMG recordings either from sEMG or needle EMG show changes in voltage potentials due to intracellular activity, either in the nerve or muscle (Preston and Shapiro, 2005). The process by which an intracellular electric potential is transmitted through the extracellular fluid and tissue is known as *volume conduction*. Volume conduction can be modelled as near-field or far-field potentials through layers of skin or fat. Recent research has focused on modeling sEMG based on volume-conduction models (Mesin and Farina, 2005, 2006; Mesin et al., 2006; Farina et al., 2000). If successful, this approach will be useful primarily in determining the decomposition of the sEMG waveform into the individual MUAP waveforms.

In near-field potentials, the recording is made as close as possible to the source of the action potential since acceptable recordings are not obtained until the electrodes are placed close to the source. The amplitude of the recording depends on the electrode proximity to the target muscle group and also the distance between the recording electrodes. The recording can be influenced by other volume-conducted near-field potentials such as compound MUAPs and sensory nerve action potentials. The single MUAP waveform produced is triphasic as the action potential approaches, passes beneath and then away from the measuring electrode (Figure 5.19). As the potential moves toward the electrode, an initial positive deflection is seen, followed by a large negative voltage spike, when the action potential is directly beneath the electrode. Finally, a smaller positive phase is observed as the potential propagates away from the electrode site. If the electrode is placed directly on the source of the action potential, the initial positive deflection may be absent and a negative spike is first observed. This results in a characteristic biphasic waveform as shown in Figure 5.19b. If the muscle action potential never passes beneath the electrode, one usually observes a small positive deflection similar to that depicted by Figure 5.19c. This situation can occur, for example, when recording a hypothenar muscle that is innervated by the median nerve. In this case, MUAPs in the thenar muscle never directly reach the hypothenar muscle. Far-field potentials quickly propagate over a wide area and two electrodes are used to record the same potential instantaneously, although one may be closer

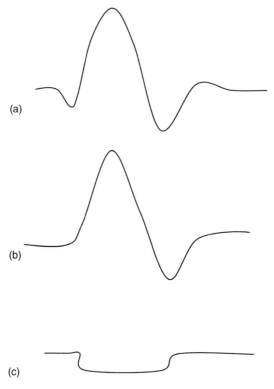

FIGURE 5.19

(a) Waveform morphology for a triphasic action potential. Initially a small
positive phase is observed as the action potential approaches, followed by a
negative spike when the potential is beneath the electrode. Lastly, a small
positive spike signals the wave travelling away from the measuring area.
(b) Waveform morphology for a biphasic action potential. The negative spike
is observed first because the measuring electrode is directly over the source
of the action potential. (c) Waveform morphology for action potential that is
far away (far-field wave). EMG records a small positive deflection.

to the source than the other. Action potentials with latencies that do not vary
with distance from the simulation site are generally far-field potentials.

Various information can be gathered by the electromyographer such as

a. *Insertional activity.* In normal muscles and nerves, initial insertion and
movement of the needle electrode results in bursts of short potentials
also known as *insertional activity.* These potentials have been observed
to have a duration of 1–3 ms and amplitudes of 0.3 mV, possibly due
to mechanical excitation of the muscles by the probing action of the
electrode.

b. *Spontaneous activity.* Abnormal or spontaneous depolarization of muscle can sometimes occur even without impulse activation. The cause

of these can be due to spontaneous depolarization of muscle cell, fasciculation potentials, myokymic discharges, and other factors (Preedy and Peters, 2002).

c. *MUAP morphology and stability.* When the initial activity has settled, the MUAP waveform and shape can then be processed. The shape of the individual MUAPs can provide clues as to the etiology of the disorder, provided their physiological generation is thoroughly understood.

d. *MUAP recruitment.* The number of motor units recruited by the CNS can also be inferred from the recorded EMG waveforms. These can measure physical strength and are useful for the detection of neuromuscular disorders.

5.4.4   Signal Amplification and Filtering

After amplification, the signal is first bandpass filtered, sampled, and then low-pass filtered again. Butterworth or Chebyshev filters are usually implemented in the instrumentation and there is seldom the need to specify the cutoff or bandpass frequencies. Some international journals such as the *IEEE Transactions on Biomedical Engineering* or the *Journal of Electromyography and Kinesiology* are, however, strict about the type of filtering and require that these values be reported. This is primarily because a major portion of power in the EMG signal is known to be contained in the 5–500 Hz frequency range of the power density spectra. Surface EMG recordings are preferred to be in the 10–350 Hz band whereas filtering within 10–150 Hz or 50–350 Hz, for example, is not favorable because portions of the signal's power above 150 Hz and below 50 Hz are eliminated. Intramuscular recording should be made with the appropriate increase of the high frequency cut-off to a minimum 450 Hz or by applying a bandpass filter of 10–450 Hz. These are ideal filter settings, which can be adjusted provided ample justification is given for doing so. Pattichis et al. (1995), for example, used an epoch of 5 s, bandpass filter with cutoffs at 3–10 kHz, and a sampling rate of 20 kHz, whereas the low-pass filter was set to 8 kHz. The filtered signal was then full or half-wave rectified, which simplified the following analysis because it was confined to positive-signal samples.

5.4.5   Signal Digitization

After the analog signal is collected, amplified, and bandpass filtered, it is digitized and stored on a computer for further processing. In digitization, sampling is done at the Nyquist rate, which is at least twice the highest frequency cutoff of the bandpass filter. For example, if a bandpass filter of 10–350 Hz is used, the minimum sampling rate should be 700 Hz (350×2) or preferably higher to improve accuracy and resolution. In some cases, a low-pass filter is used to first smooth the EMG signal such that a lower-sampling rate (50–100 Hz) can be used for digitization. The raw unfiltered EMG signal can also be stored in the computer using a sampling rate greater than 2.5 kHz. Note that the signal will be susceptible to high-frequency noise and should be bandpass filtered as described previously. The conversion of the analog EMG signal to digital signal can be done by an A/D convertor with a resolution of 8–12 bits.

Digitized signals can be further normalized, such as in force/torque experiments where normalization is with respect to the maximum voluntary contraction (MVC). This procedure requires the subjects be properly trained to avoid discrepancies and experimental inconsistencies. Normalization may also be done relative to joint angle or muscle length, velocity of shortening or elongation, or applied load.

5.4.6   Factors Influencing Electromyograph Readings

In all EMG studies, factors such as orientation over the muscle with respect to tendons, motor point, and motor fiber direction are important considerations when placing electrodes. Positional factors such as these primarily affect the collected EMG signal and are crucial when reporting experimental results. One difficulty in obtaining good EMG readings is that the MUAP originates deep within the muscle and has a very small amplitude (in the order of microvolts). This makes the signal difficult to read and also easily corrupted by background noise or muscle movements. Concentric needle EMG is used to obtain a clearer signal, but suffers from the fact that it is an invasive technique. The fact that features of the recorded EMG depend on the effort exerted by the muscle and the level of muscle fatigue increases the difficulty. For example, a patient who is uncomfortable during the EMG reading may distort the true signal reading by flexing or tensing his or her muscles. This, in turn, causes the clinician to look for a better spot to take EMG measurements, which means piercing the skin again and increasing the discomfort of the patient. From a physiological perspective, the EMG waveform depends on the properties of the particular muscle fiber itself, for example, the number of motor units firing, the firing rates of the motor units, the motor unit action duration, the propagation velocity of the action potential, and the motor unit size. In addition, the frequency of the waveforms also varies with different levels of muscle contraction.

The EMG instrumentation may also affect the appearance of the MUAP waveforms, for example, the type of electrode and the filter properties used to process the raw signal affect the detected MUAP waveform. The shape and amplitude of the EMG also depends on the distance between the muscle fiber and the recording site. A more critical factor that can adversely affect the readings is crosstalk, which is usually attributed to volume conduction–dispersion effects caused by action potentials enroute to the electrode site. As the MUAP from deep muscles propagate toward the surface, they pass through layers of muscle and fat that have different conduction properties. These layers change in waveform speed, and the waveform is then distorted or dispersed. One then observes a motor unit action potential train (MUAPT) due to the MUAP from the target muscle site being superimposed with MUAPs originating from adjacent muscle sites. Togawa et al. (1997) suggest that a tolerable crosstalk range is 2–5% of the average signal recorded on a channel, for example, a 25 mV waveform can tolerate 0.5–1.25 mV crosstalk. In addition to intermuscular effects, crosstalk can also arise from cables and switches between the electrodes and it is advisable to ensure that it does not corrupt the recorded signal. This can be achieved by selecting the appropriate electrode size, interelectrode distance, and location over the muscle prior to experimentation. Extra care should be taken when working on areas where many narrow muscles are tightly gathered (e.g., forearm) or when working with superficial/thin muscles (e.g., trapezius). For surface EMG, it is important to note specific areas with subcutaneous adipose tissue such as the abdomen, buttocks, and chest as adipose tissue enhances crosstalk (Solomonow et al., 1994). In such cases, crosstalk can be reduced by using smaller surface electrodes with decreased bipolar spacing.

Surface EMGs or sEMG are noninvasive but record instead a group of MUAPTs (Ma et al., 2001), which are a larger summation of MUAPs. It is extremely challenging to discern the individual MUAP pulses due to this time-domain superposition (Emly et al., 1992). The problem is further aggravated if sEMG readings are taken in close proximity to large organs such as the heart. In these cases, the larger EMG signal from the cardiac muscle activity tends to mask the smaller EMG signal.

5.5 Electromyograph Signal-Processing Methods

Studies of EMG signals for research and clinical purposes have been undertaken since the 1950s and were initially based on analysis of stochastic temporal characteristics and frequency domain power spectra. Prior to that, traditional methods were used with quantitative analysis performed manually by the physician or electromyographer. These methods have made only modest improvements on previous methods and are hence treated sceptically (Hassoun et al., 1994). The objective behind EMG signal processing is to extract characteristic information from the waveform that may be indicative of a particular muscle action. Highlighting special characteristics can infer neuromuscular disorders, muscle strength, and nerve health. There are three steps in preprocessing EMG signals to extract important features: segmentation of the signal, decomposition and identification of the MUAPs, and extraction of waveform characteristics. Figure 5.20 depicts a flowchart for EMG feature processing. The details of the signal-processing methods are described in the following subsections.

5.5.1 Preprocessing: Segmentation of Motor Unit Action Potentials

Since the action potential resulting in the MUAP occurs during fixed intervals of muscle contraction, it is necessary to first locate the area of interest in the

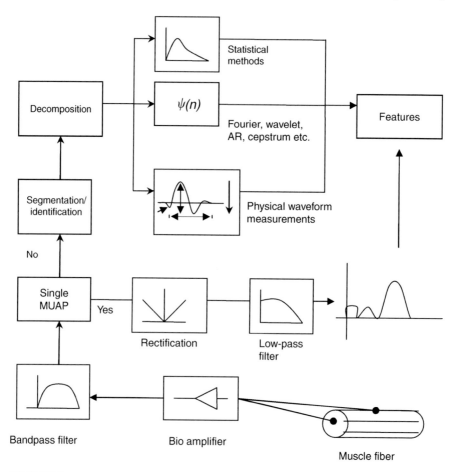

FIGURE 5.20
The major steps involved in the signal acquisition, preprocessing, and feature
extraction of EMG signals recorded from a muscle fiber.

signal. This preprocessing step reduces the amount of data to be processed
and potentially reduces the effect of noise. Noise in this context refers to
unwanted information from unimportant waveforms or background emissions
that superimpose on the required EMG signal.

EMG signals are first amplified, usually by a factor of 1000 since they
have a very small amplitude in the range of microvolts. It is preferable to
report on the characteristics of the amplification such as single, differential or
double-differential amplifiers, input impedance, common mode rejection ratio
(CMRR), and SNR because the following processing stages, such as template
matching, may require this information.

The next signal-processing method is filtering and MUAP segmentation.
In some experiments, for example, the EMG signal is high-pass filtered at
250 Hz to remove background noise (Pattichis et al., 1995). The beginning

(BEP) and end (EEP) extraction points of the MUAP are identified by sliding an extraction window of length 3 ms with width $\pm 40\,\mu V$. The BEP is found if the voltage to the left of the MUAP waveform after searching for 3 ms remains within $\pm 40\,\mu V$. The EEP is determined similarly by searching to the right. A baseline correction is then performed to obtain the area of the MUAP waveform. Some researchers have extracted MUAP information using weighted low-pass filters (Xu and Xiao, 2000) and Weiner filters (Zhou et al., 1986) to obtain the signal decomposition of surface EMGs.

5.5.2   Intermediate Processing: Decomposition and Identification of Motor Unit Action Potential Trains

Decomposition of the EMG signal refers to the separation of the complex signal, such as MUAPTs, into the individual action potentials. This intermediate processing is usually required in surface EMG because of the complexity of the MUAPTs originating from effects of the adjoining muscles. The objective of decomposition (Thompson et al., 1996) is to determine the EMG signals at all levels of contraction, separate reliably and accurately all MUAPs, and to provide an acceptable level of decomposition in the presence of noise. An inherent problem in this scheme is that superposition of MUAPs makes it difficult to determine the true number of MUAPs in the composite waveform. Besides that, movements due to surface-electrode placement distorts the composite MUAPT further, whereas discharge to discharge variability makes MUAP localization difficult.

Since the 1980s, efforts to extract single MUAP patterns from the composite MUAPTs have been attempted using basic signal-processing methods (LeFever and De Luca, 1982a,b). De Luca (1993) described a decomposition system, which employed a special quadrifilar electrode to measure three EMG channels simultaneously. The effect of superposition was reduced by bandlimiting the signals to a window of 1–10 kHz and sampling at 51.2 kHz. Next, a template-generation process was employed for 150 waveforms using a nearest-neighbor algorithm. In this process, new data were compared with templates using a least squares method, which was found to achieve up to 90% classification accuracy when decomposing 4–5 MUAPs. A competing system was the automatic decomposition electromyography (ADEMG) system developed by McGill et al. (1985), which used digital filters to transform the rising slope of MUAPs into spikes for detection using a zero-crossing detector. The authors reported an identification rate of 40–70% and argued that their method outperformed template matchingbecause sampling can be undertaken at the Nyquist frequency of 10 kHz for maximum information extraction. Work by Gerber and Studer (1984) involved partitioning the waveforms into active and inactive segments where the inactive segments were perceived to contain noise and discarded. The nearest-neighbor algorithm was used to cluster the segments based on a feature vector and superpositions were resolved using template matching. Louden (1991) in his PhD thesis studied the decomposition of needle EMG waveforms using normalization to separate nonoverlapping EMG signals. The detected waveforms were segmented based on eight features, whichincluded amplitudes, slopes, turning points, area, and duration. Active segments were identified using a dynamic time-warping method, which searched for similar sequences followed by clustering and template matching to obtain the individual MUAPs. More recently, high-order statistics have been employed to extract individual MUAP waveforms as will be seen later. Shahid et al. (2005) have attempted using second-order spectrum or bispectrum analysis to estimate the individual MUAP waveforms from needle EMG and sEMG. Their work focused on the rectus femoris and vastus lateralis muscles with estimations performed on six types of muscle contractions with subject loading. Higher-order spectrum analysis could be performed at the cost of increased complexity and slower recognition, but their added benefit has not been clearly demonstrated.

In the early 1990s, work began on applying techniques such as ANNs to the detection of individual MUAP waveforms. Hassoun et al. (1994) proposed the NNERVE algorithm in 1994 in an attempt to automate the extraction of individual MUAPs from the compound MUAPTs. They proposed that a successful system could be designed for routine clinical application but it would be required to work well over a wide range of cases. These observations were made during a time when it was widely accepted that patients presented different EMG waveform characteristics, which made general recognition systems inherently hard to design let alone function reliably. It was, however, shown that NNs could learn to recognize similarities across a wide range of waveform characteristics to high degree of accuracy and a general detection system could be devised. In EMG waveform decomposition systems, errors in processing the signal tend to propagate toward the final signals. These can be due to error in detecting the number of spikes, in the network retrieval stage, and in the interpeak interval (IPI) analysis. Several measures have been proposed to handle these errors (Hassoun et al., 1994) as an attempt to quantify them, some of which are listed in the following:

a. *Detection ratio*

$$\mathrm{DR} = \frac{I}{K} \qquad (5.2)$$

where I is the number of true spikes and K the total number of spikes from all detected MUAP classes.

b. *Classification ratio*

$$\mathrm{CR} = \frac{I_\mathrm{T}}{I} \qquad (5.3)$$

where I_T is the number of true spikes and I the total number of spikes identified by the detector.

c. *Error ratio*

$$\mathrm{ER} = \frac{I_\mathrm{E}}{I} \qquad (5.4)$$

where I_E is the number of misclassified spikes. This is the proportion of spikes incorrectly classified into one valid MUAP class.

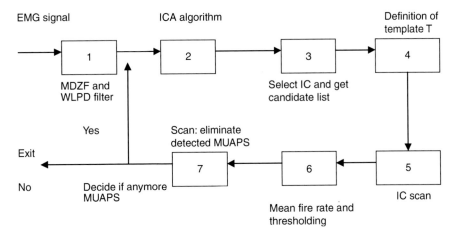

FIGURE 5.21

Example of an sEMG waveform decomposition system (Garcia et al., 2002) consisting of signal processing (1), signal decomposition (2), template matching (3–5), and postprocessing (6–7).

Comparisons of the NNERVE system were made against ADEMG (McGill et al., 1985), which used traditional signal-processing, template-matching, and pattern-firing analysis. It was found from simulation that in both systems, the number of MUAP classes detected exceeded the true number of classes. The NNERVE system was shown to have approximately 85.9% accuracy compared to ADEMG, which achieved only 52.3% using first-order filtering and 34.6% using second-order filtering.

Recently, more complex systems to decompose sEMG waveforms into their individual MUAP and MUAPTs have been proposed. The primary difficulty with identifying individual MUAPs or MUAPTs from sEMG readings is the superposition effect and the relatively low SNR of sEMG compared to waveforms obtained through intramuscular recordings or needle EMG. This is caused by the filtering effect of tissues between the electrodes and muscle fibers, which cause motion artifacts. Garcia et al. (2002) developed the decomposition system seen in Figure 5.21 consisting of preprocessing, EMG decomposition using ICA, and postprocessing of the extracted MUAPs. A modified dead-zone filter (MDZF) was used to remove MUAPs below a certain threshold and the resulting waveform was then enhanced by using a weighted low-pass differential (WLPD) filter. An ICA algorithm was then applied to the extracted individual waveforms, which were then separately matched against template waveforms to identify their origins. This process was repeated iteratively on a single MUAPT until no further MUAPs could be extracted. The authors argued that their system showed good extraction properties because the single MUAPs compared well with generally accepted values for motor unit behavior. The accuracy of this method is, however, questionable because the true number of MUAPT or MUAPs present in the

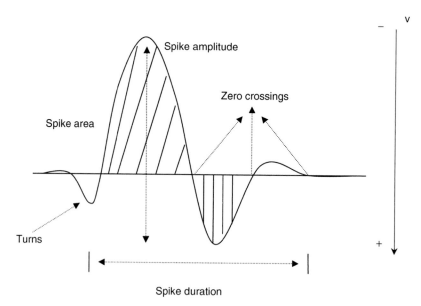

FIGURE 5.22
Example of physical measurements of a single MUAP waveform.

sEMG waveform is not known beforehand. Furthermore, the template matching phase requires eye-matching and, therefore, the presence of a human expert which is less useful in the design of a fully automated diagnostic system.

5.5.3   Postprocessing: Feature-Extraction Methods

After processing the raw EMG waveforms, further information can be extracted from the physical measurements of the waveform, statistical quantities of the signal, time- and frequency-domain characteristics, and coefficients of theoretical EMG waveform models. For physical measurements of single MUAP waveforms (Figure 5.22), the following quantities have often been used (Pattichis et al., 1995; Huang et al., 2003):

     a.   *Duration (DUR)*. The time between the beginning and end of a MUAP determined using a sliding window. For example, a window with length 3 ms and $\pm 40\,\mu$V.

     b.   *Spike duration (SPDUR)*. The time between the first and the last positive spike.

     c.   *Amplitude (AMP)*. The difference between the largest positive peak and the most negative peak gives the peak-to-peak amplitude. Other methods include measuring the difference between smallest positive peak and largest negative peak (Pattichis et al., 1995). Amplitude for sEMG can be estimated as a mean absolute value (MAV) or as a RMS of the time varying standard deviation. In this approach, the signal is first passed through a noise-rejection filter, decorrelated, demodulated, and smoothed (Clancy et al., 2002). Some researchers have also used the integrated raw EMG signal alone (Onishi et al., 2000).

d. *Area.* The area of the MUAP waveform, either the entire waveform or the rectified version of the waveform.

e. *Spike area.* This is similar to the area but it is the area under the spikes and calculated only over the spike duration.

f. *Phases.* The number of baseline crossings that exceeds a certain voltage threshold, for example, $25\,\mu V$.

g. *Turns.* The number of positive and negative peaks separated by a specified threshold voltage, for example, $\pm 25\,\mu V$.

h. *Willison amplitude (WAMP).* The number of counts for a change of signal in the EMG amplitude above a predefined threshold.

Since the EMG signal is automatically digitized by the EMG recording instrument, it can be immediately processed as a sequence of digital signals. As such these discrete signals may be assumed to be drawn from some unknown statistical distribution and standard statistical descriptors such as the mean, median, standard deviation, and variance can be calculated. These first- and second-order moments and cumulants are adequate to completely specify the EMG signal under the assumption that it is a Gaussian signal obtained from some linear measuring system. The combined MUAPT, however, is nonlinear and hardly Gaussian, which entails the computation of higher-order moments and cumulants to estimate the statistical distribution. These HOS have been applied to decompose surface EMG signals (Zazula, 1999; Zazula et al., 1998; Shahid et al., 2005). Power spectrum and higher-order spectral techniques have also been applied to extract features. Recent work by Shahid et al. (2005) looked at using statistical cumulants and comparing sEMG techniques with needle EMG to verify the accuracies of their waveform decompositions.

Although it is known that the EMG signal is nonlinear and nonstationary, several investigators, however, have attempted to model a short period of the EMG signal as a stationary Gaussian process (Mahyar et al., 1995; Graupe et al., 1982; Huang et al., 2003). Under this assumption, the AR model is used to represent the EMG signal as a time series with the following form:

$$y_k = -\sum_{i=1}^{M} a_i x_{k-i} + w_k \qquad (5.5)$$

where y_k is the kth output of the AR, x_{k-i} the $(k-i)$th sample, a_i the AR coefficient estimate, and w_k the additive white noise. The adaptive LMS algorithm is often used to compute the AR parameters given the update rule for all $i = 1, \ldots, M$ as

$$a_i(n+1) = a_i(n) - 2\beta e(n)x(n-i) \tag{5.6}$$

where β is a constant rate of convergence and $e(n)$ the difference between the nth EMG sample and the nth estimated AR parameter. The coefficients of the AR model are then used as features, where, for example, a fourth-order AR model yields four coefficients. The optimal order for the AR model can be estimated, for example, by using AIC (Akaike, 1974) where the model order minimizes some selected criterion function. The AIC criterion for selecting AR model orders was used by Coatrieux (1983), Maranzana et al. (1984), and Pattichis and Elia (1999) and depending on the application, model orders in the range 3–20 were found useful. Coatrieux (1983) applied AR models in the range 11–13 for muscle force, Maranzana (1984) used lower orders up to order 9, and Basmajian et al. (1985) found model order 6 to be adequate. Pattichis and Elia (1999) combined AR coefficients with cepstrum coefficients determined from the estimated AR coefficients, as features that were then fed into an NN.

Earlier studies had shown that EMG signals were not only white Gaussian noise but also exhibited chaotic behavior prompting some researchers (Ehtiati and Kinsner, 1999) to investigate dynamical models in an effort to investigate the chaotic origins of the EMG signal. It was found, for example, that sEMG signals of the deltoid muscle exhibited chaotic behavior and were associated with low dimensional strange attractors. An attractor is a point or set of points into which the dynamical system converges. Under this interpretation, the multifractal dimensions of the EMG signals could be extracted and used as characteristic features. These features have been shown to be related to the contribution of a specific muscle, for example, muscles supporting limb movement have smaller fractals than muscles performing the primary movement.

Fourier and wavelet decomposition analyses have also been applied as a feature extraction method for EMG signals. The advantage of using wavelet over Fourier analysis is that the technique captures information in the EMG signal pertaining to changes in both time and frequency domain whereas Fourier analysis provides only time-domain information. It should be noted though that the EMG spectrum itself lies in the range 50–70 Hz and is variable depending on fatigue and recruitment level of motor units (Ogino, 1983). In most cases, the coefficients of the wavelet decomposition are used as features to be classified. Kumar et al. (2003), for example, showed that wavelet analysis could be used to extract features that were distinctively different for cases of fatigued and nonfatigued muscles. Their experiments were based on 240 readings of surface EMG using wavelet functions such as the Haar, Daubechies, and symmetrical wavelets. The symmetrical wavelets were found to enhance the differences between the extracted features more than other wavelet functions. Karlsson et al. (1999) compared FFT and wavelet analysis in simulated EMG and recorded EMG signals (from isometric knee extensions). Khalil and Duchene (2000) processed the wavelet information and used variance and covariance matrices computed from the signal decomposition using various wavelet bases. Wavelet analysis has also been used by Zhang et al. (2000) to analyze somatosensory signals for detection of early cerebral injury. Ranniger and Akin (1997) applied wavelet decomposition to determine the mean power spectra in the EMG waveform. Features such as interscale wavelet maximum (ISWM) (Arikidis et al., 2002) have been further used to characterize the EMG interference pattern in the diagnosis of neuromuscular disease. In this method, the EMG signals are decomposed with the redundant dyadic wavelet transform and the wavelet maxima (WM) is found. A thresholding method is then applied to remove the WM resulting from noise and background activity. An efficient fine-to-coarse algorithm identifies the WM tree structure for the MUAP rising edges. The WM for each tree are summed at each scale and the largest value is the ISWM. It has been found that these ISWM values give highly significant differences between healthy, myopathic, and neuropathic subjects and thus provide very useful features for disease classification.

In cases where the full nondecomposed sEMG has been used, several methods of feature extraction have been proposed. The challenge are to extract useful information from the resulting measurements and relate them to the muscle's state. Several researchers have postulated that the full waveform may contain more accurate information about the motion and have attempted to extract useful information (Park and Lee, 1998). The following time and spectral information have been used:

a. *Integrated absolute value (IAV).* This is the estimate of the MAV of the signal taken for a segment i of a waveform. If the segment contains N samples, the IAV is defined as

$$\overline{x}_i = \frac{1}{N} \sum_{k=1}^{N} |x_k| \qquad (5.7)$$

b. *Difference absolute mean value (DAMV).* This is the MAV of the differences between two adjacent samples:

$$\Delta \overline{x}_i = \frac{1}{N-1} \sum_{k=1}^{N-1} |x_{k+1} - x_k| \qquad (5.8)$$

c. *Variance (VAR).* The variance of the discrete signal is estimated as

$$\sigma_i^2 = E(x_i^2) - E^2(x_i) \qquad (5.9)$$

where $E(x_i)$ denotes the expected value of the signal in segment i.

d. *AR model coefficients (AR).* These coefficients denote the position of the peaks in the signal spectrum.

e. *Linear cepstrum coefficients (LCC).* The cepstrum coefficients contain the spectrum information of the signal.

f. *Adaptive cepstrum vector (ACV)*. This is an enhanced algorithm, which extracts cepstrum coefficients using block- and adaptive-signal processing.

Feature extraction is an important processing step in the diagnosis of diseases or classification of muscle activity. If the extracted features accurately model certain muscular activity, then simple thresholding methods or linear discriminant analysis are adequate for classification. In these situations, the features are linearly separable (distinctively different), but this is rarely the case with EMG waveforms because many internal and external factors alter the waveform so that it no longer resembles the original waveform. Furthermore, a simple motion such as arm extension may not be described fully by information from a single MUAP waveform. We are then faced with the question of what information best characterizes a particular muscle activity or state. Although the signal-processing techniques described previously provide signal information, they do not necessarily provide critical information to fully link the waveform to the muscle state. This is not the problem of the techniques, rather it is the result of incomplete system knowledge. One way to address this problem is to first accept the knowledge limitation and attempt to estimate the generating model of the activity. In this aspect, CI techniques become useful in dealing with function estimation and estimation through learning. We will examine some of the applications in the following sections.

5.6 Classification of Neuromuscular Disease

In this section, we review CI techniques that have been applied to diagnosing neuromuscular diseases. It is often difficult to compare these techniques because the data have been privately collected and curated, and yield only quantitative information but little qualitative information as to how the techniques compare. The practical use of CI in automated diagnosis is still relatively new with the objective of designing automated diagnosis systems to aided in the diagnosis of neuromuscular diseases. Clinicians, however, are still wary of relying on automated diagnostic systems due to the fear of an unproven system providing entirely wrong diagnoses. Active research is now focused on signal-processing techniques and volume-conduction modeling to obtain good EMG signal characterization. The ensuing discussion is aimed at introducing the reader to methods and performance results that could assist in the application of CI to their research problems.

5.6.1 Artificial Neural Networks

In ANN, the popular MLP network has been applied for classification in several research ventures. A series of work in relation to diagnosis of myopathies and motor neuron disease were undertaken from the early 1990s to 2000.

TABLE 5.2
Comparison of Classification Accuracies of Myopathy, Motor Neuron
Disease, and Normal Subjects

Method	Acc. (%)	Feature Set
BP–NN (Pattichis et al., 1995)	80.0	MUAP time domain
NN-majority vote (Pattichis et al., 1995)	80.0	AR, cepstrum, time domain
SOM	80.0	MUAP time domain
SVM (Xie et al., 2003)	82.40	MUAP time domain
Neurofuzzy hybrid (Christodoulou and Pattichis, 1999)	88.58	AR, cepstrum, time domain
Hybrid SOM (Christodoulou and Pattichis, 1999)	94.81	MUAP time domain
Statistical (Christodoulou and Pattichis, 1999)	95.30	MUAP time domain
Hybrid SOM–LVQ (Christodoulou and Pattichis, 1999)	97.61	MUAP time domain
Genetics-based machine learning (Pattichis and Schizas, 1996)	≤ 95	MUAP time domain

Note: The data sets used are not the same, hence the results are just a general indication
of the potential of these techniques.

In 1990, Schizas et al. (1990) looked at the application of NN to classify
action potentials of a large group of muscles. The early results yielded accuracies of approximately 85% and simple physiological features of the EMG
waveforms were extracted. The supervised data were obtained from the decisions of two expert neurologists using the Bonsett protocol that consisted of
a series of questions used to arrive at the final diagnosis. Their work was later
extended (Schizas et al., 1992) to comparing classification algorithms using
K-means, MLP-NN, SOMs, and genetic-based classifiers. It was discovered
that the simple K-means algorithm was not suitable as it gave the lowest classification accuracy, but both NN and genetic-based models produced promising results (Table 5.2).

Pattichis et al. (1995) extended the work by Schizas et al. by applying NN
to 880 MUAP signals collected from the biceps brachii muscle. The data consisted of 14 normal subjects, 16 patients suffering from motor neuron disease,
and another 14 patients suffering from myopathy of which 24 were randomly
selected to form the training set and the remainder formed the test set. They
compared the MLP network against the K-means clustering and Kohonen's
SOM. The best performance using a K-means clustering algorithm using 3
cluster means was 86% accuracy on the training set and 80% on the test set.
The MLP network fared better giving 100% accuracy on the training data
but required a fair amount of tuning for the network gain η and momentum
α coefficients. The best model had a hidden layer with 40 neurons providing
a test accuracy of only 80–85%. The problem then was that short of trial

and error; no other way was known to select the appropriate MLP model. Further investigation revealed that the same pathological cases were always misclassified although different MLP models were used. These were subjects that suffered from a typical motor neuron disease with bulbar involvement and no serious involvement in the biceps brachii. Another commonly misclassified subject suffered from polymyositis myopathy. This suggested that the disorders possess finer EMG characteristics, which could not be easily differentiated from normal subjects. Similar results were obtained when applying Kohonen's SOM with 91% accuracy on the training set but again only 80–85% on the test set. The results on genetics-based machine learning (GML) were reported separately (Pattichis and Schizas, 1996) where the same data was classified using GML models derived by genetic learning. Better classification was achieved with less evolution indicating that frequent changes in the classifier state should be avoided. It was argued that the GML method yielded simple implementation models, were easier to train, and provided quicker classification due to parallel-rule implementation. This is, however, questionable as the number of classifiers from the evolution algorithm numbered greater than 500 on average. Comparisons with unsupervised learning methods was undertaken in 1999 by Pattichis and Elia (1999) using SOM, SOM with learning vector quantization (LVQ), and statistical methods. In this experiment, 1213 MUAPs were analyzed and classified using a proposed EMG decomposition and classification system. The EMG signal was signal thresholded using average-signal and peak-signal values and located using a 120 bit window corresponding to a sampling frequency of 6 ms or 20 kHz. The hybrid SOM–LVQ method was found to produce the highest classification accuracy of all the methods used by Schizas and Pattichis to classify between myopathies and motor neuron diseases.

NNs have also been used as decision-making tools for online classification of various neuromuscular diseases (Bodruzzaman et al., 1992). The main challenge was to design an automatic classification system based on signal-processing techniques such as AR modeling, short-time Fourier transform, the Wigner–Ville distribution, and chaos analysis to define abnormalities in diseases such as neuropathy and myopathy. It was often found that features derived from these different methods were nonseparable and, hence, difficult to classify using a single method. This was demonstrated by the probability density function of the features, which often overlapped. Ten features were extracted from the processing methods described, and feature selection was used with NN to provide at least 95% classification accuracy.

5.6.2 Support Vector Machines

The SVM classifier has also been applied to EMG signals for diagnosis of neuromuscular disorders. Xie et al. (2003) trained a multiclass SVM classifier to distinguish between healthy subjects and patients suffering from motor neuron disease or myopathies. They extracted the physical features such as EMG spike amplitude, number of turns, and spike duration. The SVM classifier was then applied to a data set consisting of 80 subjects. These were composed of 20 normal subjects, 30 suffering from motor neuron disease, and 30 from myopathies. Initial results showed improved performance over an MLP network trained using backpropagation. The multiclass SVM provided maximum accuracies of 82% whereas the MLP network achieved an accuracy of 74.8% at best.

The use of SVMs in this area of biomedical engineering is still new and many of the reported accuracies using NNs could potentially be improved using the SVM classifier instead. This area has considerable potential for further investigation to develop increasingly accurate diagnosis systems for neuromuscular diseases.

5.6.3 Fuzzy and Knowledge-Based Systems

Fuzzy techniques have also been employed for automated diagnosis of disease under the assumption that fuzzification of features will allow improved classification of differential diagnosis for certain neuromuscular diseases. In other applications, neurofuzzy controllers have been built to control the medicine dispensing and monitoring during surgery.

MUAPs recorded during routine EMG examination have been used as inputs for the assessment of neuromuscular disorders. Xie et al. (2004) proposed using a fuzzy integral to combine outputs from different NNs separately trained on time-domain measure, AR, and cepstral coefficient input data. They argued that conventional, computer-aided methods of MUAP diagnosis based on a single feature set and single NN model produce diagnosis accuracies that were not always satisfactory. Rather, they believed that using a hybrid decision support system based on a fusion of multiple NN outputs represented by a fuzzy integral would provide better results. The fuzzy integral is a nonlinear method for combining multiple sources of uncertain information by combining objective evidence for a hypothesis with prior expectation of the importance of the evidence to the hypothesis. Fuzzy measures are not necessarily additive and can be used to represent situations for which the measure of the whole is not equal to the sum of the parts. Using this measure, the authors showed that the resulting hybrid fuzzy system could produce 88% accuracy for a three class problem, which was better than the single feature single NN architecture.

Large-scale medical decision support systems have also been designed using fuzzy logic and probabilistic rules. Suojanen et al. (2001) used causal probabilistic networks to model the characteristics of the diagnostic process present in diagnostic tasks. The diagnosis is often made under uncertainty and chooses between diagnoses that have small but not insignificant prior probabilities. This is because small prior probabilities could also indicate a particular symptom frequently associated with several diseases and conversely the presence of several diseases can also aggravate the symptom. For diagnostic problems that share these characteristics, the diagnostic phase is broken down into a number of smaller phases. In the first phase, only single diseases are considered and the number of possible diseases are reduced. This is followed by pairwise comparison of diseases to eliminate those from the plausible diagnoses before larger subsets of diseases are considered together. The method was applied to the diagnosis of neuromuscular disorders, building on previous work with the MUNIN system. The results demonstrated that causal probabilistic networks gave large reductions in computation time without compromising computational accuracy making it a practical tool for deployment in large medical decision support systems.

Fuzzy logic has also been used to design neurofuzzy controllers for controlling biomedical systems, for example, to control the flow of neuromuscular block during surgical operations so that the patient remains properly sedated. A neurofuzzy controller was proposed by Shieh et al. (2004) for the delivery of a muscle relaxant (i.e., rocuronium) as general anaesthesia to control the system more efficiently, intelligently, and safely during an operation. The system was trialled on 10 ASA I or II adult patients anaesthetized with inhalational (i.e., isoflurane) anaesthesia and monitored using a Datex relaxograph. The EMG signals were pruned by a three-level hierarchical structure of filters to design the controller and controlled via a four-level hierarchical fuzzy logic controller and rule of thumb concepts. It was found that stable control could be achieved with a mean T1% error of -0.19% (SD 0.66), accommodating a good range of mean infusion rates (MIR). Similarly, good control was achieved when the system was applied to another drug, mivacurium. The hierarchical rule-based monitoring and fuzzy logic control architecture provided stable control of neuromuscular block despite the considerable variation among patients in drug requirement. Furthermore, a consistent medium coefficient variance (CV) in both drugs indicated that the controller could withstand noise, diathermy effects, artifacts, and surgical disturbances. Earlier work by Mason et al. (1994) investigated clinical trials using a fuzzy PD+I and self-organizing controllers for the delivery of atracurium.

5.7   Prostheses and Orthotics

Since the mid-1970s, research has focussed on the development of myoelectric prostheses that operate using biosignals from the muscles. Myoelectric prostheses are an advancement of the static hook-type prostheses because further motion such as grasping and lifting is available. The motivation behind them is to mimic the function of the original limbs as clearly as possible.

The concept of using myoelectric signals to power prosthesis has been recognized since the late 1970s and early 1980s (e.g., Jacobsen et al., 1982; Hudgins et al., 1994; Katutoshi et al., 1992). Later research concentrated on proper signal processing to extract motion commands or arm positions from EMG signals. Park and Meek (1995), for example, proposed adaptive filtering methods to obtain force estimations of contracting muscles, but their filter was impractical because the true signal and its derivative were not available.

To circumvent this problem, they proposed using a time-invariant second-order low-pass filter in parallel to provide the required estimates.

More recent investigations have been aimed not only toward providing mobility but also toward giving a degree of sensation to the disabled person. Research is also starting to look at using brain signals to power the prosthesis rather than muscle signals, however, many challenges remain to be addressed. A recent breakthrough reported in *Popular Science** was the first truly brain-controlled bionic arm that could move at the shoulder, elbow, and wrist. This six-motor prosthesis incorporated a 64-bit computer chip embedded in the forearm and provided the wearer with some sensation, but it was heavy and still inaccurate for practical use.

We now examine several contributions to the advancement of prostheses. We find that most of the research has reported the successful application of CI techniques, but issues such as system accuracy and speed remain fundamental research problems.

5.7.1 Kinematics and Kinetic Studies

Some fundamental research has addressed the problem of recognizing limb position using EMG readings. These studies focus on limb biomechanics with the objective of capturing the kinematic and kinetic characteristics of muscle groups and determine limb position based on the kinematic information. Ma et al. (2001) employed an NN to classify hand directions from multichannel sEMG measurements. Their EMG readings were a complex superposition of MUAPs from the palmaris longus, brachioradialis, and extensor digitorum muscles in the wrist. The integral RMS values were used because it indicated the strength of contraction of the muscle. The authors then trained an NN to classify five wrist movements based on sEMG readings of the three muscles and reported accuracies of 88% using the integral RMS as features. Similarly, Wang et al. (1997) reported preliminary results from using an ANN with features obtained using a parametric AR model to detect motion in the flexor carpi radialis and the extensor carpi ulnaris muscles. Peleg et al. (2002) applied genetic algorithms and K-means clustering to sEMG data collected from the finger flexor muscles to study finger movement and finger identification in amputees. They reported probability errors of 2%, which was a considerable improvement over earlier work (Uchida et al., 1992) with an average 15% probability of error.

Joint moments and angles have also been estimated using NN (Suryanarayanan et al., 1995; Ragupathy et al., 2004). Wang and Buchanan (2002) applied an adjusted backpropagation algorithm to train their NN since the error signal from the muscle activation was unavailable. Rather than using a complex muscle activation model, the authors argued that the complexity could be represented by a black box ANN trained on muscle activation

---

*Accessed on 28/08/06, http://www.sigmorobot.com/technology/news/toast_bionic_man.htm.

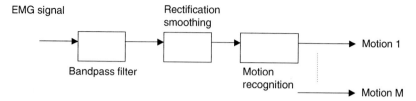

FIGURE 5.23

An example of a limb-motion detection system used to classify between the various motions captured by the EMG waveform.

information. Best prediction accuracies were 95.1% but were reduced if the limb motion was significantly different from that used in the training data.

Cheron et al. (1996) designed a recognition system based on a dynamic recurrent neural network (DRNN) to recognize more complex arm movements. The objective was to recognize the arm's trajectory during fast movements, such as drawing imaginary figures in the air. Their classification task was more complex because many muscles were involved and the system had to react quickly. EMG information from the posterior, anterior, and median deltoid, and pectoralis major superior, inferior, and latissimus dorsi were recorded. The DRNN network was shown to be more robust to signal noise and could extrapolate the muscle motion, for example, the system could correctly produce shapes such as circles and ellipses when the corresponding EMG signals were passed to it. This outcome offered some validation in that the DRNN network had captured some of the intrinsic muscle information related to the final arm movements.

5.7.2 Limb Motion in Prosthetics

The 1990s saw renewed research work into the application of NNs and other CI techniques in systems designed for limb-motion detection in myoelectric prosthetics. The basic design (Figure 5.23) of a limb-motion detection system involves first filtering, rectifying, and smoothing the EMG signal before processing the features and applying pattern-recognition techniques to determine the motion implied by the EMG input. The goal here is to develop systems that are highly accurate and robust because it is always a challenge to reduce the false detection rates whereby the user moves the prosthetic wrongly due to the system misclassifying the intention.

Work on predicting limb motion began earlier with investigations into muscle changes due to the interaction with a prosthesis. Morris and Repperger (1988) used linear-discriminant analysis to investigate changes in the antagonist tricep–bicep muscle pair due to wearing a prosthesis. Their work was an attempt to develop some measure of performance in the use of prostheses from earlier studies by Saridis and Stephanou (1977), Lee and Grimson (2002), and Graupe et al. (1982). The EMG signals for four different motions were recorded from seven healthy subjects. The integrated absolute value was used as the main feature and the mean and standard deviation of these values were computed. It was found that the RMS tracking error could be used on these values to discern the type of limb motion.

Ito et al. (1992) worked on an ultrasonic prosthetic forearm that provided 3 degrees of freedom (DOF) and contained quieter motors. Their prosthesis was an improvement over the then commonly used Russian hand, Boston arm, Otto Bock, Utah arm, and Waseda Imasen myoelectric hand (WIME) hand. The Ito forearm allowed six types of movements; wrist flexion and extension, forearm pronation and supination, hand grasping and opening. An average 90% classification accuracy was achieved using an ANN to predict the intended motion of the prosthetic wearer. The major drawback was the need to retrain the NN due to artifacts such as muscle fatigue and sweat arising from prolonged use of the prosthesis. Furthermore, only a limited number of movements were allowed.

The classification accuracy of limb motion in the upper-limb extremities continued to improve with contributions using radial basis functions (RBF) (Chaiyaratana et al., 1996), MLP NNs (Kelly et al., 1990; Karlik et al., 1994; Englehart et al., 1995), evolutionary RBF–MLP (Zalzala and Chaiyaratana, 2000) networks, and fuzzy classifiers (Boca and Park, 1994). The preferred features were the AR model or the ARMA, which had been used to model myoelectric signals that were frequently trained using RLS or SLS methods. Chan et al. (2000) recently applied fuzzy EMG classification for control of prosthesis. They demonstrated that the fuzzy logic system was superior to the standard MLP NN, with greater accuracy and potentially faster training times. The fuzzy system also provided consistent results and was resistant to overfitting of the training data.

Karlik (2003) then employed a combination of fuzzy clustering with NNs to improve classification accuracies using AR model features to classify four different classes. The primary model was an NN trained using features derived from cluster sets obtained using a fuzzy C-means algorithm (FCNN). The following weighted within-groups sum of squared errors was used as the objective function for the FCNN:

$$\min J_m(\mathbf{U}, \mathbf{V}, \mathbf{X}) = \sum_{k=1}^{N} \sum_{i=1}^{c} (u_{ik})^m \|x_k - v_i\|_{\mathbf{A}}^2 \qquad (5.10)$$

where $\mathbf{U}$ is a membership matrix, $\mathbf{V}$ the vector of cluster centers, and $\mathbf{X}$ the input data. The matrix $\mathbf{A}$ was usually set to the identity matrix so that the distance measure was Euclidean and the clusters were spherical. Comparisons were made against a 3-layered MLP NN and a conic NN with the FCNN giving the highest recognition accuracy of 98.3%. The drawback of this method, however, was that the fuzzy clusters were required to be computed prior to being used in the NN, which increased the computational overhead. Table 5.3 lists classification performances for various CI techniques in classifying motion of the upper limbs using EMG signal information.

TABLE 5.3

Classification Accuracy (Acc.) for Limb Motion Used in Prosthesis Control

Classifier	Year	Acc. (%)	Remark
Nonlinear discriminant (Graupe et al., 1982)	1982	99	Long training times limited by computer technology then. AR parameters used.
Bayes classifier (Lee and Grimson, 2002)	1984	91	Data collected from immobilized limb measurements. Zero crossing and variance features used.
Neural network (Englehart et al., 1995)	1995	93.7	Neural network model with AR features.
Fuzzy systems (Chan et al., 2000)	2000	91.2	Fuzzy inputs based on basic isodata.
Fuzzy clustering with neural network (Karlik et al., 2003)	2003	98.3	Requires two step processing, finding fuzzy clusters, and training neural network.

Recently, there has been some work toward developing a prosthesis for spinal cord-injured patients (Au and Kirsch, 2000; Hincapie et al., 2004). An adaptive NN (Hincapie et al., 2004) was used to control a neuroprosthesis for patients with C5/C6 spinal cord injury. The initial experiment employed healthy subjects to obtain kinematic data for several upper-limb muscles such as the upper trapezius, anterior deltoid, infra spinatus, and biceps. The EMG information was then fed into a finite element neuromuscular model, which simulated the actions of spinal cord–injured patients. The model performed an inverse dynamic simulation to transform the healthy data into data that mimicked spinal cord–injured patients. This was then used to train a time-delayed neural network (TDNN) to predict the muscle activations and subsequently produced a low RMS error of 0.0264. Hussein and Granat (2002) proposed a neurofuzzy EMG classifier to detect the intention of paraplegics to sit or stand. Their work applied the Gabor matching pursuit (GMP) algorithm to model the EMG waveform. A genetic algorithm was applied to increase the GMP algorithm speed whereas a neurofuzzy model was used to classify the intention of the patient to sit or stand. Results from 30 tests yielded high-classification accuracies with only one or two false-negative classifications and no false positives. False-positive classifications were when the subject performed an action, which they did not intend, whereas false negatives were not performing an intended action. In this situation, false positives would be more hazardous and highly undesirable in the design of myoelectric prostheses.

Huang et al. (2003) proposed an intelligent system to automatically evaluate the EMG features for control of a prosthetic hand. The system was built on a supervised feature-mining method, which used genetic algorithms, fuzzy measures, and domain knowledge to give an important value to the different EMG features. Several features were extracted from the raw EMG signal including the EMG integral, VAR, zero crossings, slope change, WAMP, ARM parameters, and histogram parameters. A SOM was used to convert the ARM parameters and histogram parameters to single values and it was found that ARM and EMG signal histograms gave the best discriminative feature set.

Al-Assaf (2006) had the objective of using polynomial classifiers for the control of elbow and wrist prostheses to dynamically and simultaneously detect and classify events in surface myoelectric signals. They used a dynamic cumulative sum of local generalized likelihood ratios from wavelet decomposition of the myoelectric signals to detect frequency and energy changes in the signal. The feature set encompassed multiresolution wavelet analysis and AR modeling features to detect and classify four elbow and wrist movements. The movements demonstrate on average 91% accuracy using multiresolution wavelet analysis whereas 95% accuracy was achieved with AR modeling. Classification accuracy decreased when a short prostheses response delay was needed. The polynomial classifier performed better than NNs but was only comparable in performance to the SVM. The authors claimed that parameter tuning was not required in their polynomial classifier as opposed to the SVM, however, their classifier did require selection of the polynomial degree to be used.

5.7.3   Robotic Exoskeletons

One extension to the prostheses traditionally used to replace lost limbs is the robotic exoskeleton. Initial work on exoskeletons or orthotics were mainly aimed at improving myoelectric devices introduced in the early 1980s. However, an additional characteristic of an exoskeleton is that the exoskeleton does not necessarily require an amputee to function. In actual fact, the exoskeleton can be thought of as a mechanical device that enhances limb function by supplementing human limb strength with mechanical strength. These systems are useful for assisting motion of physically weak persons such as the elderly, disabled, or injured persons and can be controlled by surface EMG signals. In rehabilitation, the exoskeleton assists patients weakened by neuromuscular disorders to regain mobility. The sensor systems in these devices usually detect EMG signals and interface with mechanical actuators. These are more advanced than the simple sensor systems, which use torque sensors or tactile pressure sensors but do have limitations such as susceptibility to noise corruption. Tanaka et al. (2003), for example, describe the development of a body suit for monitoring body action using ultrasound sensors, microelectromechanical systems (MEMS) sensors such as strain gauges and optical fiber sensors. Their proposed ultrasonic sensor disc will be embedded in a sensor suit and is expected to give better accuracy at detecting motion intention. The authors found that the square of ultrasonic transmission speed is proportional to the elasticity of the object and inversely proportional to the density. Based on this principle, they estimated the elasticity and density of the muscle as the muscle is innervated and attempted to correlate the measurements to the intention to move. Initial experimental results indicate some success but with scope for improvement.

There has also been some research to examine orthotic systems or exoskeletons as possible extensions of existing limbs. Lee et al. (1990) developed an intelligent externally powered orthotic system for the treatment of motor dysfunctions seen in the upper extremities of stroke patients. The intelligent rehabilitative orthotic system (IROS) system was an interactive and cooperative human-machine system for control of the human arm and was also used in rehabilitation where the approach was to first provide full mechanical assistance, which was decreased as the patient improved. The procedure required continuous monitoring of the sensor-integrated orthosis and a real-time graphics system to achieve the goals of therapy. The knowledge-based expert system incorporated the therapists responses while monitoring sensory information such as force, motion, and EMG signals. Treatment-related tasks were displayed on a real-time graphical system to increase interest and maintain the patient's motivation for taking part in the therapy.

Initial investigations of an exoskeleton for upper-limb motion using surface EMG resulted in a device providing 3 DOF for shoulder joint and elbow joint motion (Kiguchi et al., 2003a–c). The authors studied the effect of biarticular muscles and the application of fuzzy control to realize the sophisticated real-time control of the exoskeleton. The proposed exoskeleton (Figure 5.24) was attached directly to the lateral side of a patient's arm and assisted the elbow flexion–extension motion of the patient for rehabilitation. Kiguchi et al. (2005; 2004a; 2004c) later investigated hierarchical fuzzy controllers to control the exoskeleton. The controller was composed of input signal selection, posture region selection, and neurofuzzy control components, which allowed the system to adapt to different users despite the individual EMG waveform differences. The idea of using sophisticated real-time neurofuzzy control was also proposed to incorporate the effect of muscle motion by learning each patient's muscle activation patterns. The exoskeleton was then enhanced by including more types of movements and a 3-DOF exoskeleton system was later introduced to assist the human upper-limb motion incorporating shoulder flexion–extension motion, shoulder adduction–abduction motion, and elbow flexion–extension motion (Kiguchi and Fukuda, 2004). The next objective was to improve the exoskeleton, so that it could be worn as a haptic device and controlled using surface EMG signals. The device reported (Kiguchi et al., 2003b) was designed to restore dexterity to paralyzed hands and constructed from a lightweight, low-profile orthotic exoskeleton controlled by the user's EMG signals. Several simple strategies to control the orthotic device for a quadriplegic (C5/C6) subject were employed, for example, simple on/off strategies allowed for faster interaction with objects using contralateral arm control whereas variable control provided more precise controlled interactions especially with deformable objects. The success of the control algorithms depended largely on the type of control strategies where a more general algorithm was sought than user-specific algorithms.

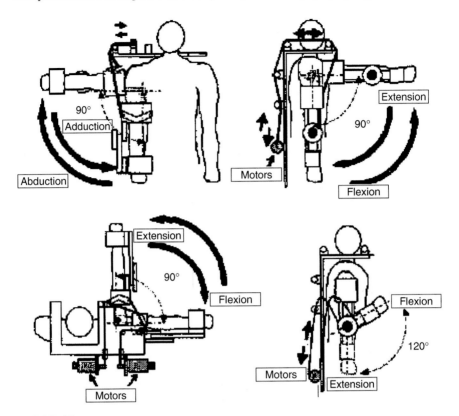

FIGURE 5.24
The exoskeleton developed by Kiguchi et al. (2004) and the neurofuzzy controller system implemented for control of the mechanical actuators. (*Continued*)

Mulas et al. (2005) explored a device for partial hand musculature loss due to stroke or spinal cord injury. Their exoskeleton (Figure 5.25) has been used primarily to detect finger motion and can be easily connected to a laptop with a graphical user interface for monitoring the patient's therapy. Initial investigation has shown the ability to detect and actuate based on the intention of a healthy volunteer to move his fingers. Andreasen et al. (2005) examined prototype development of a robotic system for upper extremity (UE) rehabilitation in individuals with neurological impairments such as cervical level spinal cord injuries (SCI), brain injuries, or stroke. Their system included programmable mechanical impedance for load variations, adjustable thresholds, and control gains and offered some user adaptability. The device was designed to provide automated rehabilitation via repeated motor practice so that neurological recovery in the UEs could be facilitated without requiring a physiotherapist. A wearable power-assisted device was developed by Naruse et al. (2005) to enable a person to lift heavier than normal objects (Figure 5.26) and was designed to support the upper body weight. This

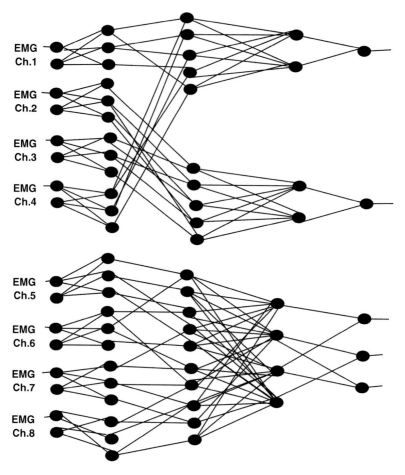

FIGURE 5.24
Continued

resulted in a reduction of compression forces to the lumbar discs, which was
known to be a major factor of lower back injury. It was observed that in lifting motions, the body was twisted and bent, which required three-dimensional
motion analysis to properly analyze the compression forces. A motion capture
system was used to measure a sequence of body positions with and without
the power-assist device in different object locations and masses. This was then
used in conjunction with surface EMG signals to model the compression force
of the lower back discs by estimating the biomechanical human body model.
It was found that when the torsion of the body was 30°, the compression
force increased by 10% but the presence of the device reduced the strain on
the spinal muscles as observed by the reduction in surface EMG magnitude.
Rosen et al. (2001) used the EMG signal as a command signal to actuate their
mechanically linked exoskeleton, which was a two-link, two-joint mechanism
corresponding to the arm. The EMG signal was fed into a Hill-based muscle

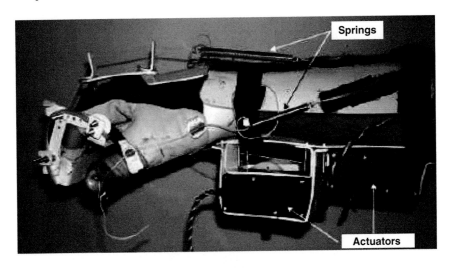

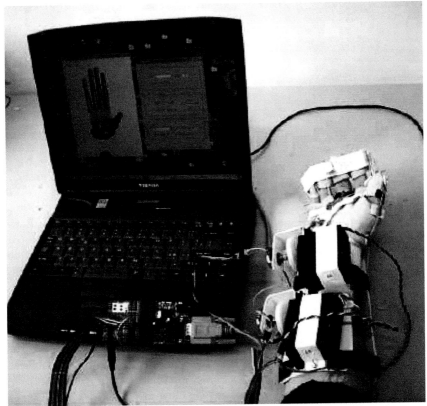

FIGURE 5.25
The hand rehabilitation exoskeleton developed by Mulas et al. (2005) at DEI, Electronic and Information Department in Italy, showing the lateral view of the device and the computer interface.

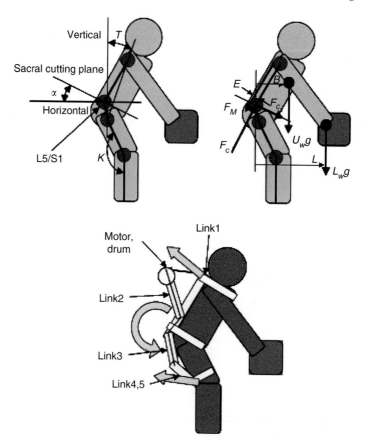

FIGURE 5.26

The prototype exoskeleton developed by Naruse et al. (2005) for assisting a person in lifting heavy weights. The top schematics show the force vectors involved in lifting weights and how the exoskeleton is designed to provide support and assistance in lifting. The prototype of the exoskeleton is also demonstrated in use. (*Continued*)

model and employed elbow movements to predict the motion of the joints. A moment-based control system integrated myoprocessor moment prediction with feedback moments measured at the human arm/exoskeleton and external load/exoskeleton interfaces. Their results suggested that synthesizing the processed EMG command signals with the external-load/human-arm moment feedback significantly improved the mechanical gain of the system. In addition, the natural human control of the system was maintained compared to other control algorithms that used only position or contact forces.

  Leg orthosis devices have also been investigated where EMG signals were used to predict the intended motion of leg joints. The device by Fleischer et al. (2005) (Figure 5.27) could be used to support the disabled or stroke patients in walking or climbing stairs. To achieve real-time processing, EMG

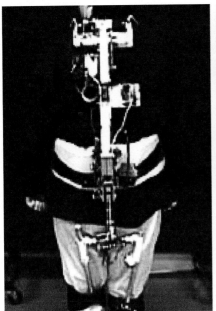

FIGURE 5.26
Continued

signal processing had to be performed with relatively small processing delay times. A model of the body was first developed to simulate physical properties and to synchronize the recorded data obtained from pose sensors whereas model animation was used in experiments aimed at calculating the intended motion of climbing a flight of stairs. The authors previously developed simulations of bones and muscles (Fleischer et al., 2004) with the aim of calibrating surface EMG signals to their force relationships. It was realized that modeling the dynamics of a single muscle was not feasible and so group muscle simulation using an online optimization algorithm was performed. The algorithm minimized the error between observed motion and motion computed from the biomechanical model. Experiments performed on leg movements in the sagittal plane with no external contacts showed good correlation between model predictions and the measured motion.

5.8 Concluding Remarks

In everyday practice, neurologists use EMG with an acoustic output to diagnose disorders. The waveform morphology cannot be used alone because amplitude and duration vary depending on the electrode position. Large variations in EMG from patients make designing a general and reliable classification

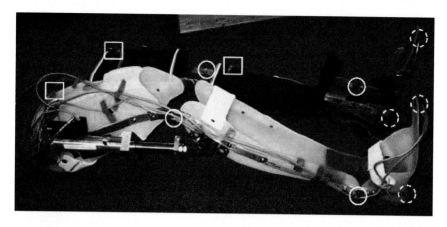

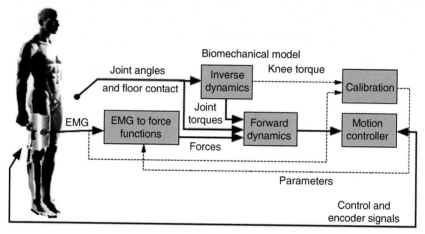

FIGURE 5.27

The leg exoskeleton developed and reported by Fleischer et al. (2005) and the system-operation flowchart.

system intricate. Signal-processing and CI techniques have addressed this problem by attempting to extract characteristic information and estimate the related muscle activity; however, further investigation is required to improve the system's performance and to extend their usefulness to other disorders, which are more difficult to diagnose.

References

Akaike, H. (1974). A new look at the statistical model identification. *IEEE Transactions on Automatic Control 19*, 716–723.

Al-Assaf, Y. (2006). Surface myoelectric signal analysis: dynamic approaches for change detection and classification. *IEEE Transactions on Biomedical Engineering 53(11)*, 2248–2256.

Andreasen, D., S. Alien, and D. Backus (2005). Exoskeleton with EMG based active assistance for rehabilitation. *9th International Conference on Rehabilitation Robotics 1*, 333–336.

Arikidis, N., E. Abel, and A. Forster (2002). Interscale wavelet maximum— a fine to coarse algorithm for wavelet analysis of the EMG interference pattern. *IEEE Transactions on Biomedical Engineering 49(4)*, 337–344.

Au, A. and R. Kirsch (2000). EMG-based prediction of shoulder and elbow kinematics in able-bodied and spinal cord injured individuals. *IEEE Transactions on Neural Systems and Rehabilitation 8(4)*, 471–480.

Basmajian, J., D. Gopal, and D. Ghista (1985). Electrodiagnostic model for motor unit action potential (MUAP) generation. *American Journal of Physics and Medicine 64*, 279–294.

Boca, D. and D. Park (1994). Myoelectric signal recognition using fuzzy clustering and artificial neural networks in real time. *IEEE International Conference on Neural Networks 1*, 3098–3103.

Bodruzzaman, M., S. Zein-Sabatto, D. Marpaka, and S. Kari (1992). Neural networkbased classification of electromyographic (EMG) signal during dynamic muscle contraction. *IEEE Proceedings on Southeastcon 1*, 99–102.

Chaiyaratana, N., A. Zalzala, and D. Datta (1996). Myoelectric signals pattern recognition for functional operation of upper-limb prosthesis. *Proceedings of ECDVRAT*, Maidenhead, UK 1, 151–160.

Chan, F., Y. S. Yang, F. Lam, Y. T. Zhang, and P. Parker (2000). Fuzzy EMG classification for prosthesis control. *IEEE Transactions on Neural Systems and Rehabilitation Engineering 8(3)*, 305–311.

Cheron, G., J. Draye, M. Bourgeios, and G. Libert (1996). A dynamic neural network identification of electromyography and arm trajectory relationship during complex movements. *IEEE Transactions on Biomedical Engineering 43(5)*, 552–558.

Christodoulou, C. and C. Pattichis (1999). Unsupervised pattern recognition for the classification of EMG signals. *IEEE Transactions on Biomedical Engineering 46(2)*, 169–178.

Clancy, E., E. Morin, and R. Merletti (2002). Sampling, noise reduction and amplitude estimation issue in surface electromyography. *Journal of Electromyography and Kinesiology 12*, 1–16.

Coatrieux, J. (1983). Interference electromyogram processing—Part II: experimental and simulated EMG AR modeling. *Electromyography Clinical Neurophysiology 23*, 481–490.

De Luca, C. (1993). Precision decomposition of EMG signals. *Methods in Clinical Neurophysiology 4*, 1–28.

Ehtiati, T. and W. Kinsner (1999). Multifractal characterization of electromyogram signals. *IEEE Canadian Conference on Electrical and Computer Engineering 2*, 792–796.

Emly, M., L. Gilmore, and S. Roy (1992). Electromyography. *IEEE Potentials 11(2)*, 25–28.

Englehart, K., B. Hudgins, M. Stevenson, and P. Parker (1995). A dynamic feedforward neural network for subset classification of myoelectric signal patterns. *19th Annual International Conference of the IEEE EMBS/CMBEC 21*. Montreal, P.Q.

Farina, D., E. Fortunato, and R. Merletti (2000). Noninvasive estimation of motor unit conduction velocity distribution using linear electrode arrays. *IEEE Transactions on Biomedical Engineering 47(3)*, 380–388.

Fleischer, C., K. Kondak, C. Reinicke, and G. Hommel (2004). Online calibration of the EMG to force relationship. *IEEE/RSJ International Conference on Intelligent Robots and Systems 2*, 1305–1310.

Fleischer, C., C. Reinicke, and G. Hommel (2005). Predicting the intended motion with EMG signals for an exoskeleton orthosis controller. *IEEE/RSJ International Conference on Intelligent Robots and Systems 1*, 2029–2034.

Garcia, J., I. Martinez, L. Sornmo, S. Olmos, A. Mur, and P. Laguna (2002). Remote processing server for ECG-based clinical diagnosis support. *IEEE Transactions on Information Technology in Biomedicine 6(4)*, 277–284.

Gerber, A. and R. Studer (1984). A new framework and computer program for quantitative EMG signal analysis. *IEEE Transactions on Biomedical Engineering 31*, 857–863.

Graupe, D., J. Salahi, and K. Kohn (1982). Multifunction prosthesis and orthosis control via micro-computer identification of temporal pattern differences in single site myoelectric signals. *Journal of Biomedical Engineering 4*, 17–22.

Hassoun, M., C. Wang, and A. Spitzer (1994). NNERVE: neural network extraction of repetitive vectors for electromyography. ii. performance analysis. *IEEE Transactions on Biomedical Engineering 41(11)*, 1053–1061.

Hincapie, J. G., D. Blana, E. Chadwick, and R. F. Kirsch (2004). Adaptive neural network controller for an upper extremity neuroprosthesis. *26th Annual International Conference of the Engineering in Medicine and Biology Society 2*, 4133–4136.

Huang, H. P., Y. H. Liu, and C. S. Wong (2003). Automatic EMG feature evaluation for controlling a prosthetic hand using supervised feature mining method: an intelligent approach. *IEEE International Conference on Robotics and Automation 1*, 220–225.

Hudgins, B., P. Parker, and R. Scott (1994). Control of artificial limbs using myoelectric pattern recognition. *Medical and Life Sciences Engineering 13*, 21–38.

Hussein, S. and M. Granat (2002). Intention detection using a neuro-fuzzy EMG classifier. *IEEE Engineering in Medicine and Biology Magazine 21(6)*, 123–129.

Ito, K., T. Tsuji, A. Kato, and M. Ito (1992). EMG pattern classification for a prosthetic forearm with three degrees of freedom. *IEEE International Workshop on Robot and Human Communication 1*, 69–74.

Jacobsen, S., D. Knutti, R. Johnson, and H. Sears (1982). Development of the Utah artifical arm. *IEEE Transactions on Biomedical Engineering 29*, 249–269.

Karlik, B., T. Osman, and M. Alci (2003). A fuzzy clustering neural network architecture for multifunction upper-limb prosthesis. *IEEE Transactions on Biomedical Engineering 50(11)*, 1255–1261.

Karlik, B., H. Pastaci, and M. Korurek (1994). Myoelectric neural neworks signal analysis. *Proceedings of the 7th Mediterranean Electrotechnical Conference 1*, 262–264.

Karlsson, S., J. Yu, and M. Akay (1999). Enhancement of spectral analysis of myoelectric signals during static contractions using wavelet methods. *IEEE Transactions on Biomedical Engineering 46*, 670–684.

Katutoshi, K., O. Koji, and T. Takao (1992). A discrimination system using neural network for EMG-controlled prosthesies. *Proceedings of the IEEE International Workshop on Robot Human Communication 1*, 63–68.

Kelly, M., P. Parker, and R. Scott (1990). The application of neural networks to myoelectric signal analysis: a preliminary study. *IEEE Transactions on Biomedical Engineering 37*, 221–227.

Khalil, M. and J. Duchene (2000). Uterine EMG analysis: a dynamic approach for change detection and classification. *IEEE Transactions on Biomedical Engineering 47(6)*, 748–756.

Kiguchi, K., R. Esaki, and T. Fukuda (2004a). Forearm motion assist with an exoskeleton: adaptation to muscle activation patterns. *IEEE/RSJ International Conference on Intelligent Robots and Systems 3*, 2948–2953.

Kiguchi, K., R. Esaki, T. Tsuruta, K. Watanabe, and T. Fukuda (2003a). An exoskeleton system for elbow joint motion rehabilitation. *IEEE/ASME International Conference on Advanced Intelligent Mechatronics 2*, 1228–1233.

Kiguchi, K. and T. Fukuda (2004). A 3 DOF exoskeleton for upper limb motion assist: consideration of the effect of bi-articular muscles. *IEEE International Conference on Robotics and Automation 3*, 2424–2429.

Kiguchi, K., T. Fukuda, K. Watanabe, and T. Fukuda (2003b). Design and control of an exoskeleton system for human upper-limb motion assist. *IEEE/ASME International Conference on Advanced Intelligent Mechatronics 2*, 926–931.

Kiguchi, K., M. Rahman, and T. Yamaguchi (2005). Adaptation strategy for the 3DOF exoskeleton for upper-limb motion assist. *Proceedings of the 2005 IEEE International Conference on Robotics and Automation 1*, 18–22.

Kiguchi, K., T. Tanaka, and T. Fukuda (2004b). Neuro-fuzzy control of a robotic exoskeleton with EMG signals. *IEEE Transactions on Fuzzy Systems 12(4)*, 481–490.

Kiguchi, K., T. Tanaka, K. Watanabe, and T. Fukuda (2003c). Exoskeleton for human upper-limb motion support. *IEEE International Conference on Robotics and Automation 2*, 2206–2211.

Kumar, D., N. Pah, and A. Bradley (2003). Wavelet analysis of surface electromyography. *IEEE Transactions on Neural Systems and Rehabilitation Engineering 11(4)*, 400–406.

Lee, S., A. Agah, and G. Bekey (1990). IROS: an intelligent rehabilitative orthotic system for cerebrovascular accident. *IEEE International Conference on Systems, Man and Cybernetics 1*, 815–819.

Lee, L. and W. Grimson (2002). Gait analysis for recognition and classification. *Proceedings of Fifth IEEE International Conference on Automatic Face and Gesture Recognition 1*, 148–155.

LeFever, R. and C. De Luca (1982a). A procedure for decomposing the myoelectric signal into its constituent action potentials I: technique, theory and implementation. *IEEE Transactions on Biomedical Engineering 29*, 149–157.

LeFever, R. and C. De Luca (1982b). A procedure for decomposing the myoelectric signal into its constituent action potentials II: execution and test accuracy. *IEEE Transactions on Biomedical Engineering 29*, 158–164.

Louden, G. (1991). Advances in knowledge based signal processing: a case study in EMG decomposition. PhD Thesis, Leicester University.

Ma, N., D. Kumar, and N. Pah (2001). Classification of hand direction using multichannel electromyography by neural network. *The Seventh Australian and New Zealand Intelligent Information Systems Conference 1*, 405–410.

Mahyar, Z., W. Bruce, B. Kambiz, and M. Reza (1995). EMG feature evaluation for movement control of upper extremity prostheses. *IEEE Transactions on Rehabilitation Engineering 3(4)*, 324–333.

Maranzana, M., R. Molinari, and G. Somma-Riva (1984). The parametrization of the electromyographic signal: an approach based on simulated EMG signals. *Electromyography Clinical Neurophysiology 24*, 47–65.

Mason, D., D. Linkens, M. Abbod, N. Edwards, and C. Reilly (1994). Automated delivery of muscle relaxants using fuzzy logic control. *IEEE Engineering in Medicine and Biology Magazine 13(5)*, 678–685.

McGill, K., K. Cummins, and L. Dorfman (1985). Automatic decomposition of the clinical electromyogram. *IEEE Transactions on Biomedical Engineering 32*, 470–477.

Mesin, L. and D. Farina (2005). A model for surface EMG generation in volume conductors with spherical inhomogeneities. *IEEE Transactions on Biomedical Engineering 52(12)*, 1984–1993.

Mesin, L. and D. Farina (2006). An analytical model for surface EMG generation in volume conductors with smooth conductivity variations. *IEEE Transactions on Biomedical Engineering 53(5)*, 773–779.

Mesin, L., M. Joubert, T. Hanekom, R. Merletti, and D. Farina (2006). A finite element model for describing the effect of muscle shortening on surface EMG. *IEEE Transactions on Biomedical Engineering 53(4)*, 593–600.

Morris, A., Jr. and D. Repperger (1988). Discriminant analysis of changes in human muscle function when interacting with an assistive aid. *IEEE Transactions on Biomedical Engineering 35(5)*, 316–322.

Mulas, M., M. Folgheraiter, and G. Gini (2005). An EMG-controlled exoskeleton for hand rehabilitation. *9th International Conference on Rehabilitation Robotics 1*, 371–374.

Naruse, K., S. Kawai, and T. Kukichi (2005). Three-dimensional lifting-up motion analysis for wearable power assist device of lower back support. *IEEE/RSJ International Conference on Intelligent Robots and Systems 1*, 2959–2964.

Ogino, K. (1983). Spectrum analysis of surface electromyogram EMG. *Proceedings of the IEEE International Conferences on Acoustics, Speech and Signal Processing*, 1114–1117.

Ohyama, M., Y. Tomita, S. Honda, H. Uchida, and N. Matsuo (1996). Active wireless electrodes for surface electromyography. *Proceedings of the 18th Annual International Conference of the IEEE Engineering in Medicine and Biology Society 1*, 295–296.

Onishi, H., R. Yagi, K. Akasaka, K. Momose, K. Ihashi, and Y. Handa (2000). Relationship between EMG signals and force in human vastus lateralis muscle using multiple bipolar wire electrode. *Journal of Electromyography and Kinesiology 10*, 59–67.

Park, S. H. and S. P. Lee (1998). EMG pattern recognition based on artificial intelligence techniques. *IEEE Transactions on Neural Systems and Rehabilitation Engineering 6(4)*, 400–405.

Park, E. and S. Meek (1995). Adaptive filtering of the electromyographic signal for prosthetic control and force estimation. *IEEE Transactions on Biomedical Engineering 42(10)*, 1048–1052.

Pattichis, C. and A. Elia (1999). Autoregressive and cepstral analyses of motor unit action potentials. *Elsevier Journal of Medical Engineering and Physics 21*, 405–419.

Pattichis, C. and C. Schizas (1996). Genetics-based machine learning for the assessment of certain neuromuscular disorders. *IEEE Transactions on Neural Networks 7(2)*, 427–439.

Pattichis, C., C. Schizas, and L. Middleton (1995). Neural network models in EMG diagnosis. *IEEE Transactions on Biomedical Engineering 42(5)*, 486–496.

Peleg, D., E. Braiman, E. Yom-Tov, and G. Inbar (2002). Classification of finger activation for use in a robotic prosthesis arm. *IEEE Transactions on Rehabilitation Engineering 10(4)*, 290–293.

Preedy, V. and T. Peters (2002). *Skeletal Muscle: Pathology, Diagnosis and Management of Disease*. U.K.: Cambridge University Press.

Preston, D. and B. Shapiro (2005). *Electromyography and Neuromuscular Disorders: Clinical-electrophysiologic Correlations*. Philadelphia, PA: Elsevier.

Ragupathy, S., D. Kumar, and B. Polus (2004). Relationship of magnitude of electromyogram of the lumbar muscles to static posture. *26th Annual International Conference of Engineering in Medicine and Biology Society 1*, 57–60.

Ranniger, C. and D. Akin (1997). EMG mean power frequency determination using wavelet analysis. *Proceedings of the 19th Annual International Conference of the IEEE Engineering in Medicine and Biology Society 4*, 1589–1592.

Reucher, H., J. Silny, and G. Rau (1987). Spatial filtering of noninvasive multielectrode EMG:part i—introduction to measuring techniques and applications. *IEEE Transactions on Biomedical Engineering 34(2)*, 98–105.

Rosen, J., M. Brand, M. Fuchs, and M. Arcan (2001). A myosignal-based powered exoskeleton system. *IEEE Transactions on Systems, Man and Cybernetics, Part A 31(3)*, 210–222.

Saridis, G. and H. Stephanou (1977). A hierarchical approach to the control of a prosthetic arm. *IEEE Transactions on Systems, Man and Cybernetics 7*, 407–420.

Schizas, C., C. Pattichis, and L. Middleton (1992). Neural networks, genetic algorithms and the k-means algorithm: in search of data classification. *International Workshop on Combinations of Genetic Algorithms and Neural Networks 1*, 201–222.

Schizas, C., C. Pattichis, I. Schofield, P. Fawcett, and L. Middleton (1990). Artificial neural nets in computer-aided macro motor unit potential classification. *IEEE Engineering in Medicine and Biology Magazine 9(3)*, 31–38.

Shahid, S., J. Walker, G. Lyons, C. Byrne, and A. Nene (2005). Application of higher order statistics techniques to EMG signals to characterize the motor unit action potential. *IEEE Transactions on Biomedical Engineering 52(7)*, 1195–1209.

Shieh, J. S., S. Z. Fan, L. W. Chang, and C. C. Liu (2004). Hierarchical rule-based monitoring and fuzzy logic control for neuromuscular block. *Journal of Clinical Monitoring and Computing 16(8)*, 583–592.

Solomonow, M., R. Baratta, M. Bernardi, B. Zhou, and Y. Lu (1994). Surface and wire EMG crosstalk in neighbouring muscles. *Journal of Electromyography and Kinesiology 4*, 131–142.

Suojanen, M., S. Andreassen, and K. Olesen (2001). A method for diagnosing multiple diseases in MUNIN. *IEEE Transactions on Biomedical Engineering 48(5)*, 522–532.

Suryanarayanan, S., N. Reddy, and V. Gupta (1995). Artificial neural networks for estimation of joint angle from EMG signals. *IEEE 17th Annual Conference in Engineering in Medicine and Biology Society 1*, 823–824.

Tanaka, T., S. Hori, R. Yamaguchi, M. Feng, and S. Moromugi (2003). Ultrasonic sensor disk for detecting muscular force. *The 12th IEEE International Workshop on Robot and Human Interactive Communication 1*, 291–295.

Thompson, B., P. Picton, and N. Jones (1996). A comparison of neural network and traditional signal processing techniques in the classification of EMG signals. *IEE Colloquium on Artificial Intelligence Methods for Biomedical Data Processing 1*, 8/1–8/5.

Togawa, T., T. Tamura, and P. A. Oberg (1997). *Biomedical Transducers and Instruments*. Boca Raton, FL: CRC Press.

Uchida, N., A. Hiraiwa, N. Sonehara, and K. Shimohara (1992). EMG pattern recognition by neural networks for multi finger control. *Proceedings of the 14th Annual International Conference of the IEEE Engineering in Medicine and Biology Society 1*, 1016–1018.

Wang, L. and T. Buchanan (2002). Prediction of joint moments using a neural network model of muscle activations from EMG signals. *IEEE Transactions on Rehabilitation Engineering 10(1)*, 30–37.

Wang, R., C. Huang, and B. Li (1997). A neural network-based surface electromyography motion pattern classifier for the control of prostheses. *Proceedings of the 19th Annual International Conference of the IEEE Engineering in Medicine and Biology Society 3*, 1275–1277.

Xie, H. B., H. Huang, and Z. Z. Wang (2004). A hybrid neuro-fuzzy system for neuromuscular disorders diagnosis. *IEEE International Workshop on Biomedical Circuits and Systems 52(9)*, S2/5–S5–8.

Xie, H. B., Z. Z. Wang, H. Huang, and C. Qing (2003). Support vector machine in computer aided clinical electromyography. *International Conference on Machine Learning and Cybernetics 2*, 1106–1108.

Xu, Z. and S. Xiao (2000). Digital filter design for peak detection of surface EMG. *Journal of Electromyography and Kinesiology 10*, 275–281.

Zalzala, A. and N. Chaiyaratana (2000). Myoelectric signal classification using evolutionary hybrid RBF-MLP networks. *Proceedings of the 2000 Congress on Evolutionary Computation 1*, 691–698.

Zazula, D. (1999). Higher-order statistics used for decomposition of s-EMGs. *Proceedings of the 12th IEEE Symposium on Computer Based Medical System*, 72–77.

Zazula, D., D. Korosec, and A. Sostaric (1998). Computer-assisted decomposition of the electromyograms. *Proceedings of the 11th IEEE Symposium on Computer Based Medical System 1*, 26–31.

Zhang, J., J. Liu, C. Zheng, W. Tao, N. Thakor, and A. Xie (2000). Noninvasive early detection of focal cerebral ischemia. *IEEE Engineering in Medicine and Biology Magazine 19*, 74–81.

Zhou, Y., R. Chellappa, and G. Bekey (1986). Estimation of intramuscular EMG signals from surface EMG signal analysis. *Proceedings of the IEEE International Conferences on Acoustics, Speech and Signal Processing 11*, 1805–1808.

6

Computational Intelligence in Electroencephalogram Analysis

6.1 Introduction

This chapter overviews the analysis of electrical brain activity for monitoring the functional status of the brain. Many control signals are initiated in the brain, which are then transmitted throughout the body to execute various bodily functions. Brain signals can be picked up by electrodes on the scalp, known as the electroencephalogram (EEG). The EEG signal arises from billions of neurons firing within the nervous system. These signals hold important information regarding the brain function and are affected by neurological conditions and other abnormalities. There are, for example, routine applications for the diagnosis and treatment of neurological disorders such as epilepsy. Recent research in EEG-based communication has also opened up exciting opportunities for restoring function in individuals with disabilities.

All biomedical signals such as EEGs are affected by noise that arises primarily due to how the signals are received from the body in addition to sources in the environment. For example, while recording EEG signals, noise may be present from surrounding EMG activities. A major challenge in biomedical signal processing is to reduce these noise artifacts prior to the signals being used to diagnose any pathological condition. CI plays a major role in the processing and analysis of EEG signals and for the detection and monitoring of various diseases. In the next sections, following an overview of the fundamentals of EEG, we outline some of the widely used CI-assisted applications for the detection and classification of abnormalities in EEG signals.

6.1.1 The Brain

The brain is the central nervous system's (CNS) control center, a highly complex structure composed of billions of interconnected neurons. Together with the spinal cord, the brain controls most of our movements. Anatomically, a human brain is approximately 17 cm anterior to posterior and includes about 50–100 compartments or functional areas (Deutsch and Deutsch, 1993). The brain's sections have unique functions, for example, movement is controlled by the frontal region, whereas vision is controlled by the back region, and some

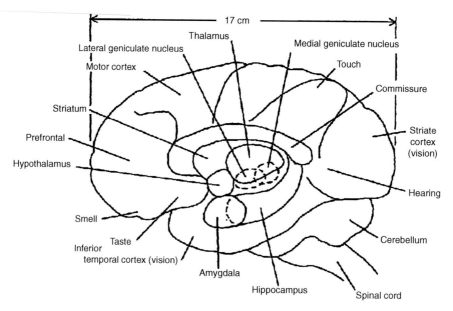

FIGURE 6.1

A cross-section of the human brain showing some of the major regions. (From Deutsch, S. and Deutsch, A., *Understanding the Nervous System: An Engineering Perspective*, IEEE Press, New York, 1993. ©IEEE.)

of these are illustrated in Figure 6.1. As seen in the figure, specific regions are linked to the sensory systems, vision, taste, touch, and smell, whereas the cerebellum controls motion. The number of neurons within a region influences the information content and functional capability such that a greater number of neurons increase the information-processing capability (Deutsch and Deutsch, 1993).

The brain constantly communicates with the various systems of the body, including organs and muscles, and the processing to control the body's actions is also within the brain. In conjunction with the spinal cord, the brain sends and receives signals through nerves. The cerebellum is in close proximity to the spinal cord, and neurons in this region control movement by muscle activation. The cerebellum is, therefore, pivotal to the recognition of EEG signals in human subjects. The motor cortex comprises a large area of the brain and this is where action potentials are initiated (Deutsch and Deutsch, 1993).

6.1.2   Electroencephalography

EEG is the recording of electrical activity from the brain using noninvasive electrodes, which are placed on the scalp. The EEG signal is the resulting waveform and represents the overall electrical activity of the brain arising from many neuronal activities. The first human EEG signal was recorded by

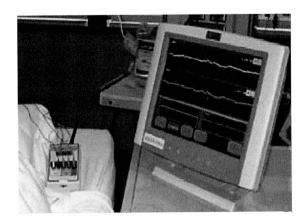

FIGURE 6.2 (See color figure following page 204.)

An EEG-monitoring system setup showing wireless EEG electrode connections to the patient's bedside and a display monitor showing EEG signals. (From Thakor, N. V., Hyun-Chool, S., Shanbao, T., et al., *IEEE Eng. Med. Biol. Mag.*, 25, 20–25, 2006. ©IEEE.)

Hans Berger in 1924, and this proved a significant breakthrough in human brain research. In this early demonstration, he connected two electrodes to a patient's skull and was able to detect a very weak current using a galvanometer. Nowadays, clinical EEG machines can be found in many clinics for routine brain electrical activity monitoring and assisting the physicians for decision-making processes. Figure 6.2 shows an EEG setup for recording brain electrical activity using wireless electrodes.

EEG signals are affected by moods such as drowsiness, excitement, and relaxation. The biopotential recorded from an electrode is a summation of the electrical activity of the nearby neurons. In clinical tests, recording electrodes are placed on the scalp using international standards EEG geometrical sites, such as the 10–20 system (Jasper, 1958). EEG recorders can accept signals from up to 256 electrodes including reference electrodes placed on the earlobes. Clean contact between the electrode and the skin is essential for good EEG recording. Conductive gel is usually used to reduce impedance between the electrode and the scalp. Each electrode is input to a differential amplifier commonly set between 1,000 and 100,000 amplification. Signal processing is a combination of low- and high-pass filtering; the low-pass filter is used to filter the slow electrogalvanic signals, whereas the high-pass filter is intended to remove any EMG or electromyographic activity from the EEG waveform.

The EEG signal from the scalp is typically characterized by an amplitude of approximately $100\,\mu V$ and a time duration of 0.01–$2\,s$ (Geva and Kerem, 1999). Rhythmic sinusoidal activities can be recognized within the EEG signal; the frequency compositions of the EEG signals commonly used for analysis

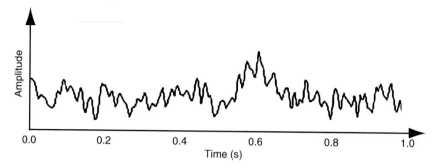

FIGURE 6.3
Time-domain information of a segment of a typical EEG raw signal.

include the following frequency bands (Bylsma et al., 1994):

- Delta (up to 4 Hz)

- Theta (4–8 Hz)

- Alpha (8–12 Hz)

- Beta (12–26 Hz)

- Gamma (26–100 Hz)

Figure 6.3 shows a 1 s recording of the raw EEG.

Waveform activity changes according to the brain function associated with physical and psychological tasks, for example, an EEG signal recorded during sleep contains a higher proportion of longer waves (delta and theta), while shorter waves (alpha and beta) dominate during awake times. As illustrated in Figures 6.4 and 6.5, the EEG frequency spectrum changes from an *eyes closed* condition to *eyes open* condition, with the *eyes closed* state containing a distinct peak in the spectrum at about 10 Hz (Felzer and Freisleben, 2003). The EEG signal is also characterized by patches of nonstationarity in the waveforms and has a number of semistationary time-dependent states. This characteristic makes the recognition task challenging, and techniques that account for both time and frequency domain information, such as wavelets, have been employed for feature extraction from the EEG waveform (Bigan, 1998).

6.1.3 Uses of EEG Signals

The EEG has proved useful for monitoring and diagnosing a number of clinical contexts that include the following:

- Epilepsy

- Alzheimer disease (AD)

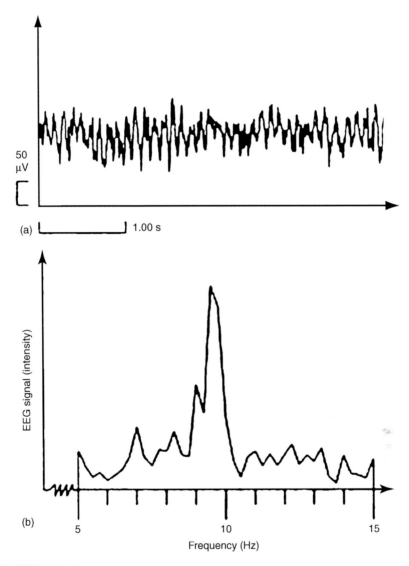

FIGURE 6.4
(a) Time-domain information of an EEG record while eyes are closed.
(b) The corresponding frequency spectrum showing dominant frequency in
the signal. (Adapted from Felzer, T. and Freisieben, B., *IEEE Trans. Neural
Syst. Rehabil. Eng.*, 11, 361–371, 2003. ©IEEE.)

- Huntington's disease (HD)

- Sleep disorders

EEG is used by neuroscientists to study brain function and with the
widespread use of CI techniques and mathematical modeling, EEG analysis

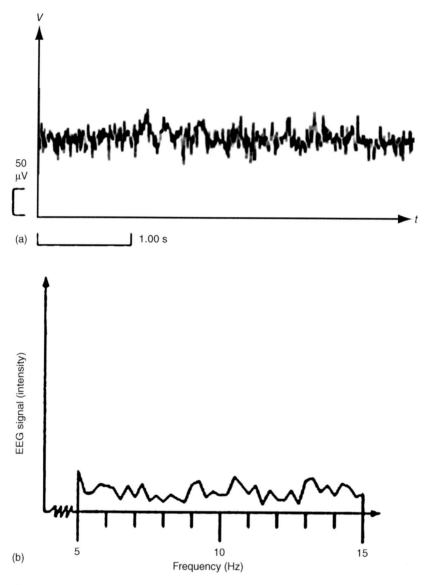

FIGURE 6.5

(a) Time-domain information of an EEG record while eyes are open.
(b) The corresponding frequency spectrum. (Adapted from Felzer, T. and
Freisleben, B., *IEEE Trans. Neural Syst. Rehabil. Eng.*, 11, 361–371, 2003.
©IEEE.)

has advanced significantly. CI tools have also assisted in the identification of
patterns embedded within the EEG to improve recognition of. The application of brain computer interface (BCI) has been used recently to help the
disabled restore day-to-day activities. It has been demonstrated, for example,

that BCI can facilitate the communication of physically handicapped individuals with the help of a computer by using EEG signal characteristics. Other applications of EEG are in evoked potentials or evoked responses that record the brain's responses to sound, touch, and light, which is useful for evaluating a number of neurological conditions.

6.2 Computational Intelligence in EEG Analysis

6.2.1 Computational Intelligence in the Diagnosis of Neurological Disorders

In conjunction with CI techniques, EEG signals can be used to detect disorders of brain function and underlying pathologies. There have been many studies of the EEG signal characteristics associated with neurological diseases; in the following discussion, however, we will focus on three common pathologies, epilepsy, HD, and AD.

6.2.1.1 Epilepsy Detection

Epilepsy is a neurological disorder, which affects over 50 million people worldwide. It is regarded as the second most common neurological disorder (D'Alessandro et al., 2003) after stroke. Epilepsy is characterized by excessive electrical discharge from brain cells, which usually leads to seizures and abnormal movements. Imaging techniques such as MRI can be used to diagnose the brain's structural disorders, but EEG has become the routine procedure for examining brain function in epilepsy.

Seizure-related EEG signals have characteristic patterns that enable health professionals to distinguish them from normal (nonseizure) EEG signals, these have been defined as "repetitive high-amplitude activities" with fast, slow, or a combination of spikes and slow waves (SSW) (Geva and Kerem, 1998). The visual recognition of epileptic seizures from EEG signals can be time-consuming and problematic due to lack of clear differences in EEG activity between epileptic and nonepileptic seizures (Bigan, 1998). Automated-recognition techniques have, therefore, been trialled to enable faster and more accurate identification of pathological EEG waveforms with links to epileptic seizures, and several techniques have been proposed to detect spikes in the EEG to predict epileptic events. Pang et al. (2003), for example, compared the performance of ANN-based classifiers and concluded that ANNs trained with features selected using feature-selection algorithms and a raw EEG signal fed directly to the ANNs could yield similar results with accuracies ranging from 82.83% to 86.61%. Bigan (1998) used time–frequency (wavelet) techniques for extracting features from EEG recordings to develop a system for the automatic analysis and identification of seizure events. The EEG characteristics for epileptic seizures exhibited more frequency changes than the EEG segments of nonepileptic seizures. An NN with a multilayer perceptron was proposed for the automatic analysis of the seizure-related EEG waveforms.

Fuzzy clustering techniques have been applied to the detection of epileptic events. Geva and Kerem (1998) used rats to trial an EEG-based brain-state identification technique, which allowed neural behavior prediction to estimate the occurrence of epileptic seizures. The rats were exposed to epileptic conditions from which the features leading up to and during the seizure were analyzed using the wavelet transform method. These extracted features were then used as input to the unsupervised optimal fuzzy clustering (UOFC) algorithm from which a classification of the data was undertaken. The classification was successful in determining the unique behavioral states as well as the features associated with the seizure. Figure 6.6 illustrates a flow diagram of the steps followed by Geva and Kerem (1998). Although the results hold some uncertainty due to the fluctuation in the data, the applications of this technique have promise for recognizing seizures.

A number of researchers have applied hybrid approaches to improve epilepsy detection. Harikumar and Narayanan (2003) applied fuzzy logic techniques through genetic algorithms (GA) to optimize the EEG output signals for the classification of patient risk associated with epilepsy. The risk factor was determined by inserting data regarding energy, variance, peaks, sharp and spike waves, duration, events, and covariance measurements into a binary GA and continuous GA. The performance index measures and quality value measures were also evaluated. The authors claimed over 90% accuracy in detecting the epileptic seizures using their technique, and from the results it appears that the continuous GA may provide the most accurate risk assessment.

A hybrid CI approach based on genetic search techniques has been applied by D'Alessandro et al. (2003) to identify patient-specific features and electrode sites for optimum epileptic seizure prediction. The genetic search algorithm was trained using features extracted from a subset of baseline and preictal data, the trained algorithms were then validated on the remaining data sets. A three-level feature-extraction process was adopted by the authors as illustrated in Figure 6.7. Feature selection was applied to identify the most important features contributing to the prediction task. Synchronized video and EEG signals were recorded from four patients at 200 Hz, which included information regarding 46 preictal states and 160 h of baseline data. A 60-Hz notch filter was applied to remove the power line frequency noise. The training data constitutes 70% of these data with testing carried out on the remaining data sets; a block diagram of the steps in the prediction of epileptic seizures is illustrated in Figure 6.7. EEG signals were collected from multiple implanted intracranial electrodes and quantitative features derived from these signals. Features representing information in the time domain, frequency domain, and nonlinear dynamics were used in the study. By using an intelligent genetic search process and a three-step approach (Figure 6.7), important features were selected for use with the classifiers. A probabilistic NN classifier was used to classify "preseizure" and "no preseizure" classes; and the predicted performance, based on four patients implanted with intracranial electrodes, was reported to

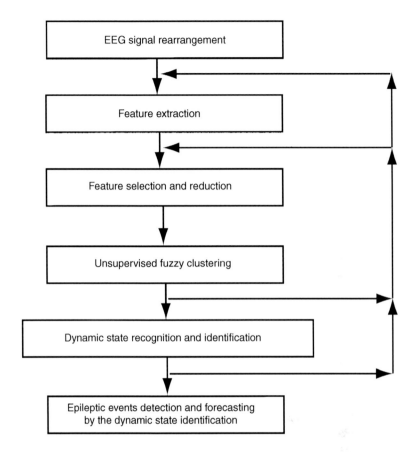

FIGURE 6.6
Block diagram of the fuzzy clustering technique used by Geva and Kerem (1998) to identify epileptic events using the EEG waveforms. Features were extracted using techniques such as statistical moments calculations, correlation analysis, spectrum estimation, and time–frequency decomposition. Features were reduced using principal component analysis (PCA). (Adapted from Geva, A. B. and Kerem, D. H., *IEEE Trans. Biomed. Eng.*, 45, 1205–1216, 1998. ©IEEE.)

have a sensitivity of 62.5% with 90.5% specificity. The authors highlighted the importance of using individualized data for developing classifiers for epileptic seizure prediction.

6.2.1.2   Huntington's Disease

HD is a genetically inherited neurological condition, which affects a significant proportion of the population. The disease is caused by a constant expansion in the Huntingtin gene that leads to cell death in regions of the brain

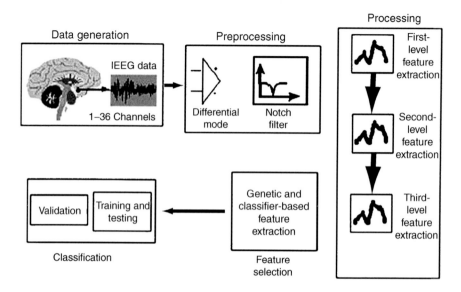

FIGURE 6.7

Block diagram illustrating the various steps used to predict epileptic seizure using hybrid genetic and classifier-based feature selection. A notch filter was applied to eliminate power line frequency and three levels of features were extracted. (Adapted from D'Alessandro, M., Esteller, R., Vachtsevanos, G., et al., *IEEE Trans. Biomed. Eng.*, 50, 603–615, 2003. ©IEEE.)

that control movements and cognitive functions. Physical, cognitive, and psychiatric abnormalities are symptoms of HD and all children of an affected individual have a high probability (50%) of developing the disease. Medical specialists use both genetic techniques and neurological tests to diagnose HD.

It is important to have procedures for the diagnosis and monitoring of HD patients prior to the symptoms becoming apparent. Several studies have used EEG waveform characteristics to identify potential at-risk HD patients. Bylsma et al. (1994), for example, reported that EEG recordings from HD patients show substantial reduction in the alpha rhythm or waveform. De Tommaso et al. (2003) explored EEG differences between HD patients and healthy subjects, features were extracted from their EEG signals, and then tested using Fisher's linear discriminant (FLD) method, a likelihood ratio method (LRM), and an MLP NN (with a single hidden layer). The NN classified the two groups with high accuracy (11 of 13 patients and 12 of 13 healthy subjects), and these procedures outperformed the other classifiers, such as FLD and LRM.

Jervis et al. (1992) explored cognitive negative variation (CNV), the evoked response potential within the EEG signal, which is modified due to HD. The voltage magnitude and timing of CNV samples were completed for 11 HD patients, 21 at-risk subjects, and an equal number of age and sex-matched controls. These CNV data combined with the voltage response values were then used as data sources for Kohonen unsupervised ANNs and ART in an attempt to classify the groups. The recognition results of HD versus normal healthy controls were reported to be similar for the two classifiers.

6.2.1.3   Alzheimer's Disease

AD is a common form of dementia and the causes are uncertain although it is thought that genetic factors are highly related. The most serious effect is on memory and long-term locomotive and lingual skills. Alzheimer's patients usually demonstrate a steady decline in such abilities due to the neurological consequences, which has no cure at present. Treatment usually includes acetylcholinesterase inhibitors, psychosocial interventions, various therapies, and vaccines. With early diagnosis essential to the efficient treatment of Alzheimer's patients, EEG techniques are being developed to provide early detection.

Owing to the increasing prevalence of Alzheimer's, as indicated earlier, there is a need for techniques to predict onset so that appropriate care can be planned. In addition to methods such as MRI, EEG is regarded as being potentially effective in identifying the development of this disease. Some bands of the EEG waveform are affected by the onset of AD, for example, the development of AD was found to correlate well with an increase in the delta and theta power band (Kamath et al., 2006). Bennys et al. (2001) experimented with several spectral indices calculated from the EEG to discover whether these indices are affected in the AD population, their study included EEG recordings from 35 AD patients and 35 controls with no history of neurological disorders. The spectral indices were calculated from the quantitative EEG recordings and statistical analysis was conducted to evaluate the discriminatory power of these indices between the two groups. The high-frequency (alpha + beta)/low-frequency (delta + theta) ratio was found to differentiate well between the control and AD patients suggesting that an increase in delta and theta power in AD patients provides a good discriminating variable.

The early detection of AD is critical for effective treatment and this can slow the rate of progression. Yagneswaran et al. (2002), for example, compared the power frequency and wavelet characteristics of an EEG segment between patients with and without Alzheimer's; LVQ-based NNs were then trained to distinguish between individuals with and without AD.

Although the detection of Alzheimer's from an EEG signal is generally suitable, various "abnormalities"within the symptoms of Alzheimer's patients can cause such models to be less efficient. Petrosian et al. (1999), therefore, focused on the development of a computerized method for the recognition of Alzheimer's patients from EEG where they applied trained recurrent neural networks (RNNs) pooled with wavelet processing for the determination of early-Alzheimer's patients as opposed to non-Alzheimer's. Cho et al. (2003) investigated an Alzheimer's recognition method involving ANNs and GA. The EEG from early AD patients and non-Alzheimer's patients were analyzed to find features that could be used for recognition. Moreover, ANNs and GA were implemented to find dominant features prior to this information being used as a data source for an ANN.

6.2.2 Evoked Potentials

Evoked potentials (EVP), also referred to as evoked responses, record the brain's responses to various stimuli, such as sound, touch, and light. These tests often help in evaluating neurological conditions. For each test, electrodes are attached to the patient's scalp and, in some cases, to the earlobes, neck, shoulders, and back.

> *Auditory evoked potential.* This test records the brain's electrical potential through EEG signals in response to stimulation of the auditory sensory mechanism (Hoppe et al., 2001; Davey et al., 2006). Such tests have many clinical applications, such as obtaining an objective audiogram for auditory organ evaluation (Hoppe et al., 2001). Pattern-recognition techniques such as NNs can be applied to classify the feature vector extracted from the EEG signal.

> *Somatosensory evoked potential.* The aim of this test is to monitor through responses reflected in EEG signals when mild electrical currents are passed through body through electrodes. During the test, electrodes are located on the scalp, neck, shoulders, wrist, and ankle and applied electrical currents cause a mild twitch of the thumb or big toe (Davey et al., 2006).

> *Visual evoked potential.* This test involves EEG response analysis while visual cues are presented, for example, the patient may be instructed to sit in a chair and watch a television screen displaying a checkerboard pattern while recording his/her brain EEG potentials (Davey et al., 2006).

6.3 Brain–Computer Interfaces

A BCI uses noninvasively recorded EEG signals from the brain to analyze an individual's intended action and then translates the information into an external action. Such a system can provide disabled persons with a means of communicating with the physical environment such as for controlling a wheelchair, a computer mouse, and other everyday appliances. A BCI system includes EEG signal acquisition, signal-processing, feature extraction and selection, and finally, pattern recognition. Pattern recognition is utilized to categorize or classify the intention from the signal characteristics embedded within a segment of the EEG waveform. There has been much recent interest in the implementation of BCI for a variety of tasks, and to help classify the tasks several machine-learning tools have been applied including NNs, fuzzy logic, and SVMs.

Several studies have been reported of NN applications in an EEG-based BCI system design and for categorizing an EEG segment into movement tasks (Pfurtscheller et al., 1998; Deng and He, 2003; Felzer and Freisieben, 2003; Wang et al., 2004). Pfurtscheller et al. (1998) applied an NN to EEG-based controls for moving a cursor on a computer screen. EEG signals were recorded by two bipolar electrodes over the left and right central areas. Data were collected in two sessions with the first needed for acquiring EEG data for subject-specific classifier development. In the second session, the classifier was used to analyze frequency bands and to classify EEG while the subject was provided with a feedback stimulus and asked to imagine right- or left-hand movements. Four subjects participated in the evaluation test and the average error was $21.9 \pm 6.4\%$ for the online classification of hand movements. This system would particularly benefit those patients who are aware of their environment but are unable to interact with it. The eventual event of this research would be to obtain signals from the brain and transform them into control signals.

Wang et al. (2004) proposed an algorithm (see Figure 6.8) to classify a single-trial EEG for left- and right-finger movements. Fisher discriminant analysis was applied to extract features and a perceptron NN was used to classify the movements. Test results yielded an accuracy of 84%.

Fuzzy classifiers have been proposed as an element in the design of BCI systems to classify mental tasks. Palaniappan et al. (2002) reported a BCI system based on a fuzzy ARTMAP (FA) NN; the data acquisition for this study involved recording EEG signals from locations C3, C4, P3, P4, O1, and O2 (see Figure 6.9a). EEG signals were recorded for 10 s for each of five mental tasks, that is, baseline tasks when the participants were asked to relax and not think about anything; a math task involving multiplication; geometric figure rotation, in which participants were asked to study a 3D object within 30 s;

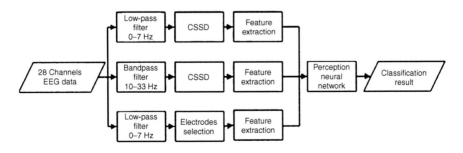

FIGURE 6.8

Block diagram of the steps undertaken to classify the movement from EEG data (28 channels of EEG data were recorded at 100 Hz); CSSD—common spatial subspace decomposition. (Adapted from Wang, Y., Zhang, Z., Li, Y., et al., *IEEE Trans. Biomed. Eng.*, 51, 1081–1086, 2004. ©IEEE.)

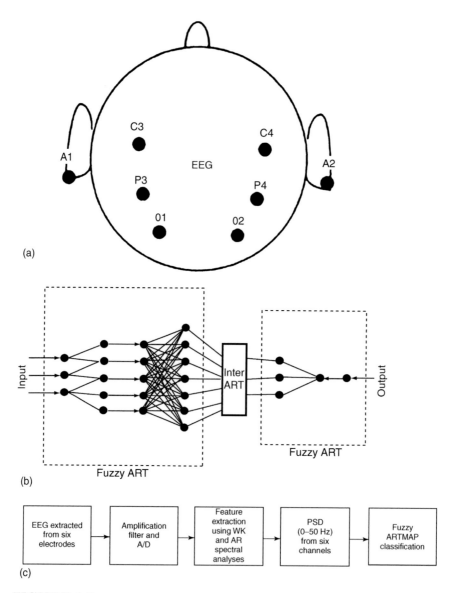

FIGURE 6.9
(a) Diagram showing electrode placement for EEG data acquisition; (b) fuzzy
ARTMAP structure; and (c) BCI–FA design proposed by Palaniappan et al.
(Adapted from Palaniappan, R., Paramesran, R., Nishida, S., et al., *IEEE
Trans. Neural Syst. Rehabil. Eng.*, 10, 140–148, 2002. ©IEEE.)

mental letter composition tasks requiring participants to mentally compose a letter to their friends; and finally, visual counting tasks visualizing numbers written sequentially on a blackboard. During these conditions, participants were instructed to visualize tasks and not to verbalize either the numbers or texts. A total of 300 power spectral density (PSD) values were extracted from EEG segments using Wiener–Khinchine (WK) methods (Palaniappan et al., 2002). The FA classifier was trained to classify three mental tasks using the 300 PSD values. Figures 6.9b and 6.9c illustrate the operation of the BCI system and the steps followed using the two fuzzy ART modules. The evaluation results reported an average error of $<6\%$ with regard to the system's capability in recognizing the mental tasks.

In addition to NNs and fuzzy logic, EEG-based BCI systems have been evaluated using other CI approaches. The major tasks for the CI in these systems were selection of important features from the EEG and classification of the intended mental tasks. Some of this work has been reported recently, for example, SVMs for feature and channel selection (Garrett et al., 2003; Lal et al., 2004) and HMM and Mahalanobis distance as BCI feature classifiers (Millan et al., 2000).

6.4   Concluding Remarks

EEG signals provide an important source of information for studying brain activity and associated functional status. An EEG recording usually results in a large amount of data, thus, data management and processing present complex analysis problems. The use of CI techniques in processing this data has, however, helped significantly in extracting useful information for assisting in decision making. EEG analysis is facilitating the diagnosis and treatment of neurological diseases in clinical settings as well as assisting researchers in translating mental tasks and intentions to a computer to form a communication channel between cognition and actions. Recent years have seen research advances involving BCI and EEG pattern recognition designed to assist the physically disabled to link with a computer to control external devices and appliances using the power of thought.

References

Bennys, K., G. Rondouin, C. Vergnes, and J. Touchon (2001). Diagnostic value of quantitative EEG in Alzheimer's disease. *Neurophysiologie Clinique 31(3)*, 153–160.

Bigan, C. (1998). A recursive time-frequency processing method for neural networks recognition of EEG seizures. In *Neural Networks and Expert*

Systems in Medicine and Healthcare, E. C. Ifeachor, A. Sperduti, and A. Starita (Eds), Singapore: World Scientific.

Bylsma, F. W., C. Peyser, S. E. Folstein, et al. (1994). EEG power spectra in Huntington's disease: clinical and neuropsychological correlates. *Neuropsychologia 32(2)*, 137–150.

Cho, S. Y., B. Y. Kim, E. H. Park, et al. (2003). Automatic recognition of Alzheimer's disease with single channel EEG recording. *Proceedings of the 25th Annual International Conference of the IEEE Engineering in Medicine and Biology Society, 3*, 2655–2658.

D'Alessandro, M., R. Esteller, G. Vachtsevanos, et al. (2003). Epileptic seizure prediction using hybrid feature selection over multiple intracranial EEG electrode contacts: a report of four patients. *IEEE Transactions on Biomedical Engineering 50*, 603–615.

Davey, R. T., P. J. McCullagh, H. G. McAllister, et al. (2006). The use of artificial neural networks for objective determination of hearing threshold using the auditory brainstem response. In *Neural Networks in Healthcare: Potential and Challenges*, R. Begg, J. Kamruzzaman, and R. Sarker (Eds), pp. 195–216, Hershey: Idea Group Publishing.

Deng, J. and B. He (2003). Classification of imaginary tasks from three channels of EEG by using an artificial neural network. *Proceedings of the 25th Annual International Conference of the IEEE 3*, 2289–2291.

De Tommaso, M. M., F. De Carlo, O. Difruscolo, et al. (2003). Detection of subclinical brain electrical activity changes in Huntington's disease using artificial neural networks. *Clinical Neurophysiology 114(7)*, 1237–1245.

Deutsch, S. and A. Deutsch (1993). *Understanding the Nervous System: An Engineering Perspective*. New York: IEEE Press.

Felzer, T. and B. Freisieben (2003). Analyzing EEG signals using the probability estimating guarded neural classifier. *IEEE Transactions on Neural Systems and Rehabilitation Engineering 11*, 361–371.

Garrett, D., D. A. Peterson, C. W. Anderson, et al. (2003). Comparison of linear, nonlinear, and feature selection methods for EEG signal classification. *IEEE Transactions on Neural Systems and Rehabilitation Engineering 11*, 141–144.

Geva, A. B. and D. H. Kerem (1998). Forecasting generalized epileptic seizures from the EEG signal by wavelet analysis and dynamic unsupervised fuzzy clustering, *IEEE Transactions on Biomedical Engineering 45*, 1205–1216.

Geva A. B. and D. H. Kerem (1999). Brain state identification and forecasting of acute pathology using unsupervised fuzzy clustering of EEG temporal patterns. In *Fuzzy and Neuro-Fuzzy Systems in Medicine*, H. N. Teodorescu, A. Kandel, and L. C. Jain (Eds), Boca Raton, FL: CRC Press.

Harikumar, H. and B. S. Narayanan (2003). Fuzzy techniques for classification of epilepsy risk level from EEG signals. *Conference on Convergent Technologies for Asia-Pacific Region TENCON, 1(15–17 Oct.)*, 209–213.

Hoppe, U., S. Weiss, R. W. Stewart, et al. (2001). An automatic sequential recognition method for cortical auditory evoked potentials. *IEEE Transactions on Biomedical Engineering 48*, 154–164.

Jasper, H. (1958). The ten twenty electrode system of the international federation. *Electroencephalograph Clinical Neurophysiology 10*, 371–375.

Jervis, B. W., M. R. Saatchi, A. Lacey, et al. (1992). The application of unsupervised artificial neural networks to the sub-classification of subjects at-risk of Huntington's disease. *IEE Colloquium on Intelligent Decision Support Systems and Medicine 5*, 1–9.

Kamath, M., A. R Upton, J. W. McMaster, et al. (2006). Artificial neural networks in EEG analysis. In *Neural Networks in Healthcare: Potential and Challenges*, R. Begg, J. Kamruzzaman, and R. Sarker (Eds), pp. 177–194, Hershey: Idea Group Publishing.

Lal, T. N., M. Schroder, T. Hinterberger, et al. (2004). Support vector channel selection in BCI. *IEEE Transactions on Biomedical Engineering 51*, 1003–1010.

Millan, J., J. Mourino, F. Cincotti, et al. (2000). Neural networks for robust classification of mental tasks. *Proceedings of the 22nd Annual International Conference of the IEEE Engineering in Medicine and Biology Society 2*, 1380–1382.

Palaniappan, R., R. Paramesran, S. Nishida, et al. (2002). A new brain-computer interface design using fuzzy ARTMAP. *IEEE Transactions on Neural Systems and Rehabilitation Engineering 10*, 140–148.

Pang, C. C. C., A. R. M. Upton, G. Shine, et al. (2003). A comparison of algorithms for detection of spikes in the electroencephalogram. *IEEE Transactions on Biomedical Engineering 50*, 521–526.

Petrosian, A., D. Prokhorov, and R. Schiffer (1999). Recurrent neural network and wavelet transform based distinction between Alzheimer and control EEG. *Proceedings of the 21st Annual Conference and the Annual Fall Meeting of the Biomedical Engineering Society BMES/EMBS Conference, 2 (13–16 Oct.)*, 1185.

Pfurtscheller, G., C. Neuper, A. Schlogl, et al. (1998). Separability of EEG signals recorded during right and left motor imagery using adaptive auto-regressive parameters. *IEEE Transactions on Rehabilitation Engineering 6*, 316–325.

Thakor, N. V., S. Hyun-Chool, T. Shanbao, et al. (2006). Quantitative EEG assessment. *IEEE Engineering in Medicine and Biology Magazine 25*, 20–25.

Wang, Y., Z. Zhang, Y. Li, et al. (2004). BCI competition 2003-data set IV: an algorithm based on CSSD and FDA for classifying single-trial EEG. *IEEE Transactions on Biomedical Engineering 51*, 1081–1086.

Yagneswaran, S., M. Baker, and A. Petrosian (2002). Power frequency and wavelet characteristics in differentiating between normal and Alzheimer EEG. *Proceedings of the 24th Annual Conference and the Annual Fall Meeting of the Biomedical Engineering Society EMBS/BMES 1*, 46–47.

7

Computational Intelligence in Gait and Movement Pattern Analysis

7.1 Introduction

Analysis of gait or human movement offers insights into information that can help in the diagnosis and treatment of walking and movement disorders. In this chapter, we begin by providing an overview of the commonly used biomechanical and motion analysis techniques used in gait analysis and highlight the key features extracted from graphs for characterizing movement patterns. The applications involving motion analysis are extensive, ranging from clinical applications, rehabilitation and health, to technique analysis and performance enhancement in sport. Many features and parameters are used to characterize movement patterns, and the type of gait features in such analysis tasks range from directly measurable variables to parameters that require significant data processing. When analyzing movement, in addition to statistical techniques, approaches based on machine learning have been used to create better models for linking the inputs and outputs in the assessment of movement patterns. We briefly discuss some areas of applications that have benefited from using CI methods.

7.1.1 Biomechanics, Gait, and Movement Analysis

Biomechanics involves the study of the mechanics of living systems. Aristotle (384–322 BC) is regarded as the father of biomechanics, his book *De Motu Animalium* or *On the Movement of Animals* is considered the first biomechanics book, in which animals were observed as mechanical systems. This concept inspired him to explore questions relating to his observations. Biomechanics has evolved such that there are now many applications including clinical, rehabilitative, and sports-related fields. A most interesting aspect of the field is that many disciplines use biomechanics concepts in their specific areas of interest. Some examples of biomechanics include the exploration of forces and moments acting on human joints, limbs and segments, the aerodynamics of insect flight, and the characteristics of locomotion in many nonhuman species.

Movement in humans occurs due to complex interactions between the CNS, the peripheral nervous system, and the musculoskeletal system. As illustrated in Figure 7.1, Vaughan et al. (1992) identified the following sequence of events

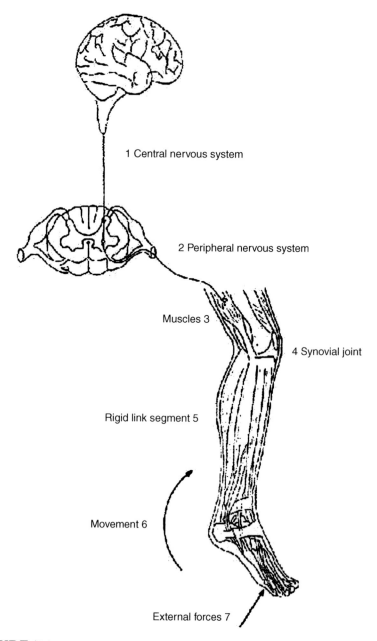

FIGURE 7.1

Various components of the locomotor system that contribute to the overall movement. (From Vaughan, C. L., Davis, B. L., and O'Connor, J. C., *Dynamics of Human Gait*, Human Kinetics, Champaign, 1992. With permission.)

that leads to the generation of movement. These include:

1.   Registration and activation of movement pattern signals in the CNS.

2.   Communication of signals to the peripheral nervous system.

3.   Contraction of muscles to develop tension.

4.   Generation of joint forces and moments.

5.   Regulation of joint forces and moments by the rigid link system.

6.   Movement or motion of the segments.

7.   Production of ground reaction forces.

Disruption to any of these components can result in abnormal movement patterns, for example, cerebral palsy (CP) patients show movement-related problems due to abnormalities of the central nervous system, often caused by brain damage during birth.

Gait is the most common and essential activity that humans undertake, such as walking and running. Although apparently a very easy task, gait is one of the most complex and totally integrated human movements, involving coordinated interactions of the neural and musculoskeletal systems (Winter, 1991). There are many challenges to the task of walking. First, walking requires integration of the CNS with peripheral sensory systems to control muscle activation in the correct sequence within the dynamic skeletal system. Second, the human mechanical system operates in a gravitational environment on two small bases of support (the feet) with the center of mass about two-thirds of body height from the ground. The body is mostly in a single-foot support phase ($\sim$80% of the time), which implies a state of continuous imbalance. Moreover, there is the need for safe foot trajectory during locomotion, which not only requires adequate ground clearance to avoid tripping but also a gentle heel contact (Winter, 1991). The regulation of such a system requires the neural control of well-defined total limb synergies that are sufficiently flexible to accommodate perturbations and are able to anticipate changes well in advance (Winter, 1991). Many investigations of human gait have sought to understand the processes of movement control, quantify changes due to diseases, and determine the effects of interventions and assistive devices.

7.1.2   The Gait Cycle

Gait is a periodic phenomenon and the gait cycle is defined as the interval between two successive events (usually heel contact) of the same foot. Figure 7.2 depicts one gait cycle or stride from heel contact of the right foot. It is characterized by a stance phase (60% of the total gait cycle), where at least one foot is in contact with the ground, and a swing phase (40% of the total gait cycle) during which one limb swings through to the next heel contact.

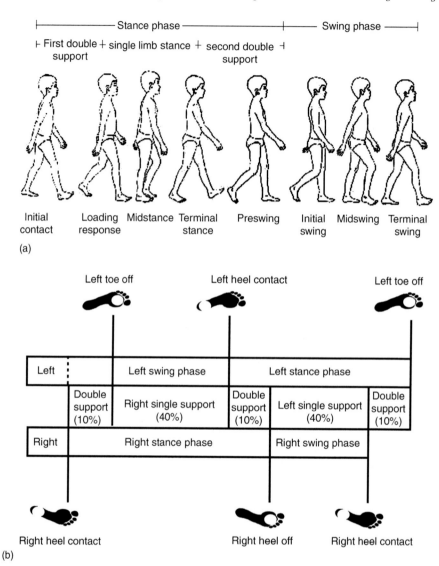

FIGURE 7.2

(a) One full gait cycle illustrating the various gait phases and events. There
are two major phases: stance and swing. These are further divided into a
number of subphases and major events. (From Vaughan, C. L., Davis, B.L.,
and O'Connor, J. C., *Dynamics of Human Gait,* Human Kinetics, Champaign,
1992. With permission.) (b) *Gait cycle.* The period of time from one event
(usually initial contact) of one foot to the subsequent occurrence of initial
contact of the same foot.

Absolute stance and swing times can be quite different between individuals but when normalized to a percentage of the gait cycle they maintain close similarity. For healthy young adults, the stance phase is estimated to be within 58–62%, but these durations can change with pathology, age, and variations in walking speed. The stance phase includes heel contact (initial contact), foot flat (loading response), midstance, heel off (terminal stance), and toe off (preswing); these events are illustrated in Figure 7.2a. The swing phase begins at toe off, continues through midswing and ends with heel contact of the same foot, as illustrated in Figure 7.2. In gait analysis and assessment, deviations or abnormalities are frequently referred to in terms of these events (Perry, 1992).

Gait analysis includes measurement, analysis, and assessment of the biomechanical features of walking. Technical and theoretical progress has been made in the area due to advances in computing and the development of sophisticated movement-recording systems. Improved computing speeds and algorithms have inspired increasingly complex and innovative data-analysis techniques.

7.2   Gait Measurement and Analysis Techniques

Typically, gait measurement involves obtaining a number of biomechanical variables using different measurement techniques (Best and Begg, 2006). Biomechanical analysis of human movement requires information about motion (kinematics) and internal and external forces (kinetics) on the limbs. Detailed laboratory-based gait analysis can be achieved using (1) a motion-recording system to measure the motion of body segments and joints, (2) a force platform measuring the forces exerted on the ground by the foot, and (3) an electromyography (EMG) data-acquisition system for recording the muscle activity during movements.

Other measurement techniques can also be used in gait analysis, for example, foot pressure distributions (Figure 7.3a), which can be applied to diagnose and treat foot disorders (such as diabetic neuropathy) and recording body accelerations using accelerometers and gyroscopes (Aminian et al., 2002). Researchers have extracted many parameters from the gait cycle to characterize the gait pattern. Most gait parameters can be categorized under the following headings:

1. *Basic time–distance or temporospatial* measures such as cadence, walking velocity, step/stride lengths, and stance/swing times

2. *Kinematic* measures such as joint/segment angles and angular range of motion (ROM), toe trajectory and clearance over the walking surface, and velocity/acceleration of joints and body segments

3. *Kinetic* measures that include force and pressure distribution under the foot, and joint/muscle moments and forces

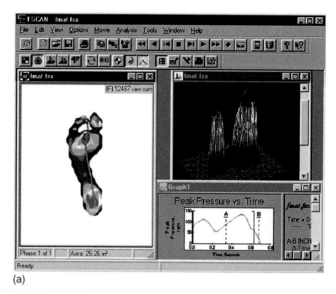

(a)

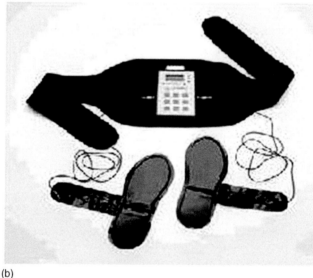

(b)

FIGURE 7.3 (See color figure following page 204.)

(a) Examples of foot pressure distribution using the F-scan system (www.tekscan.com). (b) Footswitches insoles and stride analyzer system for recording stride phase data (B & L Engineering: www.bleng.com). (c) Distance measures such as stride length, step length/width, and foot angles. (From Vaughan, C. L., Davis, B. L., and O'Connor J. C., *Dynamics of Human Gait*, Human kinetics, Champaign, 1992. Reproduced with permission.) (*Continued*)

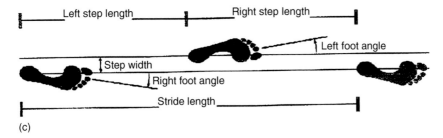

(c)

FIGURE 7.3
Continued

4. *Electromyography* data, which relate to electrical activity measures (timing and amplitude) of the contracting muscles

5. *Energy* expenditure (potential and kinetic) of the limb segments and also the whole body while performing the walking task.

7.2.1 Temporospatial Data

This type of gait analysis is often undertaken prior to a more detailed and complex analysis. It involves measurement of quantities such as walking speed, stride and step lengths, step width, stance and swing times, and cadence. Such measurements can be obtained using unsophisticated equipment such as a stopwatch, tape measure, and footswitches (Figure 7.3b). This straightforward analysis allows key parameters to be measured providing a general picture of the gait pattern but no detailed information. Commercial systems are available for temporospatial gait analysis; Figure 7.4 shows a typical output from the GAITRite system involving various time–distance measures.

> *Timing data.* Footswitches (see Figure 7.3b) are often used to measure the timing of foot placement or temporal parameters of a gait cycle, such as stance, swing, and single- and double-support time, as illustrated in Figure 7.2.
>
> *Cadence.* Cadence is the number of steps per unit time often expressed in steps per minute. Cadence can be calculated simply by counting the number of steps in a fixed time.
>
> *Velocity.* Velocity is the average horizontal velocity (displacement over time) of the body in the plane of progression, usually measured over multiple strides.
>
> *Spatial data.* Spatial data includes step length, stride length, and step width (see Figure 7.3c), which can be measured using instrumented mats (e.g., GAITRite) or simply from footprints.

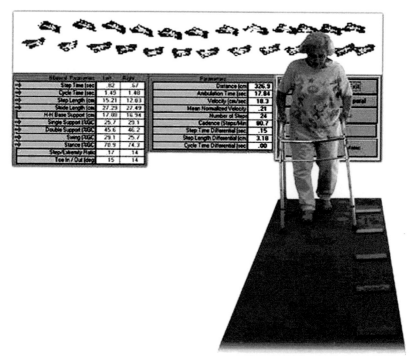

FIGURE 7.4 (See color figure following page 204.)

GAITRite walking analysis system showing portable 14 in. carpet with embedded sensors for recording temporospatial parameters and an output from GAITRite system (electronic footprint and time/distance data) (http://www.gaitrite.com/).

7.2.2 Kinematics

Kinematics involves the measurement and analysis of motion of body segments, joints, anatomical, or other landmarks of interest. It includes measures of displacement, velocity, and acceleration, either linear or angular. These data may be one-dimensional but sometimes information in all three dimensions is required for describing and analyzing complex activities.

Markers are usually attached to the body to identify anatomical points of interest and are normally placed on bony landmarks. A variety of motion-capture systems have been developed but most can be classified as one of two types of systems based on marker-recognition principles (Best and Begg, 2006) (Figure 7.5):

1. Reflective or passive marker systems (e.g., VICON and Peak) use reflective markers attached to the participant. Cameras record reflections from the markers to automatically capture their position and then present 3-D position–time representation of the markers.

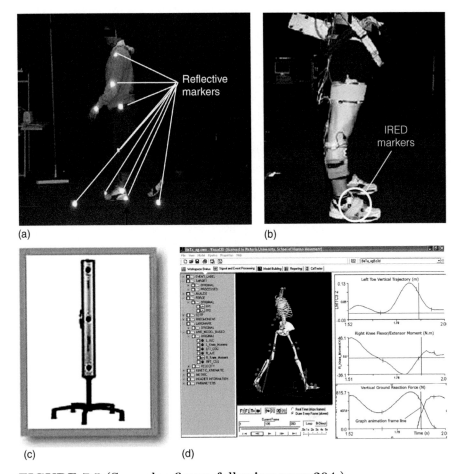

FIGURE 7.5 (See color figure following page 204.)
(a) Retro-reflective and (b) IRED active markers (arrows) are used to identify location of interest on the body. (c) Optotrak Certus position sensor (Northern Digital Inc, Waterloo, CA.). (d) Typical output from the Optotrak system and C-Motion Visual3D software package (www.c-motion.com. showing skeleton model, plots of knee joint angle, joint moment and vertical ground reaction forces from VU Biomechanics Lab).

2. Active marker systems (e.g., Optotrak and CODA) are optoelectronic systems that use active (light emitting diodes [LED]) markers to locate points of interest. Three dimensional position–time data is captured automatically in real time because the markers are individually identifiable (Best and Begg, 2006).

Position–time data can be further processed to minimize noise. Noise reduction is important prior to calculations of velocity and acceleration because noise is highly amplified, especially, in acceleration data due to double differentiation. Most of the noise in movement data is in the high-frequency region, thus, a low-pass digital filter with a low cutoff frequency (6–8 Hz) is sufficient to effectively reduce noise.

7.2.3   Kinetics

Kinetics is the study of motion that includes consideration of forces as the cause of movement.

Forces and torques are the two most widely used kinetic parameters in gait investigations. Ground reaction forces (GRF) result from contact between the foot and the ground and are frequently used when analyzing human gait, balance, and sports technique. There are various types of force platforms, each based on different principles. The two most widely used are the Kistler (http://www.kistler.com/), which is based on piezoelectric transducers, and the AMTI (http://www.amtiweb.com/), which uses strain gages mounted on precision strain elements.

Outputs (Figure 7.6) from these force platforms provide information regarding three orthogonal force components (F_x, F_y, F_z) and three moments (M_x, M_y, M_z). F_x is the horizontal medio-lateral force or M/L shear component, F_y is the horizontal anterior–posterior force or A/P shear component, and F_z represents the vertical force component.

Force platforms usually have standard software that provide information on variables such as component forces and directions, resultant forces, force vectors, center of pressure, and moments. As shown in Figure 7.6, the vertical (F_z) force has two distinct maxima (max1 and max2) during weight acceptance and push off and one minimum (min) during midstance. In healthy adults, both maxima usually exceed the bodyweight with the minimum below bodyweight, both maxima increase, whereas the minimum decreases with increased walking velocity. The anterior–posterior (F_y) force has a braking phase followed by a propulsive phase. The medio-lateral (F_x) is the smallest force out of the three and is highly variable across trials.

7.2.4   Electromyography

Human movement is produced by muscle contractions. As seen in Chapter 5, the electromyogram measures the electrical activity associated with muscle contractions. EMG signals can be used to identify which muscles are working during an activity and can also reveal the intensity or level of activation. Most applications involving EMG and gait have focused on monitoring the on/off timing of muscles to ensure that the correct muscles are activated at the required magnitude.

Surface or indwelling electrodes record motor unit action potentials (MUAPs) associated with muscle activation. In gait analysis, many factors can influence the MUAP and the EMG signal, including electrode location on the muscle, electrode size and shape, spacing between electrodes, fat or other tissue over the muscle, muscle area, and cross talk from other muscles.

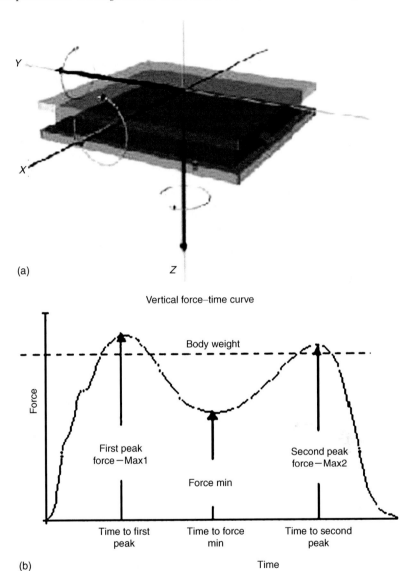

(a)

Vertical force–time curve

(b) Time

FIGURE 7.6
(a) AMTI force platform to measure multiaxis force/moment (http://www.amtiweb.com/transducerhome.htm.). (b) Typical vertical (Fz) force–time graph showing major force and time features. (c) Typical anterior–posterior (Fy) force–time graph and the key features. (d) Typical medio-lateral (Fx) force–time graph and the key features that are used for the analysis of gait. (*Continued*)

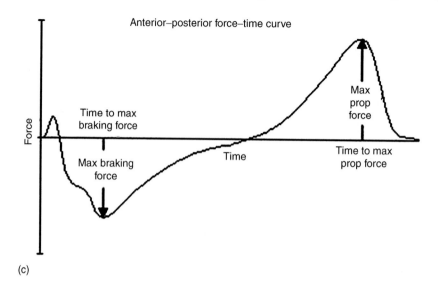

(c)

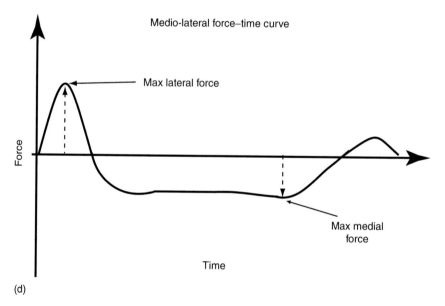

(d)

FIGURE 7.6
Continued

As shown in Figures 7.7b and 7.7c, surface electrodes are disc electrodes and are more commonly used in gait analysis because they are less invasive and easier to use. Nevertheless, surface electrodes are not suitable for recording EMG activity from the deep muscles because of cross talk or electrical activity from neighboring muscles; so in such cases, wire electrodes are used.

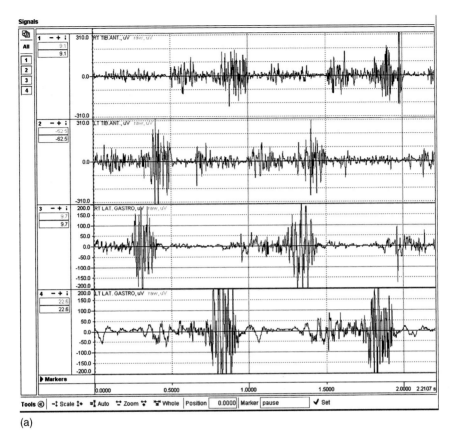

(a)

(b) (c)

FIGURE 7.7
(a) Raw EMG signal recorded (from tibialis anterior and gastrocnemicis muscles) during gait using the Noraxon EMG system at VU Biomechanics Lab.
(b) Surface electrode. (c) Wire/needle electrode (http://www.electrodestore.com/EMG).

Commercial systems are available for both tethered and wireless surface and fine-wire electromyography recording. Although raw EMG signals (Figure 7.7a) can show approximate on/off timing and activity levels, these signals are usually further processed to gain more information prior to their application. This includes minimizing both low- and high-frequency noise using digital filters (low and high pass) and rectification to convert negative signals to positive values. Low-pass filters allow for high-frequency noise to be eliminated, such as that from electrical equipment, whereas high-pass filters minimize the low-frequency noise due to sources such as movement between the electrode and the skin or movement of the cable connecting the electrode to the amplifier. Refer to Chapter 5 for details of EMG signal processing.

A low-pass filtered EMG signal is known as a *linear envelope*; this closely approximates the muscle forces and is frequently used in gait applications. Figure 7.8 illustrates the activity of major lower-limb muscle groups over the gait cycle. Within the human body, there are both two-joint and single-joint muscles. Single-joint muscles affect motion at one joint only, thus, *tibialis anterior* activity at the start of the gait cycle keeps the foot dorsiflexed and ensures proper heel contact (see Figure 7.8). A two-joint muscle can have two functions at the two joints. For example, the *gastrocnemius* may work either as an ankle plantarflexors during the push-off phase or as a knee flexor (see Figure 7.8).

7.2.5   Combined Output

Motion data are frequently collected in conjunction with force and EMG data, and postprocessing is often undertaken to derive useful parameters. Motion and force data, for example, can be combined using anthropometric measures to calculate internal joint moments and the energy and power around ankle, knee, and hip joints using *inverse dynamics* (Winter, 1990). Figure 7.9 and 7.5d illustrate typical 3-D output from a commercial motion analysis system (Optotrak motion analysis system and Visual3D software) displaying kinematic, kinetic, and muscle activity information during gait.

There are also techniques to superimpose force, force vectors derived from the GRF data, and EMG information onto the video image in real time. This type of analysis can assist the study of force vectors relative to lower-limb joint positions. There have been attempts to combine information from two or more biomechanical measurement systems into an integrated output to allow visualization of the motion and associated forces (Begg et al., 1991; Roberts et al., 1995; Rowe, 1996). Figure 7.10 illustrates output from the system developed by Begg et al. (1991), which uses a combination of video and force platform information (force vector). One advantage is that this output can reveal effects of the foot–ground reaction forces on the individual lower-limb joints. This can be achieved by observing the type (flexing/extending) of joint moment. Moments can be estimated by studying the moment arms and magnitude of the resultant GRF. Such real-time analysis can be helpful for the assessment of orthotic and prosthetic interventions. From the two sagittal plane outputs in Figure 7.10, although Figure 7.10a illustrates a flexing

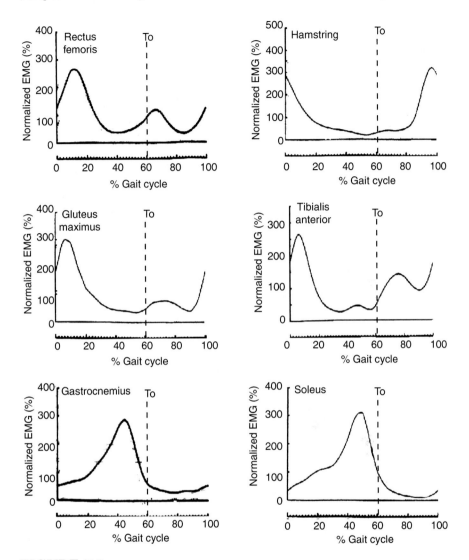

FIGURE 7.8

EMG activity from six major lower-limb muscle groups over one full gait cycle (stance phase is 60% and swing phase is 40% of the gait cycle), To-Toe-off event. (Adapted from Winter, D.A., *The Biomechanics and Motor Control of Human Gait: Normal, Elderly, and Pathological*, Waterloo Press, Waterloo, 1991.)

demand moment created by the GRF in a normal healthy subject at the knee joint, a different type of joint moment (extending) can be seen in the cerebral palsy patient (Figure 7.10b). This type of output can be used to quickly check GRF effects on the joints during an activity. They hold limited quantitative information regarding the actual joint moments and are not highly accurate

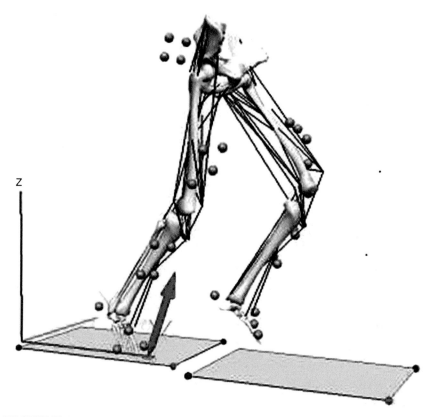

FIGURE 7.9
Typical output from Optotrak system and Visual 3-D software showing major
segments, marker locations, muscle connections, and force vector measured
concurrently by two force plates (www.c-motion.com).

when estimating moments around more proximal joint centers such as the hip.
Accuracy will decrease as we move from knee to the hip joints and also as the
walking speed increases because of the contribution from inertial components.

7.3 Gait in Various Populations

7.3.1 Normal Adult Gait

Most gait research concerns normal healthy gait (Whittle, 1991, 2002; Winter,
1991; Best and Begg, 2006) because it is important to first understand the
characteristics of normal gait such as the mean, variability, and ranges of
the parameters, before attempting to identify deviations due to a pathology.
Normative information is, therefore, important.

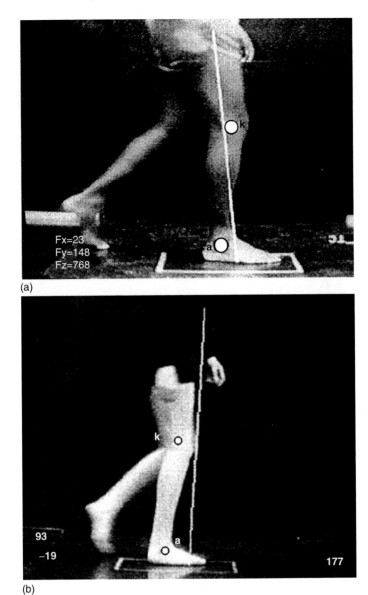

(a)

(b)

FIGURE 7.10
Output from the video vector system for (a) a healthy young gait; and
(b) a cerebral palsy (CP) patient gait. Note, two different types of moments
are evident around the knee joint for the two subjects due to location of the
force vector with respect to the joint center, where a = ankle joint and k = knee
joint. (From Begg, R. K., Wytch, R., J. M. S. Hutchison, et al., *Clin. Bomech.*,
6, 168–172, 1991.)

7.3.2 Gait in Children

A few studies have explored gait in young children. Characteristics of immature walking in healthy children are important for understanding pathological gait changes in this population. With regard to basic time–distance data (velocity, cadence, stride and step lengths, and stance and swing times), research has indicated that these parameters mature quickly with only a few minor differences from the adult gait by the age of 4 years (Sutherland, 1997). Children's gait has been characterized by higher cadence, lower walking velocity, and slight increases in pelvic and hip-joint rotation (Sutherland, 1997). Single-stance time increases from 32 (1 year) to 38% (7 years) compared to the adult single-stance time of approximately 40% of the gait cycle. Walking velocity increases progressively with age, with a rapid increase up to 3.5 years (0.64–1.14 m/s by year 7; adult values are 1.46 m/s for male and 1.3 m/s for female). Cadence decreases with age (176–144 steps/min by year 7; adult values are 113 and 118 for males and females, respectively). Other major deviations include reduced hip extension, lack of full knee extension, flexed knee heel contact, and reduced push-off force (Sutherland, 1997).

7.3.3 Gait in Older Adults

In recent years, considerable research has been directed toward describing and understanding age-related changes to gait. The primary objective has been to develop diagnostic techniques based on gait degeneration, due to ageing. The affected gait features may also be used for the identification of individuals' balance problems and the risk of falls. Falls in elderly individuals are a major public health issue because of the financial cost of surgery and rehabilitation and the human cost of associated pain and disability. In one study, 32% of a sample of community-dwelling elderly persons were shown to have sustained a fall at least once a year, with 24% of them sustaining serious injuries (Tinetti and Speechley, 1989). Furthermore, approximately 30% of people aged 65 years or more and approximately 50% of individuals over the age of 75 years sustain a fall sufficiently serious to cause injury, at least once a year (Snow, 1999). With life expectancy projected to rise and the associated ageing population increasing, the incidence of falls and associated costs are expected to rise unless effective preventative techniques are implemented.

It has been well documented that with age, balance-control mechanisms and gait function change, which affect the locomotor system. The reported declines in gait measures include, basic spatial–temporal parameters such as stride length, walking speed, and stance and swing times (Winter, 1991). In addition, there are changes to joint angular excursions at the hip, knee and ankle joints (Öberg et al., 1994), and kinetic parameters as reflected in the foot–ground reaction force data, such as the vertical and horizontal peak forces (Judge et al., 1996). Previous studies have reported that walking speed and stride length decreases with age (Ostrosky et al., 1994; Blanke and Hageman, 1989; Winter et al., 1990), whereas stance and double support time increase (Whittle, 1991). The range of joint movements (ROM) over the gait cycle has also been examined to show how coordination of the thigh, shank, and foot are affected by age. Olney and Colborne (1991) reported that the total knee ROM in older adults over the gait cycle was significantly reduced compared to young adults. Peak flexion and extension angles during the stance and swing phases have revealed reduced maximum ankle plantarflexion (Winter, 1991; Olney and Colborne, 1991) and significantly reduced maximum knee extension in elderly subjects (Ostrosky et al., 1994).

Begg and Sparrow (2006) explored whether joint angles at critical gait events and during major energy-generation and absorption phases of the gait cycle would reliably discriminate age-related degeneration during unobstructed walking. The gaits of 24 healthy adults (12 young and 12 elderly) were analyzed. The elderly participants showed significantly greater single (60.3 versus 62.3%) and double-support times, reduced knee flexion (47.7° versus 43.0°), and ankle plantarflexion (16.8° compared to 3.3°) at toe off, reduced knee flexion during push off, and reduced ankle dorsiflexion (16.8° compared to 22.0°), during the swing phase. The plantarflexing ankle joint motion during the stance to swing-phase transition for the young group (31.3°) was about twice that of the elderly (16.9°). The reduced knee extension range of motion may suggest that the elderly favored a flexed-knee gait to assist in weight acceptance. The reduced ankle dorsiflexion during the swing phase suggests increased risk of toe contact with obstacles. Given that tripping has been identified as the leading cause of falls in older individuals, the reduced vertical toe clearance and the associated increased potential for contact with over ground obstacles can be attributed to restricted ankle and knee flexion.

7.3.4   Pathological Gait

Movement patterns are changed by disease and pathological conditions, which affect the neuro-muscular system. Gait analysis is routinely used in the clinical settings for the assessment of walking performance. One important application of gait analysis is for the assessment of CP patients; in a recent study of the treatment of 102 patients with CP, for example, gait analysis was found to play a significant role in the decision making and planning for surgical outcomes (Cook et al., 2003). Identification of CP and other pathological gait patterns from characteristic features has considerable potential for gait assessment and the evaluation of treatment outcomes.

Some commonly observed gait abnormalities that affect the human musculoskeletal locomotor system (Whittle, 1991) are listed as follows:

> *Amputee.* Artificial limb, for example, below knee or above knee prosthesis.
>
> *Anterior trunk bending.* Compensation for weak knee extensors.
>
> *Cerebral palsy.* Neurological (brain) disorders, mainly seen in children that arise due to brain damage during birth. This condition affects communication between brain and various muscles leading to uncoordinated movement patterns.

Diplegia. This is a type of CP that affects both sides.

Hemiplegia. This is the mildest form of CP that affects only one side of the body. An individual with hemiplegic type of CP will have only a few limitations while performing his/her daily activities.

Increased lumbar lordosis. Lordosis is used to aid walking.

Lateral trunk bending. Painful/abnormal hip joint, wide walking base, and unequal leg length.

Parkinsonism. Degeneration of the basal ganglia of the brain that often results in shuffling gait.

Posterior trunk bending. Compensation for weak hip extensors early in the stance phase.

Quadriplegia. Involvement of the four limbs, the trunk, and the head.

7.4   CI Techniques in Gait Analysis

7.4.1   Artificial Neural Networks

CI techniques including NNs, fuzzy logic, genetic algorithms, and SVMs have been applied to model gait data and to classify gait types from their features (Chau, 2001; Schöllhorn, 2004; Begg et al., 2005b). Of these techniques, NN applications are the most widespread and most of these applications focus on gait pattern classification and recognition. There have also been attempts to use CI techniques for gait data modelling, for example, Sepulveda et al. (1993) successfully applied NNs to map lower-limb joint angles and joint moments onto electromyographic activity patterns.

A number of investigations have used NNs for the automatic identification of gait types, such as normal, abnormal, or pathological, using features from gait recordings. Figure 7.11 illustrates steps in a gait classification task based on NNs. Such an automated model could lead to useful applications, for example, in the diagnosis of gait abnormalities and assessment of gait following treatments and interventions.

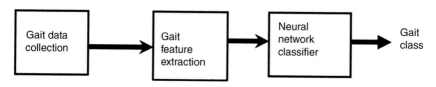

FIGURE 7.11

Main steps in a gait classification task using neural networks.

7.4.1.1 Neural Networks in the Recognition of Simulated Gait

Barton and Lees (1997) applied NNs to identify three gait conditions: normal walking, leg length difference, and leg weight difference from joint kinematics data. They used hip and knee joint angles over the gait cycle along with transformed Fourier coefficients as inputs to the classifier, and a small sample of five male and three female subjects was used. The kinematic data were collected using reflective markers at the neck, hip, knee, and ankle joints, which allowed the determination of hip and knee angles. The knee and hip angle data were then normalized to 128 time intervals and the corresponding FFT coefficients were determined. The first 15 Fourier coefficients (8 real and 7 imaginary) were used for NN training and testing to evaluate NN's performance in recognizing the three simulated gaits. A BP-learning algorithm was used to train the network. The input layer had 30 nodes for accepting FFT coefficients concerning hip and knee angle information, the two hidden layers consisted of nine nodes, and the output layer had three output nodes that represented the three simulated gait types. The training data set included six subjects, whereas the test set had information on two subjects. The correct recognition ratio of the NN was reported to be 83.3%. Although a small sample was used and accuracy was moderate, the results suggest that a trained NN is powerful enough to recognize changes in simulated gait kinematics; joint angles in this case.

7.4.1.2 Neural Networks in the Recognition of Pathological Gait

Several studies have reported the application of NNs to recognizing healthy and pathological gait patterns, one of the first was by Bekey et al. (1977), who showed that an EMG-based NN model could distinguish normal from pathological gaits. Holzreiter and Köhle (1993) applied NNs to differentiate healthy and pathological gaits using vertical ground-reaction forces recorded from two force plates; this experiment employed 131 patients and 94 healthy individuals. The vertical force–time information was divided into 128 constant time intervals and all the force components were normalized to the subject's body weight. FFT coefficients of the resulting data were calculated, of which the 30 low (real and imaginary) coefficients were used as inputs to the classifiers. The NN architecture included a three-layer feedforward topology with 30 input nodes and a single-classification output node corresponding to the healthy or pathological class. The network was trained 80 times using a BP algorithm with random weight initialization. A different proportion of the data set (20–80%) was used for training and testing the network performance. The maximum success rate for the training set that involved 80% of the data was reported to be ~95%. Wu and Su (2000) also used models based on NNs and foot–ground reaction forces to successfully classify gait patterns of patients with ankle arthrodesis and healthy individuals.

There have been attempts to compare the classification performance from NNs with that obtained using statistical techniques. Wu et al. (1998), for example, applied linear discriminant analysis (LDA) for gait classification of healthy and pathological gait patterns and compared them to the NN results. The gait recognition rate of NNs was reported to be superior to predictions based on LDA in a group consisting of both normal and pathological cases (98.7% versus 91.5%). Recently, Kamruzzaman and Begg (2006) compared classification outcomes obtained using LDA, NNs, and SVMs, which the classifiers were tested to recognizing healthy and CP gait patterns based on cadence and stride length. Superior classification was obtained using either an ANN or an SVM compared to the LDA. These results demonstrate considerable potential for NNs as a gait diagnostic tool. Similar studies of NNs as a gait classifier can be found in review articles (e.g., Chau, 2001; Schöllhorn, 2004; Uthmann and Dauscher, 2005).

7.4.1.3   Neural Networks in the Recognition of Ageing and Falls-Risk Gait

Gait changes occur across the lifespan as reflected by changes in kinematic, kinetic, and other features. Some of these changes in ageing may be detrimental due to modifications to balance control, which may lead to trips, slips, and falls. NNs have been used for the recognition of ageing effects on gait, specifically, to test whether NNs can effectively differentiate between the gait of young and elderly individuals. Begg and Kamruzzaman (2006) applied three algorithms (standard BP, scaled conjugate gradient, and Bayesian regularization) and measures of joint angles, foot–ground reaction forces, and stride-phase variables, to test whether a NN algorithm could detect age-related changes in gait patterns. The model test results were evaluated for percentage accuracy and measures were extracted from the ROC curves, such as sensitivity, specificity, and area under the ROC plots. Analysis was conducted on 12 healthy young adults and 12 healthy older adults during normal walking on a laboratory walkway. Kinetic measures were captured using a force platform and gait kinematics was recorded using a motion analysis system. Lower-limb movement was recorded using reflective markers attached to the hip, knee, ankle, heel, and toe. Gait features used for training and testing the classifiers included stride cycle time–distance data (walking speed, stance, swing and double-stance times, and their corresponding normalized data and stride length), lower-limb joint angles, and angular ROM. The angular data included knee and ankle joint angles at key events (heel contact and toe off), joint angular displacement during stance and swing, and stance-to-swing transition phases of the gait cycle. Maximum and minimum forces during key phases of the gait cycle were normalized to body weight. The specific kinetic features included peak vertical forces during weight acceptance, midstance and push-off forces, and horizontal peak forces during the braking and propulsive phases (see Section 7.2.3).

Test results indicated that all three NN algorithms provided good classification accuracy (83%); however, the Bayesian regularization gave superior classification performance in ROC area, sensitivity, and specificity (Table 7.1). Performance of the Bayesian regularization was significantly enhanced when feature selection was applied, that is, a subset of features was selected based on their relative contribution to the classification tasks. The advantage of feature selection is illustrated by the increased ROC area as shown in Figure 7.12.

TABLE 7.1

Comparison of Performance of Neural Network Algorithms for the Classification of Young–Old Gait Using ROC Area and Sensitivity Results

Neural Network Algorithms	Back Propagation	Scaled Conjugate Gradient	Bayesian Regularization
ROC area	0.82	0.84	0.90
Sensitivity at specificity of 0.9	0.60	0.67	0.73
and 0.75	0.75	0.79	0.86

Source: Adapted from Begg, R. K. and Kamruzzaman, J., *Aus. Phy. Eng. Sci. Med.*, 29(2), 188–195, 2006.

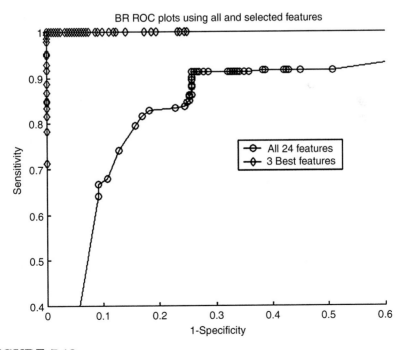

FIGURE 7.12

ROC plots of BR classifier using 24 features and 3 key features selected by the feature selection algorithm. (Adapted from Begg, R. K. and Kamruzzaman, J., *Aus. Phy. Eng. Sci. Med.*, 29(2), 188–195, 2006.)

These results demonstrated that an NN is able to successfully map relationships between ageing effects and gait characteristics.

To test the effectiveness of NNs in differentiating healthy and falls-prone elderly individuals, Begg et al. (2005a) applied NNs to measures of minimum toe clearance (MTC) during the gait cycle swing phase. During locomotion, foot clearance plays a critical role in the successful negotiation of obstacles and uneven terrain. MTC time series data were recorded from 10 healthy elderly and 10 elderly individuals with reported balance problems and tripping falls history, while they walked for 10–20 min on a treadmill. During the swing phase in normal gait, MTC above the ground is quite low and is lower in older people than in a young population (young MTC = 1.56 cm and elderly MTC = 1.48 cm). Figure 7.13 shows a histogram plot of MTC during the swing phase during an extended period of continuous treadmill walking. The histogram provides considerably more detailed information about the locomotor system than has previously been achieved using a small number of gait trials; it identifies the central tendency and variability in the data and suggests the associated control strategies to avoid foot–ground contact. The MTC event is considered an important parameter in understanding falls, specifically, falls resulting from a trip. Features extracted from the histograms such as mean, standard deviation, and skewness were used as inputs to a three-layer NN model with a BP error-correction algorithm to establish links between the MTC features and the healthy or balance-impaired categories. Cross-validation techniques were utilized for training the models and, subsequently, testing the models. To evaluate the diagnostic model accuracy rates, ROC plots and sensitivity and specificity results were used. For this MTC histogram data set, NN architecture with a single hidden layer of eight neurons provided optimal performance. Generalization performance of the diagnostic model with respect to classifying healthy and balance-impaired gait patterns was found to be >91% (ROC_{area} >0.94), with sensitivity >0.90 and specificity >0.71 (Table 7.2). The results suggest that NN models are effective for the detection of gait changes due to falling behavior.

7.4.2   Fuzzy Logic

Gait pattern recognition and classification has also been undertaken with the help of fuzzy-clustering techniques. Su et al. (2001) used a fuzzy cluster paradigm to separate the walking patterns of 10 healthy individuals and 10 patients with ankle arthrodesis. The features extracted were based on Euler angles from foot-segment kinematics, and the various clusters represented distinct walking patterns for the normal and pathological subjects.

O'Malley et al. (1997) demonstrated the usefulness of fuzzy-clustering technique in differentiating the gait of normal healthy and CP children using two time–distance data (stride length and cadence). Clusters were used to represent the walking strategies of the children with CP and healthy children. Using gait data from 88 spastic diplegia CP children and from 68 normal children, each child's gait function was represented from the membership values

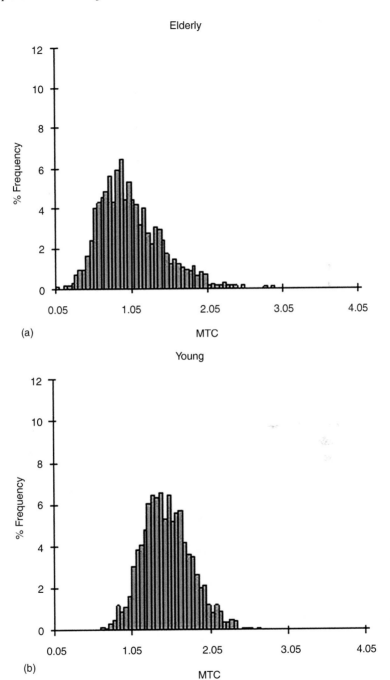

FIGURE 7.13
MTC histogram during continuous treadmill walking (a) an elderly subject
and (b) a young subject.

TABLE 7.2

Neural Networks Model Classification Performance Results

	Cross-Validation Test	Leave One Out
Accuracy (%)	91.70	95.00
ROC area	0.95	0.94
Sensitivity	0.95	0.90
Specificity	0.71	0.90

Source: Adapted from Begg, R. K., Hassan, R., Taylor, S., et al., *Proceedings of the International Conference of Intelligent Sensing and Information Processing*, 2005a, IEEE Press, Chennai, India ©IEEE.

TABLE 7.3

Classification Performance of the Fuzzy System

	Cross-Validation Test	Leave One Out
Accuracy (%)	89.30	95.00
ROC area	0.93	0.87

Source: Adapted from Hassan, R., Begg, R.K., and Taylor, S., *Proceedings of the IEEE-EMBS Conference*, 2005, China ©IEEE.

for each of the five clusters. The cluster results in addition to changes in the cluster membership were also used to show improvement due to the surgical outcomes in the children with CP.

Despite the demonstrated success of fuzzy systems in clinical gait classification, there has been limited research exploring its potential for dealing with falls-risk gait. Hassan et al. (2005) applied fuzzy logic to map a relationship between MTC features in fall-prone individuals and healthy participants with no-falls history. MTC data sets included 10 healthy elderly individuals and 10 elderly with a balance and tripping falls history; all participants were over 65 years. The MTC histogram characteristics were used as inputs for the set of fuzzy rules and features were extracted by estimation of the clusters in the data. Each cluster corresponded to a new fuzzy rule, which was then applied to associate the input space to an output region. The gradient descent method was used to optimize the rule parameters. Both cross-validation and leave one out techniques were utilized for training the models and subsequently for testing performance of the optimized fuzzy model. Test results indicated a maximum of 95% accuracy (Table 7.3) in discriminating the healthy and balance-impaired gait patterns.

Lauer et al. (2005) applied an adaptive neurofuzzy system (ANFIS) to help detect seven gait events in the CP patients from EMG recordings. Footswitches are not only traditionally used to sense events within the gait cycle but they are also used for regulating muscle stimulation in CP patients. Lauer et al. (2005) have, therefore, developed a fuzzy inference system for predicting gait events using EMG activity from the quadriceps muscles. This CI-based technique for identifying gait events was tested on eight CP children and the predicted accuracy in gait event detection was determined from 3-D motion analysis data. Accuracy ranged from 95.3% to 98.6% when compared to the gait events obtained from the motion analysis system.

7.4.3   Genetic Algorithms

GA are based on the "survival of the fittest" principle in natural systems and they have been applied to gait pattern recognition. The GA technique can be used in conjunction with NNs to determine an optimal number of hidden layer neurons and for selecting input features that contribute most to the classification. Su and Wu (2000) have used GA for searches in appropriate spaces to determine optimal NN parameters for gait recognition in both healthy and pathological (ankle arthrodesis) populations. To differentiate normal and pathological gait patterns ground-reaction forces have been recorded using force platforms with 10 healthy and 10 pathological subjects (Su and Wu, 2000). The results showed that a GA-based NN performed better than a traditional NN (98.7% compared to 89.7% using a traditional NN). These findings indicate the advantage of first optimizing parameters in the classifiers and, second, selecting the more important features for a classification task. So far, there has been little investigation into the application of GA to gait pattern-recognition but future gait pattern-recognition applications could benefit from using GA as a feature-selection technique.

7.4.4   Support Vector Machine in Gait Analysis

An SVM is a machine-learning tool, which has emerged as a powerful technique for solving various classification and regression problems (Chapter 3). SVMs originated from Vapnik's (1995) statistical learning theory; and the key advantage over NNs is that they formulate the learning problem as a quadratic optimization problem with an error surface that is free of local minima and has a global optimum (Kecman, 2002). It has been particularly effective for binary classification in biomedical problems. In gait research, SVM has been applied to the classification of young and elderly gait patterns (Begg et al., 2005b) and also used in gender-classification tasks from video gait sequence data (Lee and Grimson, 2002).

In developing SVM-based gait classification models (Begg et al., 2005b), MTC data during the swing phase of the gait cycle was recorded over multiple gait cycles. The MTC time series data were transformed into *histogram* and *Poincaré* plots for feature extraction and the features were then used as inputs to the SVM model for training and testing. The test results suggested that MTC data could provide useful information regarding steady-state walking characteristics of individuals. The features could be used to train a machine-learning tool for the automated recognition of gait changes due to ageing.

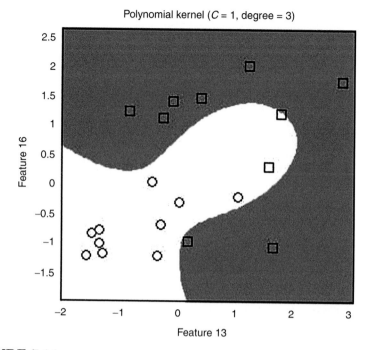

FIGURE 7.14

Two-dimensional plot showing decision surface for two features, that is, push-off force and knee range of motion. (From Begg. R. K., Palaniswami, M., and Owen, B., *IEEE Trans. Biomed. Eng.*, 52, 828–838, 2005b. ©IEEE.)

Figure 7.14 illustrates a 2-D plot of the decision surface for the young and elderly participants using two gait features from the MTC histograms.

Early detection of locomotor impairments using such supervised pattern-recognition techniques could be used to identify at-risk individuals. Corrective measures can then be implemented to reduce the falls risk. Owing to superior performance of the SVM for gait classification (Begg et al., 2005b) and in other biomedical applications, SVMs hold much promise for automated gait diagnosis.

The generalization performance of a classifier depends on a number of factors but primarily on the selection of good features, utilizing input features that allow maximal separation between classes. A feature-selection algorithm was employed by Begg et al. (2005b), which iteratively searched for features that improved the identification results. The results, from a number of training and test samples, clearly showed that a small subset of selected features is more effective in discriminating gait than using all the features (see Figure 7.15).

Classification performance of the SVM depends on the value of the regularization parameter, that must be carefully selected through trial and error.

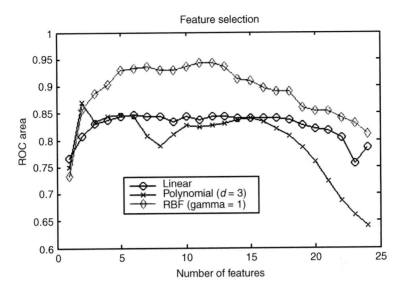

FIGURE 7.15

SVM classifier performance as a function of selected features. Note, higher performance for all three SVM classifiers for a few selected features compared to the use of a full feature set. (From Begg, R. K., Palaniswami, M., and Owen, B., *IEEE Trans. Biomed. Eng.*, 52, 828–838, 2005b. ©IEEE.)

Nonoptimal selection can lead to significant reduction in classifier performance, but both histogram and Poincaré plot features were effective in discriminating the two age groups. This suggests that gait changes with age are reflected in these plots, future research could, however, include other groups (such as people with prior falls history) to develop gait models for other populations. Other types of gait features, such as kinetics and EMG data can also be included in future classification models involving SVM, which may provide useful additional information and improve prediction.

SVM has also been applied to help differentiate healthy and CP gait patterns. Accurate identification of CP gait is important for both diagnosis and treatment evaluation. Kamruzzaman and Begg (2006) explored the use of SVMs for the automated detection and classification of children with CP using stride length and cadence as input features. The SVM classifier identified the groups with an overall accuracy of 79.3% (sensitivity = 74.5%, specificity = 85.2%, ROC area = 0.86). Classification accuracy improved significantly when gait parameters were normalized to leg-length, providing 95.9% accuracy. The effect of normalization is illustrated in the scatter plot of Figure 7.16. This normalization also resulted in an improvement of ROC area (0.99) and the accuracy was higher than with an LDA and an NN-based classifier. Among the four SVM kernel functions studied, the polynomial kernel outperformed the others. The enhanced classification accuracy of the SVM using only two

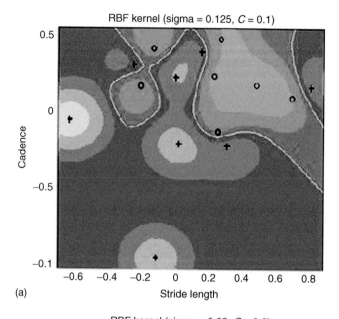

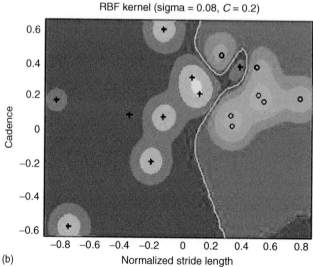

FIGURE 7.16 (See color figure following page 204.)
Scatter plots showing the test data and separating surfaces using raw and normalized gait data by the SVM classifier with RBF kernel. (a) With raw stride length and cadence and (b) with normalized stride length and cadence. (From Kamruzzaman, J. and Begg. R. K., *IEEE Trans. Biomed. Eng.*, 53, 2479–2490, 2006. ©IEEE.)

basic gait parameters makes it an attractive alternative for identifying CP children and evaluating the effectiveness of treatment.

7.5 Concluding Remarks

In this chapter, we reviewed the gait pattern recognition and classification abilities of CI techniques. Gait features used in training and testing the machine classifiers are usually extracted from data recorded using standard techniques. Gait features employed in studies to date include kinematic, kinetic, and EMG variables. Past research supports the use of NNs, fuzzy logic, GAs, SVMs, or a combination of these (e.g., neurofuzzy) for the recognition of gait types (healthy, ageing, pathological, or faller). When differentiating gaits, performance can be expressed using cross-validation accuracy rates and ROC plots (ROC area, sensitivity and specificity measures). Classification performance has been found to be significantly enhanced when a subset of selected features are used to train the classifiers. Feature selection appears, therefore, to be important when dealing with gait data classification; this can also reduce the dimensionality of input data with less computational complexity.

CI offers benefits for gait data modelling and gait pattern prediction. In most applications, the reported classification accuracy is reasonably high, suggesting excellent prediction abilities and suitability for modelling the underlying gait data structures. Furthermore, performance using CI techniques, for example, NNs and SVM, is superior to an LDA-based classifier. In studies of both healthy and pathological populations (Wu et al., 1998; Kamruzzaman and Begg, 2006), improved performance using CI tools also suggested the importance of nonlinear approaches for modelling the relationships between disease or ageing and gait characteristics.

Based on this review, it is reasonable to conclude that further advances in movement pattern recognition may be achieved through the incorporation of two approaches. The first is *feature selection*, including additional feature-selection algorithms, such as backward elimination (Chan et al., 2002). GAs appear to have much potential (Su and Wu, 2000; Yom-Tov and Inbar, 2002) for separating relevant features to improve gait detection and classification. The second approach which has considerable potential is *hybridization*, that is, incorporating more than one artificial intelligence technique (Lauer et al., 2005) into the classification process. Some examples are combining an NN with fuzzy logic (neurofuzzy scheme) or evolutionary algorithms (Yao and Liu, 2004).

Good progress has been made over the past decade in the application of CI to recognizing gait movement abnormalities, and also establishing CI as a valuable diagnostic tool in biomedical science. Further research is needed before these techniques become routine diagnostic tools within clinical settings.

However, there is no doubt that the future looks promising for the applications of CI techniques on movement diagnostics.

References

Aminian, K., B. Najafi, C. Bula, et al. (2002). Spatio-temporal parameters of gait measured by an ambulatory system using miniature gyroscopes. *Journal of Biomechanics 35*, 689–699.

Barton, J. G. and A. Lees (1997). An application of neural networks for distinguishing gait patterns on the basis of hip-knee joint angle diagrams. *Gait and Posture 5*, 28–33.

Begg, R. K., R. Hassan, S. Taylor, et al. (2005a). Artificial neural network models in the assessment of balance impairments. *Proceedings of the International Conference on Intelligent Sensing and Information Processing*, January, Chennai, India: IEEE Press.

Begg, R. K., M. Palaniswami, and B. Owen (2005b). Support vector machines for automated gait recognition. *IEEE Transactions on Biomedical Engineering 52*, 828–838.

Begg, R. K. and W.A. Sparrow (2006). Ageing effects on knee and ankle angles during major events and phases of the gait cycle. *Journal of Medical Engineering and Technology 30*, 382–389.

Begg, R. K. and J. Kamruzzaman (2006). Artificial neural networks for movement recognition. *Australian Physical and Engineering Sciences in Medicine 29(2)*, 188–195.

Begg, R. K., R. Wytch, J. M. S. Hutchison, et al. (1991). Microcomputer-based system for clinical gait studies. *Clinical Biomechanics 6*, 168–172.

Bekey, G. A., C. W. Chang, J. Perry, et al. (1977). Pattern recognition of multiple EMG signals applied to the description of human gait. *Proceedings of the IEEE 65*, 674–681.

Best, R. J. and R. K. Begg (2006). Overview of movement techniques and gait features. In *Computational Intelligence for Movement Sciences*, R. Begg and M. Palaniswami (Eds), Chapter 1, pp. 1–69, Hershey: IGI Publishing.

Blanke, D. J. and P. A. Hageman (1989). Comparison of gait of young men and elderly men. *Physical Therapy 69*, 144–148.

Chan, K., T. W. Lee, P. A. Sample, et al. (2002). Comparison of Machine Learning and Traditional Classifiers in Glaucoma Diagnosis. *IEEE Transaction on Biomedical Engineering 49*, 963–974.

Chau, T. (2001). A review of analytical techniques for gait data. Part 2: neural network and wavelet methods. *Gait and Posture 13*, 102–120.

Cook, R. E., I. Schneider, M.E. Hazlewood, et al. (2003). Gait analysis alters decision-making in cerebral palsy. *Journal of Paediatric Orthopaedics 23*, 292–295.

Hassan, R., R. K. Begg, and S. Taylor (2005). Fuzzy logic-based recognition of gait changes due to trip-related falls. *Proceedings of the IEEE-EMBS Conference*, China, September 1–3.

Holzreiter, S. H. and M. E. Köhle (1993). Assessment of gait patterns using neural networks. *Journal of Biomechanics 26*, 645–651.

Judge, J. O., R. B. Davis, and S. Õunpuu (1996). Step length reductions in advanced age: the role of ankle and hip kinetics. *Journal of Gerontology: Medical Science 51*, 303–312.

Kamruzzaman, J. and R. K. Begg (2006). Support vector machines and other approaches to the recognition of cerebral palsy gait. *IEEE Transactions on Biomedical Engineering 53*, 2479–2490.

Kecman, V. (2002). *Learning and Soft Computing: Support Vector Machines, Neural Networks and Fuzzy Logic Models*. Cambridge, MA: IEEE, MIT Press.

Lauer, R. T., B. T. Smith, and R. Betz (2005). Application of a neuro-fuzzy Network for gait event detection using electromyography in the child with cerebral palsy. *IEEE Transactions on Biomedical Engineering 52*, 1532–1540.

Lee, L. and W. E. L. Grimson (2002). Gait analysis for recognition and classification. *Proceedings of the Fifth International Conference on Automatic Face and Gesture Recognition*, IEEE Computer Society.

Öberg, T., A. Karsznia, and K. Öberg (1994). Joint angle parameters in gait: reference data for normal subjects, 10–79 years of age. *Journal of Rehabilitation Research and Development 31*, 199–213.

Olney, S. J. and G. R. Colborne (1991). Assessment and treatment of gait dysfunction in the geriatric stroke patient. *Topics in Geriatric Rehabilitation 7*, 70–78.

O'Malley, M. J., M. F. Abel, and D. L. Damiano, et al. (1997). Fuzzy clustering of children with cerebral palsy based on temporal-distance gait parameters. *IEEE Transactions on Rehabilitation Engineering 5*, 300–309.

Ostrosky, K. M., J. M. VanSwearingen, R.G. Burdett, et al. (1994). A comparison of gait characteristics in young and old subjects. *Physical Therapy 74*, 637–646.

Perry, J. (1992). *Gait Analysis. Normal and Pathological Function*, NJ: Slack Inc.

Roberts, G., R. Best, R. K. Begg, et al. (1995). A real-time ground reaction force vector visualisation system. *Proceedings of Australian Conference of Science and Medicine in Sport*, pp. 230–231 (October) Hobart, Australia.

Rowe, P. J. (1996). Development of a low-cost video vector for the display of ground reaction forces during gait. *Medical Engineering and Physics 18*, 591–595.

Schöllhorn, W. I. (2004). Applications of artificial neural nets in clinical biomechanics. *Clinical Biomechanics 19*, 876–898.

Sepulveda, F., D. M. Wells, et al. (1993). A neural network representation of electromyography and joint dynamics in human gait. *Journal of Biomechanics 26*, 101–109.

Snow, C. M. (1999). Exercise effects on falls in frail elderly: focus on strength. *Journal of Applied Biomechanics 15*, 84–91.

Su, F. C. and W. L. Wu (2000). Design and testing of a genetic algorithm neural network in the assessment of gait patterns. *Medical Engineering and Physics 22*, 67–74.

Su, F. C., W. L. Wu, Y. M. Cheng, et al. (2001). Fuzzy clustering of gait patterns of patients after ankle arthrodesis based on kinematic parameters. *Medical Engineering and Physics 23*, 83–90.

Sutherland, D. H. (1997). The development of mature gait. *Gait and Posture 6*, 163–170.

Tinetti, M. E. and M. Speechley (1989). Prevention of falls among the elderly. *New England Journal of Medicine 320*, 1055–1059.

Uthmann, T. and P. Dauscher (2005). Analysis of motor control and behavior in multi agent systems by means of artificial neural networks. *Clinical Biomechanics 20*, 119–125.

Vapnik, V. N. (1995). *The Nature of Statistical Learning Theory*. New York: Springer.

Vaughan, C. L., B. L. Davis, and J. C. O'Connor (1992). *Dynamics of Human Gait*. Champaign; IL Human Kinetics.

Whittle, M. W. (1991). *Gait Analysis: An Introduction*. Oxford: Butterworth-Heinemann.

Whittle M. W. (2002). *Gait Analysis: An Introduction (Third Edition)*. Oxford: Butterworth-Heinemann.

Winter, D. A. (1990). *The Biomechanics and Motor Control of Human Movement* (2nd ed.) New York: Wiley.

Winter, D. A. (1991). *The Biomechanics and Motor Control of Human Gait: Normal, Elderly, and Pathological*. Waterloo: Waterloo Press.

Winter, D. A., A. E. Patla, J. S. Frank, et al. (1990). Biomechanical walking pattern changes in the fit and healthy elderly. *Physical Therapy 70*, 340–347.

Wu, W. L. and F. C. Su (2000). Potential of the back propagation neural network in the assessment of gait patterns in ankle arthrodesis. *Clinical Biomechanics 15*, 143–145.

Wu, W. L., F. C. Su, and C. K. Chou (1998). Potential of the back propagation neural networks in the assessment of gait patterns in ankle arthrodesis. In *Neural Networks and Expert Systems in Medicine and Health Care*, E.C. Ifeachor, A. Sperduti, and A. Starita (Eds), pp. 92–100, Singapore: World Scientific.

Yao, X. and Y. Liu (2004). Evolving neural network ensembles by minimization of mutual information. *International Journal of Hybrid Intelligent Systems 1*, 12–21.

Yom-Tov, E. and G. F. Inbar (2002). Feature selection for the classification of movements from single movement-related potentials. *IEEE Transactions on Neural Systems 10*, 170–177.

8

Summary and Future Trends

8.1 Overview of Progress

Computational intelligence (CI) has only recently emerged as a discipline and is becoming more effective in solving biomedical-engineering problems, specifically those in which human intervention can be replaced by some form of automated response. A major part of this book has been devoted to reviewing everyday biomedical problems that have benefited from recent CI applications. Until now, most CI research into biomedical-engineering problems has focused on classifying disorders based on pathological characteristics. The motivation is to first replicate the diagnostic skill of the medical specialist and if possible provide more reliable diagnosis. In this endeavor, two trends in the application of CI have emerged; the first is classification of the pathology or disease and the second is the modeling of biological systems from where the pathology arises. The former has enjoyed cautionary acceptance within the clinical environment because physicians easily recognize the concepts of classification and pattern recognition. The latter is, however, less well accepted by clinicians because CI techniques tend to present as "black box" solutions, whose success depends on the practitioner's knowledge. It is, therefore, important to recognize that more research is needed to bring CI to the forefront of medical technology and to quell the remaining fears. For now, let us briefly review what has been achieved so far to summarize the book.

One of the most successful applications of CI has been the diagnosis of cardiovascular diseases (CVDs). This is due to several factors, firstly, CI applications here have focused primarily on the QRS wave complex, which is easily measured from the recorded ECG. This wave is periodic and can occasionally be corrupted by electrode noise or superposition of action potentials from other sources. The average peak-to-peak voltage of the QRS complex is, however, considerably larger than sources of noise thereby making signal processing ideal for retrieving the original waveform. In addition, the heart mechanics that produce the QRS waveform are well understood and existing QRS-detection algorithms are sufficiently accurate for practical application. Obtaining a clear and clean QRS wave is no longer a major problem as is information loss due to preprocessing.

Cardiologists examine the QRS waveform for abnormalities and when used in conjunction with CI techniques, physical measurements such as ST segment length, peak amplitudes, number of waveform turns, and peak-to-peak intervals have been effective in discriminating classifier features. More advanced processing methods such as wavelet transforms, AR, and principal components analysis (PCA) have also been applied in anticipation of extracting further information. Classification of cardiovascular disorders such as arrhythmias, myocardial infarctions, and ischemia has now become highly effective with most CI techniques being applied. Tests on a variety of publicly available databases have yielded high accuracies on average, which is very promising.

The remaining challenge is to design and build heart-monitoring devices, which incorporate these forms of CI. The key concern in transferring such technologies to commercial devices is reliability. CI techniques such as neural networks (NNs) rely heavily on the training data and also on correct parameter tuning for any derived diagnostic model. High accuracies in computer simulations are reassuring but clinical trials are needed to verify the usefulness of CI techniques.

The classification of neuromuscular diseases has not been straightforward. Unlike the QRS waveform, the MUAP waveforms measured via EMG have been composites due to the superposition of action potentials emanating from several muscle fibers. Even direct measurement of a single MUAP using needle EMG does not yield clear isolated waveforms. This is mainly due to motor units in the muscle fibers producing comparably smaller action potentials (-35 mV to -70 mV) making them susceptible to noise from electrode motion artifacts and from the electrical activity of larger organs. In short, although muscle mechanics and action potentials are well understood, motor-neuron firing is still thought to be random and nondeterministic. Research has examined the physical measurement of a single MUAP such as the number of terms and peak-to-peak amplitude similar to QRS analysis. This could be adequate provided one can obtain a single clean MUAP waveform, but in practice, this is rarely possible with superimposed waveforms. Autoregressive models (AR) models and independent component analysis (ICA) have garnered some success. The first is based on a linear-filter system and generates a wide-based stationary random signal and does not attempt to decompose the MUAPTs into single MUAP waveforms. ICA, however performs decomposition but is highly dependent on the final mixing matrix.

Both of these processing methods have been used to extract features for classification. The AR model performs better with higher model orders, whereas ICA has been used in conjunction with template-matching systems for the decomposition of MUAPTs. Other methods such as the popular wavelet decomposition have also been applied, but it requires correct wavelet selection to extract meaningful information, which entails considerable algorithmic adjustment. Much research is focused on finding more accurate signal-processing techniques to interpret the MUAPTs. Another issue is the type of features that should be extracted given that it may not be possible to recover the single MUAPs. In terms of classification, the literature has reported success in using NNs, whereas SVM classifiers and hybrid models remain to be researched with few current successful results. The emphasis for now is on using feature-extraction methods to determine neuromuscular disorders. Most research has, however, primarily investigated differences between myopathy, neuropathy, and nonpathology using CI techniques. Further research is, therefore, required to use CI to capture NMJ disorders and specific neuropathies with many differential diagnoses.

In addition to neuromuscular diseases, CI techniques have been applied to the development of prostheses and orthotics. Early myoelectric prostheses possessed very simple logic, few degrees of movement and depended on detecting a muscle position after the user had flexed and held the position for some time. This approach was inadequate for situations requiring instant responses due to the slow prosthesis response times. The problem was further exacerbated by the time required for training an amputee to use the prosthesis. Recent research has concentrated on recognizing limb position in isometric contraction with very few studies of muscle activity when the limb is moving. There is considerable scope for further investigation of CI techniques for intention detection.

A more advanced prosthesis is the exoskeleton aimed at providing mechanical strength either for everyday use or rehabilitative purposes. More work is needed here on translating and predicting the user's intention to move the exoskeleton. One hurdle is the need for a high degree of accuracy. In prostheses and exoskeletons, the error of not detecting an intended motion can be tolerated to a certain degree. The error of detecting an unintended motion can, however, lead to injuries. In view of this, effective application of CI techniques and optimal feature selection remain as challenging research topics.

EEG signals from the brain provide important information regarding brain function and cognitive performance but they are affected by various neurological conditions. The EEG has, therefore, found many routine applications in the diagnosis and treatment of neurological disorders. CI techniques are proving useful in biomedical diagnostics for classifying the EEG as either normal or pathological. Because these signals are low in magnitude, the EEGs are also affected by noise due to the sensing techniques and also from the surrounding environment. Noise reduction has, therefore, been a major concern even before application to the identification of pathological conditions. CI plays a major role in EEG processing and analysis, and the detection and monitoring of a disease. Another area that has received recent attention is the development of EEG-based communication links such as the brain–computer interface (BCI). This is a particularly exciting new area that is growing quickly and has wide applications. Interestingly, the BCI technology offers new opportunities for helping persons with disability to restore functional activities. CI techniques are proving very useful in classifying mental tasks from features embedded in the EEG signals, and subsequently for the control of external devices and actuators.

Steady progress has been made over the last few years in CI applications for recognizing gait abnormalities. Research has demonstrated the use of NNs, fuzzy logic, genetic algorithms, SVMs, or a combination of these (e.g., neurofuzzy) for the recognition of gait types such as healthy, aged, pathological, or faller. Recent work involving classier output as a scale for grading gait abnormalities holds promise both for diagnosis and evaluating treatment. Numerous features can be recorded in gait analysis, and feature selection is important in reducing the dimensionality of input data for gait-data classification and recognition. The uptake of CI approaches in gait analysis has not kept pace with other biomedical areas, and further research is needed before these techniques become routine in clinical and rehabilitative settings, but the future looks promising.

8.2 Future Challenges and Research Areas

To bring CI closer to the forefront of healthcare technology, fundamental challenges remain, some of which will require contributions from the machine-learning community because they are regarded as critical to driving these techniques. Further work in electrical engineering and telecommunications will also be invaluable. We are on the brink of a new healthcare paradigm, one that allows more of the physician's responsibilities to be shifted toward automated-healthcare systems powered by some type of CI.

The greatest difficulty in designing an automated diagnostic system is accuracy. As seen in previous chapters, CI techniques are sensitive to the input features that hold discriminative information of a particular disorder, motion, or state of a biomechanical system. The issue is what optimal set of features should we use; should they be entirely the representatives of pathology or consist of implicit information? Even when the optimal set is known, obtaining it may be too expensive and, therefore, impractical for diagnostic systems. At present, no general rule exists for all cases, resulting in CI practitioners being required to experiment with medical data to ascertain the proper feature set. There are several methods for doing this such as hill climbing, fuzzy integrals, ROC areas, and other information criterion measures, which can be used to investigate the separability of features. Whichever technique is used, feature selection remains a design issue that depends on the diagnostic application.

In the past, NNs have been a popular CI technique due to the extensive literature, but more recently SVMs have emerged as a contender to the NN paradigm. The SVM is considered to be an improved machine-learning paradigm based on Vapnik's structural risk minimization principle. Rather than selecting the classifier based on error minimization as in NNs, the SVMs choose a hyperplane from a subset of possible functions. This is known as capacity control, which is enforced by minimizing a regularized function; intuitively it is equivalent to maximizing the distance between the two classes. The advantage of SVMs, however, is that they generalize well in most cases, even if the training set is small. In contrast, NNs require a large training dataset to work well and the solution is dependent on the training algorithm initial conditions. A drawback to SVMs is that parameter tuning is needed for them to work well and they remain useful primarily for two class problems (binary classifier). If the problem is multiclass (three or more), NNs fare better in terms of ease of use, but research is ongoing to extend the SVM formulation to multiclass problems. A further issue is training speed, while training time for NNs scale linearly with training set size, this is not the case for SVMs, which have an almost quadratic slowing due to the quadratic programming formulation. Nevertheless, research into decomposition techniques for more efficient training of SVMs is proceeding.

When the measurements in a feature set are uncertain, rule-based methods incorporating fuzzy logic have been applied, but this area of interest is relatively new to the biomedical field and deserves further investigation. Hybrid or expert systems for medical diagnostics have been previously proposed, but there is an avenue for further improvement, especially in designing the best rules or heuristics. These systems have proven useful as educational tools for medical students and also as aids to physicians; again the research emphasis should be the accuracy of such systems. It could be beneficial to design specific systems, rather than generalized medical systems, to take maximum advantage of the selected CI technique and specific feature set.

As seen in the previous chapters, most core CI technologies have been applied to diagnostic medicine to classify as healthy rather than pathology. Efforts toward this goal have been successful, though some pathologies, for example, neuromuscular disorders, require further investigation. The next step is to apply these technologies to prognosis prediction and rehabilitation monitoring where in prognosis prediction we envisage an automated CI system to predict the probability of recovery. For example, if a patient had been diagnosed with cardiac arrhythmia, an important issue would be the chance of survival and how the disease would progress if untreated. Clinicians generally base their prognosis on experience and there is left, therefore, considerable scope for the application of CI techniques to improve the accuracy. In rehabilitation-rate monitoring, we envisage the application of CI techniques to estimate recovery rate due to a treatment, in an exercise intervention, for example, to regain balance following a fall. An automated system capable of achieving this would be useful in determining the effectiveness of drug treatment or therapy.

Figure 8.1 outlines three areas in which we foresee research in CI applications flourishing. There are many challenges to be addressed but substantial research is being undertaken in diagnosis and classification of pathologies. In prognosis prediction, we require CI techniques related to time-series prediction to estimate how well or how quickly a patient will recover. Research in this direction will require physicians to record their patient's recovery to generate examples of the required time-series data. Based on these data, it should be possible to use CI techniques to predict a patient's future condition. Fundamental questions that require attention are the accuracy of prediction, the amount of data required, methods of performing the prediction, and index measures or confidence values, which intuitively indicate the severity of disease. Accurate prediction would open up new ways of dispensing medication,

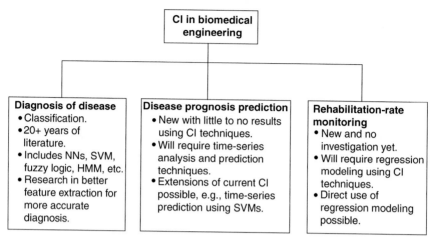

FIGURE 8.1

Some of the current and future research areas for CI techniques in biomedical engineering.

for example, dosages could be predicted and pharmacies better prepared for the demand.

CI prediction could also be used, as an early warning system, for example, using a time-series analysis of MTC variables, it may soon be possible to predict the risk of falling (see Section 7.4.1.3). Research in this direction is relatively new but a successful outcome requires a suitable application of CI technology. In rehabilitation-rate monitoring a similar prediction objective might be pursued, the idea here is to see how quickly a patient can regain normal function. This will require regression CI techniques because we are monitoring the recovery process, but similar problems are present regarding accuracy, index measures, and the quantity of data. Both prognosis prediction and rehabilitation-rate monitoring are important applications in the treatment of a patient recently diagnosed with a debilitating disease. These phases will benefit considerably from CI techniques, which have now been introduced to medicine.

A further emerging trend in healthcare is the development of automated healthcare systems also known as healthcare information technology (IT). Some of these exciting new technologies will be considered in the following sections and there is a broad avenue of possibilities for the implementation of CI techniques in the healthcare IT. In addition to research into the interconnectivity of more advanced monitoring systems, there is a need to examine the monitoring devices themselves. There are now many portable devices that can monitor the body's vital signs such as portable ECG machines. More research is required to miniaturize these devices and provide them with wireless capabilities, so that monitoring can be done not only in hospitals but also outdoors and at home. These possibilities introduce further research questions, ranging from the efficiency of the communication networks, the use of biomedical sensors, secure data transfer systems, reliability, and patient privacy. Consistent with the book's theme, there is an emerging requirement for CI techniques at almost every stage of these healthcare systems.

In the following sections, we consider some of the latest CI technologies and indicate new areas of application. There are many issues to be considered, including efficient communication networks, challenges with biomedical sensors, data transfer devices, and sensor networks.

8.3 Emerging and Future Technologies in Healthcare

8.3.1 Wireless Healthcare Systems

One emerging technology in healthcare is wireless communications to link monitoring devices and systems to the patient (Figure 8.2). The fundamental concept is to deploy wireless networks of medical devices to reduce healthcare costs and address the shortage of medical staff in industrialized countries

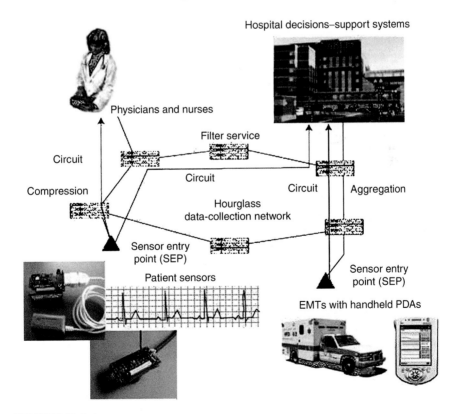

FIGURE 8.2
A wireless sensor network for healthcare. (From Gaynor, M., Moulton, S. L., Welsh, M., et al., *IEEE Internet Computing,* 8(4), 32–39, 2004. ©IEEE.)

(Noorzaie, 2006). The great advantage of a true wireless architecture is that there is little to no requirement for static infrastructure such as wires, cables, access points, and transmitting antennae to modems and routers. Wireless networks and communications can be deployed using portable devices, which communicate via radio frequency transmission and minimal infrastructure is required. This technology is new to the healthcare sector and so we will review current research and industrial applications. We will also refer to areas in which CI can be deployed to enhance the technology.

8.3.1.1 Wireless Standards and Protocols

Several wireless technologies are now used in everyday devices such as the wireless LAN (IEEE 802.11) and Bluetooth standards usually used in personal computers, personal digital assistants (PDAs), and mobile phones. Wireless LAN has been used in hospitals for Internet access and to transfer patient data between departments, and now enjoys a widespread acceptance due to its extensive coverage and ease of use. The weaknesses of this standard include high power consumption, large-packet overhead, and security issues. Bluetooth, however was designed as a short-range protocol for communication between a small group of devices and is ideal for office spaces where data can be transmitted to devices within a 10–50 m radius. It was envisaged that Bluetooth would be used to setup wireless personal area networks (WPANs) in workplaces but was later found to be susceptible to degradation due to noise and multipath effects, which reduced its effective range. Nevertheless, this protocol features prominently in mobile handsets and is sufficient for the local exchange of music, pictures, and data files. These two protocols, however, are not suitable for the design of wireless healthcare monitoring systems.

The key difference between wireless-computing networks and sensor networks is that small, low computing–powered devices (sensors) are equipped with wireless capabilities to form a network. These devices are intended to collect data from the body in our case, and transmit it to a processing centre, (such as a PC) usually called a mother node. Because the sensor nodes are small, they do not usually have substantial on-board energy for transmission, which restricts them to smaller volumes of data (low-data rates) over shorter distances. Other protocols, such as the IEEE 802.15.4 or Zigbee, have been proposed to solve those problems. Zigbee was conceived in 1998 and completed in 2003 and is primarily designed for low-powered devices over short ranges. This type of protocol is well suited to wireless-monitoring systems for healthcare (Golmie et al., 2004, 2005), for example, a wireless ECG monitor will allow a patient to move around the hospital while still being monitored. Rapid action can be taken if complications arise while providing more mobility and privacy for the patient. These networks are also known as body area networks (BANs) comprising unobtrusive devices to monitor health signs such as heart rate or blood pressure (Jovanov et al., 2005).

Another emerging technology is radio frequency identification (RFID), a labeling technology, which has become increasingly deployed. This technology is used primarily in the retail sector for the speedy handling of goods along a supply chain. RFID tags enable identification of goods from a greater distance than with standard bar code labeling. RFID is also capable of encoding more information than standard bar codes, such as the manufacturer, product type, and even measurements of temperature. It appears that RFID chips combine identification and sensor technology, which enhance the supply-chain process allowing the practice of just in time (JIT) policies (Roy, 2006). When combined with databases and globally linked systems, RFID can pinpoint a unit as it travels the supply chain from the manufacturing plant to the retail outlet. These features are the driving force behind continued research into RFID technology and it is finding more applications such as in hospitals. A hospital could, for example, use RFID technology to tag blood samples. There are two types of tags, active and passive, the former requires power whereas the latter uses electromagnetic technology (near and far fields). The passive tag is more easily used, it is cheaper, and can be printed by RFID printers, such as in Figure 8.3.

FIGURE 8.3
The I-class RFID printer produced by Datamax. (www.datamaxcorp.com.)

8.3.1.2   Research in Wireless Healthcare

One of the major technological developments in the twenty-first century will be wireless sensor networks. Microsensors with inbuilt intelligent devices, networked through wireless communications, have unprecedented applications (Figure 8.4). There are many potential applications in environment and habitat monitoring, and defence infrastructure and security monitoring (Chong et al., 2003).

Several projects have already commenced on using wireless technologies in healthcare, they focus on core technologies for the deployment of wireless networks, ranging from the sensors to the underlying transmission protocols. Some projects have been undertaken by universities in collaboration with hospitals.

| TRSS Node | Crossbow | Ember | Sensoria | Dust Inc. |

(a)

	Yesterday (1980s–1990s)	Today (2000–2003)	Tomorrow (2010)
Manufacturer	Custom contractors	Commercial, e.g., Crossbow Technology	Dust Inc. and others to be formed
Size	Large shoe box and up	Pack of cards to small shoe box	Dust particle
Weight	Kilograms	Grams	Negligible
Node architecture	Separate sensing, processing, and communication	Integrated sensing, processing, and communication	Integrated sensing, processing, and communication
Topology	Point to point, star	Client server, peer to peer	Peer to peer
Power supply: lifetime	Large batteries: hours, days, and longer	AA batteries: days to weeks	Solar: months to years
Deployment	Vehicle placed or air-drop single sensors	Hand emplaced	Embedded

(b)

FIGURE 8.4

(a) Emergence of dust sensors; (b) three generation of sensor networks. (Adapted from Chong, C. Y., Srikanta, N. D., and Kumar, P., *Proc. IEEE*, 91(8), 2003. ©IEEE.)

8.3.1.2.1 CodeBlue: Wireless Sensor Networks for Medical Care
Harvard University has initiated a wireless network project for healthcare
called Codeblue (http://www.eecs.harvard.edu/~mdw/proj/codeblue/). This
project is in collaboration with hospitals (such as the Boston Medical Center)
with support from national institutions such as the National Science Foundation (NSF), National Institutes of Health (NIH), and also companies such
as Microsoft. The aim of the project, which began in 2003, is to explore
applications of the wireless sensor network technology to a range of medical
applications, including prehospital and in-hospital emergency care, disaster
response, and stroke patient rehabilitation.

The Codeblue group has developed several small sensor devices with wire-
less capabilities such as a wireless pulse oximeter (Figure 8.5a) and 2-lead
ECG (Figure 8.5b) based on the Mica and Telos sensor motes, developed by
Crossbow and the University of Berkeley, respectively. Other devices include

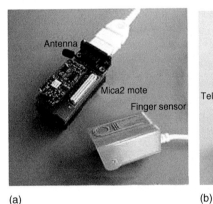

(a)   (b)

(c)

FIGURE 8.5
(a) The wireless pulse oximeter; (b) 2-lead ECG devices; and (c) a combined accelerometer, gyroscope, and electromyogram (EMG) sensors for stroke
patient monitoring, developed by the Codeblue project. (http://www.eecs.
harvard.edu/~mdw/proj/codeblue/, accessed 15/4/2007.)

(a)   (b)

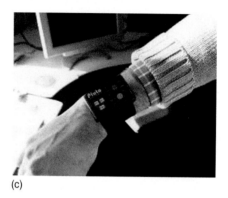

(c)

FIGURE 8.6

(a) The Pluto mote designed at Harvard. The architecture is based on the design of the Telos mote from the University of Berkeley and uses rechargeable lithium ion batteries. (b) The Pluto mote can be mounted in a case, which makes it easy for wearable wrist devices (c). (http://www.eecs. harvard.edu/~mdw/proj/codeblue/, accessed 4/4/2007.)

modified Telos mote known as the Pluto mote (Figure 8.6a); this has been equipped with accelerometers, gyroscopes, and EMG circuitry. The mote can be worn as a wrist device for monitoring vital health statistics such as heart rate and a Codeblue-software platform has also been developed for connecting wireless devices. The software is a general scalable platform, which aims to provide routing, naming, discovery, and security for wireless medical sensors, PDAs, PCs, and other devices for healthcare monitoring (Lorincz et al., 2004). A subsystem to this platform is MoteTrack, used for tracking the location of individual patient devices, it can undertake both indoor and outdoor tracking using radio signal from radio beacons fixed in the surroundings. The Codeblue group is focused on integration of medical sensors with low-power wireless networks, wireless routing, security, 3-D location tracking, and adaptive resource management; all problems faced in using sensor networks. The network architecture for emergency response is depicted in Figure 8.7a.

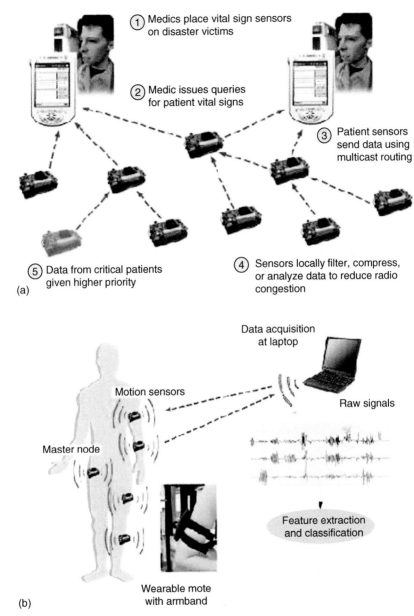

(1) Medics place vital sign sensors on disaster victims

(2) Medic issues queries for patient vital signs

(3) Patient sensors send data using multicast routing

(4) Sensors locally filter, compress, or analyze data to reduce radio congestion

(5) Data from critical patients given higher priority

(a)

Data acquisition at laptop

Motion sensors

Raw signals

Master node

Feature extraction and classification

Wearable mote with armband

(b)

FIGURE 8.7 (See color figure following page 204.)

(a) The Codeblue architecture consisting of Telos or Pluto motes used for emergency response. (b) Limb motion monitoring for rehabilitation in stroke patient. (Photo courtesy of Codeblue homepage; http://www.eecs. harvard.edu/~mdw/proj/codeblue/, accessed 15/4/2007.)

The Codeblue group has collaborations with other research centers such as the Motion Analysis Laboratory in the Spaulding Rehabilitation Hospital (http://spauldingrehab.org/) and the AID-N project at the Johns Hopkins Applied Physics Laboratory (http://www.aid-n.org/), which is investigating a range of technologies for disaster response. The AID-N wireless sensors employ the Codeblue software and include devices such as an electronic "triage tag" with pulse oximeter, LCD, and LEDs for indicating a patient's status. There is also a packaged version of the 2-lead ECG mote developed by the Harvard group and also a wireless blood pressure cuff. Other collaborations include a wide-area event delivery infrastructure for medical care, with Harvard University, and a project integrating vital dust sensors into a PDA-based patient record database (iRevive). The iRevive system provides real-time monitoring of vital signs, which will allow seamless access to a patient's data by emergency department personnel and clinicians; this could be achieved via a variety of interfaces such as a handheld PDA or Web-based computers.

8.3.1.2.2   Mobihealth: Body Area Networks using Wireless Mobile Phone Technology   Mobihealth (http://www.mobihealth.org/) is a collaborative project, initiated in 2002, based on an European initiative to create a generic platform for home healthcare. This platform uses BAN-based sensors and low bandwidth GPRS/UMTS wireless telephony for communication. The project has 14 partners in universities, hospitals, and industries. The project aims to develop medical devices that can be worn by the patient and form BANs for monitoring their vital signs. A schematic of the BAN's architecture is depicted in Figure 8.8. The devices are designed to be low powered, customizable, and to provide a service to the healthcare sector.

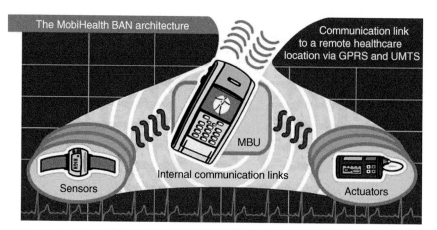

FIGURE 8.8 (See color figure following page 204.)

Mobihealth BAN architecture using GPRS/UMTS wireless protocol. (http://www.mobihealth.org/.)

The goals of this project include new services in remote monitoring of the chronically ill patients (in hospital and at home), remote assistance in emergency and disaster situations, and the remote management of clinical trials (research augmentation). It is hoped that these developments will reduce healthcare costs while maximizing response times.

From the perspective of CI, Mobihealth presents exciting new possibilities and challenges. To date, the project designers have focused on developing a generic platform to support plug-and-play devices. One of their objectives is to incorporate an early warning if a patient's vital signs fall below a threshold, for example, heart rate measured by ECG. This presents an opportunity for the application of CI techniques, not only for the detection of abnormalities in the heart rate but also for predicting the onset disorders such as myocardial infarction and ventricular tachycardia. The detection and prediction models can be placed on-board the sensor devices and increase their advantages. In short, not only can BANs be used for monitoring a patient but they can also be incorporated with intelligent systems to predict the risk or onset of a disorder, then contact emergency personnel and direct them to the patient to improve response times.

8.3.1.2.3 Other Research Initiatives Smaller research initiatives have been investigating the deployment of wireless technologies in healthcare with a primary focus on the connectivity of devices and infrastructure, as described in the previous section. The UPDATA project (http://www.sscnet.ucla.edu/sissr/da/PDA/) was initiated by the biomedical library at the University of California Los Angeles (UCLA), with the objective of incorporating statistical analysis in medical data via a PDA. The project is attempting to create a central server with customizable searches, calculation tools (such as ratios and dosages), and a database of links to major healthcare sources, all of which can be accessed by researchers using PDAs.

Another project (Jovanov et al., 2005) has investigated wireless sensor technology for ambulatory and implantable psycho–physiological applications. These devices are capable of monitoring the functioning of the heart, long-time motion of prosthetic joints, and for monitoring other vital signs. Their main focus is wearable and implantable (or subcutaneous) biosensors, biomaterial technologies, modeling, and wireless communication. There is also scope for using CI as backend software either for assisting diagnosis or for mining the collected data.

Research initiatives involving personalized and nonhospital-based care are also receiving significant attention. It has been estimated that two-thirds of deaths due to cardiac disease occur outside hospital environments (Rubel et al., 2004). This suggests the need for personalized diagnostic devices that can monitor ECG waveforms in the community and warn of any forthcoming cardiac events. Such a system will require intelligent information processing to allow the acquired signals to be analyzed for useful decision making and then to alert the care providers. Rubel et al. (2004) reported the development of an intelligent personal ECG monitor (PEM) for early detection of cardiac events. The system has in-built advanced decision making and can generate alarm levels and messages for care providers via wireless communication. This type of device demonstrates how effectively pervasive computing and CI-based intelligent information processing, combined with personalized and wearable ubiquitous devices, could improve healthcare.

8.3.1.3   Telerehabilitation

Another area that is expected to flourish with technological development is telerehabilitation, the remote delivery of neuromuscular rehabilitation (Silvennan, 2003). This area will arise due to improvements in biomedical informatics, such as expert systems and CI, in addition to developments in high-speed communication networking technology. Silvennan (2003) argued for using artificial intelligence techniques such as NNs to carry out routine rehabilitation work allowing the physical therapist to attend to other duties. This would also provide healthcare services to a broader range of patients. Such a system will collect the necessary biological signals from the patient and transmit the information in real-time over the Internet with relevant figures and diagrams. This information can be processed and analyzed by a CI tool to examine the extent of abnormality using pattern recognition, for example, in the rehabilitation needs and services of a stroke patient. Individual muscles and their EMG signals would be properly assessed and the appropriate rehabilitation procedure administered. Figure 8.9 illustrates such a rehabilitation system, with the interaction of software agents and elements, and patient feedback. The NN can be trained to estimate the deficit in neuromuscular function,

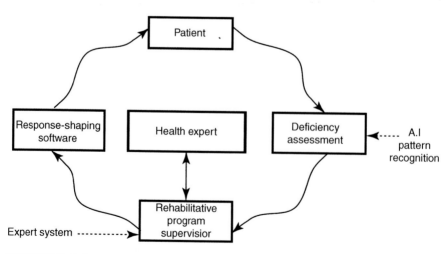

FIGURE 8.9

An overview of various elements of the telerehabilitation system. (Adapted from Silvennan, G., *First Int. IEEE EMBS Conf. Neural Eng.*, 664–667, 2003. ©IEEE.)

including EMG (e.g., biceps activity), and joint motion. This information can be used to devise goals that are transmitted to the patient's screen to restore functional status. Feedback to the patient consists of all the graphical data and therapist's feedback, including videotaped records and the rehabilitation tasks to be followed.

References

Chong, C.-Y., N.D. Srikanta, and P. Kumar (2003). Sensor networks: evolution, opportunities, and challenges. *Proceedings of the IEEE, Vol. 91(8),* 1247–1256, (August) (invited paper).

Gaynor, M., S.L. Moulton, M. Welsh, et al. (2004). Integrating wireless sensor networks with the grid. *IEEE Internet Computing 8(4)*, 32–39.

Golmie, N., D. Cypher, and O. Rebala (2004). Performance evaluation of low rate WPANs for sensors and medical applications. *Military Communications Conference (MILCOM).* http://w3.antd.nist.gov/pubs/milcom04_golmie.pdf (accessed 14/4/2007).

Golmie, N., D. Cypher, and O. Rebala (2005). Performance analysis of low rate wireless technologies for medical applications. *Computer Communications 28(10)*, 1255–1275. http://w3.antd.nist.gov/pubs/com04_golmie.pdf.

Jovanov, E., A. Milenkovic, and C. Otto (2005). A WBAN system for ambulatory monitoring of physical activity and health status: applications and challenges. *Proceedings of the 27th Annual International Conference of the IEEE Engineering in Medicine and Biology Society (EMBS)*, Shanghai, China, September. http://www.ece.uah.edu/~jovanov/papers/embs05_wban.pdf.

Lorincz, K., D.J. Malan, T.R.F. Fulford-Jones, et al. (2004). Sensor networks for emergency response: challenges and opportunities. *IEEE Pervasive Computing (Special Issue on Pervasive Computing for First Response) 4*, 16–23.

Noorzaie, I. (2006). *Survey Paper: Medical Applications of Wireless Networks*, http://www.cs.wustl.edu/~jain/cse574-06/ftp/medical_wireless.pdf (accessed 15/4/2007).

Roy, W. (2006). An introduction to RFID technology. *IEEE Pervasive Computing 5(1)*, 25–33.

Rubel, P., J. Fayn, L. Simon-Chautemps, et al. (2004). New paradigms in telemedicine: ambient intelligence, wearable, pervasive and personalized. In *Wearable eHealth Systems for Personalised Health Management: State of the Art and Future Challenges*, A. Lymberis and D. de Rossi (Eds), Amsterdam: IOS Press, pp. 123–132.

Silvennan, G. (2003). The role of neural networks in telerehabilitation. *First International IEEE EMBS Conference on Neural Engineering (20–22 March)*, Manhattan, NY, 664–667.

Index

A

accuracy issues, computational intelligence, 328

acute inflammatory demyelinating neuropathy (AIDP), pathophysiology, 227

adaptive cepstrum vector (ACV), electromyography feature extraction, 248

adaptive filters
basic properties, 63–64
LMS algorithm, 64–67
QRS complex (ECG) detection, 179
RLS algorithm, 67–69

adaptive linear network (ADALINE) system, maternal/fetal heartbeat analysis, 192–193

adaptive neurofuzzy system (ANFIS), gait analysis, 316–317

adenosine diphosphate (ADP), muscle contraction, 216–218

adenosine triphosphate (ATP) molecules, muscle contraction, 214–218

adult population, gait analysis, 306

afferent nerves, motor units, 219

AID-N wireless sensors, computational intelligence applications in, 338

Aikaike information criterion (AIC)
autoregressive model, 39

electromyography feature extraction, 246–248
moving average model, 39–40

aliasing, analog signal sampling and reconstruction, 20

Alzheimer disease (AD), electroencephalography signals, 276–277
computational intelligence with, 283–284

American Heart Association (AHA) database, cardiovascular disease, computational intelligence sources, 186

amplitude algorithms
electromyographic feature extraction, 244–248
QRS complex (ECG), 177–178

amputees, gait analysis in, 309

amyotrophic lateral sclerosis (ALS), pathophysiology, 227–229

analog filters
basic properties, 40–42
Butterworth filters, 46–48
Chebyshev filter, 48–49
resistor-capacitor filter, 42–44
RL high-pass filter, 45

analog signal
basic properties, 16
sampling and reconstruction, 19–20

Ann Arbor Electrogram Libraries, cardiovascular disease, computational intelligence sources, 186

344